GAS CHROMATOGRAPHY AND LIPIDS

A Practical Guide

by

William W. Christie

The Hannah Research Institute, Ayr, Scotland KA6 5HL

THE OILY PRESS

AYR, SCOTLAND

The Oily Press Ltd.
36 Woodend Road
Alloway
Ayr
Scotland KA7 4QR.

First edition 1989

British Library Cataloguing in Publication Data

Gas Chromatography and Lipids : a practical guide
1. Lipids, chemical analysis I. Title 547.7'7046

ISBN 0-9514171-0-X

By the same author

CHRISTIE

Lipid Analysis, 2nd Edition (Pergamon Press)
Lipid Metabolism in Ruminant Animals
High-Performance Liquid Chromatography and Lipids

Phototypesetting by The Publishing & Visual Presentation Dept.
The West of Scotland College, Auchincruive, Ayr
Printed in Great Britain by Bell and Bain Ltd., Glasgow

Preface

Gas chromatography was first developed by lipid analysts, and lipid analysts have been at the forefront in the development of the technique. In recent years, there has been a remarkable improvement in the resolution attainable by gas chromatography thanks to the availability of capillary columns fabricated from fused silica. The stability and inertness of such columns has simultaneously increased the range of applications. Related developments in mass spectrometry have meant that the combined technique of gas chromatography-mass spectrometry is less costly, more versatile and more accessible than formerly. I, therefore, felt that a book on "Gas Chromatography and Lipids" was timely and would complement my book on "High-performance Liquid Chromatography and Lipids", published at the end of 1987. There are still many gaps in our knowledge that remain to be filled, and many published procedures could be improved. Hopefully, this book will stimulate further endeavours. In the Preface to my previous books, I stated that I hoped they would "remain on the laboratory bench, not on the library shelf". That comment is equally apposite here.

I am grateful to the Director and Council of the Hannah Research Institute and to the Department of Agriculture and Fisheries for Scotland for permission to write this book. Dr. John H. Shand read and criticized each of the chapters as they were written, and his assistance is gratefully acknowledged. Similarly, D. Cooney and S. Armour helped greatly with the preparation of the figures. My final thanks must go to my wife, Norma, for her assistance throughout the project, including a great deal of the typing.

This is the first book from a new publishing company, "The Oily Press Ltd", whose objective will be to serve lipid chemists and biochemists by issuing compact, practical and readable texts on all aspects of lipid research.

William W. Christie

Contents

PART 1

AN INTRODUCTION TO LIPIDS AND GAS CHROMATOGRAPHY

CHAPTER 1

INTRODUCTION AND SUMMARY

A. A HISTORICAL PERSPECTIVE ON GAS CHROMATOGRAPHY AND LIPIDS

The initial stages in the development of gas chromatography were in the hands of lipid analysts, and this technique probably above all others so facilitated the separation of lipids, and especially their fatty acid components, that it must be considered one of the major factors in the explosive growth of knowledge about these natural products over the last thirty years. In the hands of a well-organised analyst, it is now possible to convert a lipid sample of a fraction of a milligram in size to the methyl ester derivatives, separate these by gas chromatography, and have a quantitative result in under one hour. This was not always so. In the 1940s and early 1950s, with even simple mixtures as in seed oils, it was necessary to start with a sample of 20 to 200g of the methyl esters. These were subjected to a preliminary fractionation by low-temperature crystallisation, steam distillation or as the urea adducts, then the individual fractions were subjected to fractional distillation under high vacuum. Further chemical reactions, many calculations and perhaps several weeks later, an approximate result was obtained [317]. For a brief period, a form of column partition chromatography in the reversed phase mode came into use that would appear rather crude to the modern observer, but it reduced the time required to complete an analysis to one or two days.

This technique was based on the classic paper by Martin and Synge of 1941 [598], in which liquid-liquid column chromatography was used to separate amino acids. The paper contained the significant phrase - "the mobile phase need not be a liquid but may be a vapour". However, the idea was not pursued until A.T. James joined Martin in 1949. James has recently provided an entertaining account of their joint collaboration [431]. In the initial work, they could only attempt to separate those compounds which could be detected and quantified by titration. The first essays at separations were made with a column packed with celite coated with liquid paraffin as the liquid phase and with short-chain fatty acids as the solute, apparently for the reason simply that Popjak in an adjacent laboratory was working on the biosynthesis of these compounds in mammary gland. Success eluded them initially, but no complications were experienced with low molecular weight amines and the first successful application was to the resolution of mono- and dimethylamine.

They soon realised that the difficulty with the fatty acids was because of the occurrence of dimerisation, and when a long-chain fatty acid was incorporated into the liquid phase the problem was resolved. In fact, the application to the resolution of volatile fatty acids (formic to dodecanoic) was chosen for illustrative purposes, when the first paper on the new technique was published in 1952 [432]. From the theoretical standpoint, they showed that a series of straight-chain homologues had a constant increment in retention volume for each additional methylene group.

The instrumentation became increasingly sophisticated with the addition of provision to heat the column, an automatic titrating device and then the first successful sensitive and universal detector, the gas-density balance, all of which required to be hand-crafted [597]. In a companion paper to the description of the detector, the technique was demonstrated with the first separations of methyl ester derivatives of fatty acids [433]. It appears that the initial contact with the problem of the resolution of fatty acids must have inspired James to continue with the study of lipids, and in 1957 he independently published a paper on the determination of the structures of longer-chain fatty acids using gas chromatography and microchemical methods [434]. He subsequently went on to a distinguished career in lipid biochemistry.

The scientific instrument companies rapidly took up the technique of gas chromatography, bringing it to the high state of development described in this book. The technique began with a lipid problem, and lipid analysts have been at the forefront at every stage. Now, all but those lipids of the highest polarity or molecular weight can be separated by the technique, and even these may succumb before long to the new procedure of supercritical-fluid chromatography, which in effect is a hybrid of gas chromatography and high-performance liquid chromatography.

Of course, many other instrumental procedures are utilised by lipid analysts, and the author has recently published a monograph on "High-Performance Liquid Chromatography and Lipids" [168]. Perhaps the most important technique to be used with gas chromatography directly, however, is mass spectrometry. It may come as a surprise to many readers, but mass spectrometry is in fact a much older technique than gas chromatography, the basic principle and applications to the separation of atomic masses first having been demonstrated towards the end of the last century. Methods for separating the isotopes of the elements dominated the development of the technique after the First World War, but it was the demands of the petroleum industry during the Second World War together with improvements in electronics that led to the construction of instruments we would recognise today. When it became possible to couple the mass spectrometer to the outlet of the column of a gas chromatograph, lipid analysts were not slow to recognise the possibilities that lay before them. Because of further progress in computerisation and micro electronics, together with the introduction of wall-coated open-tubular columns of fused silica that greatly reduces the problem of interfacing, the price of gas chromatography-mass spectrometry systems has plummeted, and their reliability and ease of use have increased immeasurably. They have become the routine research and analytical tools for many, rather than the prized possessions of a few, and this trend can only accelerate.

B. A SUMMARY

In Chapter 2, the structures, chemistry and compositions of lipids in animals, plants and microorganisms are summarised, since a knowledge of the lipid types, which may be encountered in an analysis, might assist in determining the best procedure to adopt. The first step in the analysis of lipids generally involves the preparation of a lipid fraction, relatively free of non-lipid contaminants, by means of solvent extraction of the tissue. Methods of achieving this are described here. Unwanted degradation of lipids can occur during the storage and handling of tissues and lipid samples, and autoxidation of unsaturated fatty acids can be especially troublesome. Methods of avoiding these difficulties during extraction are described, that are applicable to all stages of analytical procedures, and they are relevant to each of the subsequent chapters. Some preliminary fractionation of lipid samples into simple lipid, glycolipid and phospholipid groups may then be desirable to facilitate their analysis or the isolation of single lipid classes on a small scale. The last objective can be accomplished by high-performance liquid, thin-layer or ion-exchange chromatography.

In order to make the best use of the technique, it helps to have some knowledge of the theory of gas chromatography, and this is outlined in Chapter 3. Many important aspects of chromatography are governed by a few simple equations, so a highly detailed account of the mathematics and physics of the subject is not necessary here. Novices on the subject or newcomers to capillary gas chromatography are confronted by a bewildering array of different columns, injection systems and detectors, and these are discussed systematically in relation to specific analytical problems. Various precautions are described to extend column life, and to ensure reproducibility in quantification, and details of these obviously cannot be repeated in each of the subsequent Chapters.

The important complementary technique of mass spectrometry is discussed briefly here, but applications are dealt with at length in later Chapters. Some modern mass spectrometry techniques are not directly compatible with gas chromatographic separations, requiring direct probe inlet systems, and these are not treated in depth in this book.

Part 2 of the book deals with the analysis of fatty acids. This has grown to be a rather large subject, and for convenience has been divided here into four chapters. However, aspects of the subject matter in each may be helpful for particular problems, and it is never safe to rely on evidence from a single analytical technique.

The first of these chapters is concerned with the preparation of volatile non-polar derivatives of fatty acids. Methyl ester derivatives are used almost universally for fatty acid analysis, although many analysts continue to employ older cumbersome procedures for their synthesis rather than the simple safe ones now available. It should be noted that no single method can be used in all circumstances. In addition, fatty acids with a wide range of different substituent groups in the aliphatic chain occur in nature, and their analysis by means of gas chromatography can also be facilitated by the preparation of specific derivatives.

Gas chromatography on stationary phases consisting of polar polyesters of various kinds has become the standard technique for the separation of the common range of fatty acids encountered in most animal and plant tissues (Chapter 5). Packed columns afford adequate resolution for most purposes, but the high resolving power, stability and inertness of wall-coated open-tubular (capillary) columns made from fused silica have made these the standard in increasing numbers of laboratories. Fatty acid derivatives differing according to degree of unsaturation, that have been separated by the technique, can frequently be identified with a reasonable degree of certainty by their relative retention times alone, provided that these are measured with care. When other functional groups, such as hydroxyl, branched or cyclic moieties, are present in the alkyl chain, such methods can also be used provided that a sufficient range of model compounds are available for comparison of their chromatographic properties. Modern flame ionisation detectors have sufficient range and sensitivity to cope with most problems of quantification, provided that they are set up correctly and calibrated with care.

There are many circumstances when it is not possible to rely on gas chromatographic retention data for the identification of a fatty acid, for example in the analysis of material from a new natural source or when the interpretation of a metabolic experiment hinges on the identity of some specific component. It is then necessary to isolate the fatty acids by an appropriate method, nowadays high-performance liquid chromatography in the reversed-phase or silver ion modes is chosen most often, and then determine the structures by chemical and/or spectroscopic procedures (Chapter 6). Chemical degradative methods are available for locating double bonds amd many other functional groups in fatty acids on a microscale. If the material can be isolated in sufficient amount, then spectroscopic techniques, especially proton and carbon-13 nuclear magnetic resonance spectroscopy, can be substantial non-destructive aids to structure determination.

Gas chromatography and mass spectrometry in combination make a particularly powerful technique for the identification of fatty acids, and this is the subject of Chapter 7. In some instances, the methyl ester derivatives alone give adequate mass spectra for identification purposes. More often, it is preferable to prepare pyrrolidides or picolinyl esters, since these give characteristic fragmentations that permit the location of many functional groups, including double bonds and methyl branches, in aliphatic chains. Sometimes it is necessary to prepare derivatives of other functional groups in order to facilitate chromatography and to ensure that interpretable mass spectra are obtained.

Lipid classes do not occur in nature as single pure entities, but rather they are a complex mixture of molecular species in which the fatty acids and other aliphatic moieties are present in different combinations. One excellent method of resolving this complexity is to use high-temperature gas chromatography, and this is the subject of Chapter 8. With lipids of high molecular weight, such as triacylglycerols, it is possible to obtain fractions differing in size by one or two carbon atoms on short packed columns containing non-polar methylsilicone phases. By using wall-coated open-tubular columns of glass or fused silica coated with similar phases or even, as in recent work, with more polar phases, greatly improved resolution is obtained, including components differing in degree of unsaturation. This methodology is perhaps stretching current technology to its limits, and it is a much simpler task to resolve molecular species of phosphoglycerides, after they have been hydrolysed to diacylglycerols with the enzyme phospholipase C and then chemically derivatised. Similarly, gas chromatography is an excellent procedure for the separation of molecular species of monoacylglycerols, cholesterol esters and wax esters. Glycosyldiacylglycerols can be analysed by this methodology, but it has not been used a great deal for the purpose. Ceramides derived from sphingomyelin are easily resolved by high-temperature gas chromatography, but glycosphingolipids are not readily converted to ceramides and the technique has been little used for these compounds for which high-performance liquid chromatography is now generally favoured. In all such work, gas chromatography coupled with mass spectrometry has proved invaluable for identifying fractions, and also for the quantification of unresolved components in a single chromatographic peak. The last section of this Chapter deals with methods for obtaining a profile of all the main lipid classes in body fluids, such as plasma, by gas chromatography.

Gas chromatographic methods are only capable of separating a limited number of molecular species in complicated lipid samples, but greatly improved resolution can be obtained if they are used in conjunction with a complementary technique, such as adsorption or silver ion chromatography, as described in Chapter 9. Adsorption chromatography is of most value for the separations of species containing fatty acids with polar substituents. Silver ion chromatography, which has historically been associated with thin-layer chromatography although high-performance liquid chromatographic methods are now being actively developed, gives separations of triacylglycerols, diacylglycerol acetates and even of intact phospholipids by degree of unsaturation. On the other hand, high-performance liquid chromatography in the reversed-phase mode is best regarded as an alternative separatory procedure to high-temperature gas chromatography and recent developments are summarised to indicate the wider choice available to the analyst.

In the final Chapter (Ten), some miscellaneous but none the less important topics, which do not conveniently fit into other Chapters, are gathered together. Fatty alcohols, for example, are major components of waxes in addition to having other biological functions, and they are best analysed by gas chromatography in the form of various non-polar derivatives. Most glycerolipids exist in alkyl and alkenyl forms in addition to having esterified fatty acids; each of these ether-linked residues is conveniently analysed by gas chromatography in the form of alkylglycerol and aldehyde derivatives respectively. Platelet-activating factor is a special case of an ether lipid with vital metabolic properties, and the analysis of this compound is described here also. In addition to the fatty acid constituents, which are considered earlier, sphingolipids contain a range of different sphingoid bases. Procedures for the isolation of these, and for their analysis by gas chromatography are considered here. With all of these compounds, gas chromatography-mass spectrometry is of immense value for identification purposes. Cholesterol and glycerol are important lipids or components of lipids, and gas chromatographic methods for their analysis are likewise described in this Chapter.

C. ABBREVIATIONS

The following abbreviations are employed at various points in the subsequent text:

amu	atomic mass units
BDMS	*tert*-butyldimethylsilyl
BHT	2,6-di-*tert*-butyl-*p*-cresol
CI	chemical ionisation
DNP	dinitrophenyl
ECL	equivalent chain-length
ECN	equivalent carbon number
EI	electron-impact ionisation
FCL	fractional chain-length
GC	gas chromatography
GLC	gas-liquid chromatography
HPLC	high-performance liquid chromatography
IR	infrared
MS	mass spectrometry

NMR	nuclear magnetic resonance
PAF	platelet-activating factor
ODS	octadecylsilyl
TLC	thin-layer chromatography
TMS	trimethylsilyl
UV	ultraviolet

CHAPTER 2

FATTY ACIDS AND LIPIDS: STRUCTURES, EXTRACTION AND FRACTIONATION INTO CLASSES

A. DEFINITIONS

There is no satisfactory universally-accepted definition of a "lipid", although most chemists and biochemists who work with these fascinating natural products have a firm intuitive understanding of the term. Most general text books describe lipids rather loosely as a group of compounds, which have in common a ready solubility in organic solvents such as chloroform, ethers and alcohols - a definition which encompasses steroids, carotenoids, terpenes and bile acids (generic terms in their own right!) in addition to fatty acids and glycerolipids, for example. The author has recently criticised this concept of a lipid as unnecessarily broad [168], and has proposed a definition which harkens back to the origins of the term, i.e.

"*Lipids* are fatty acids and their derivatives, and substances related biosynthetically or functionally to these compounds."

The *fatty acids* are compounds synthesised in nature via condensation of malonyl-coenzyme A units by a fatty acid synthetase complex. By such definitions, cholesterol (but not steroid hormones) can be considered a lipid, as are phospholipids and glycolipids. Gangliosides, which are acidic glycolipids, are soluble in water and would not be treated as a lipid if some of the looser definitions were accepted. In the subsequent text of this book, the strict definition of lipid given above was followed in selecting the subject matter.

Some further terms which have proved their worth, especially in discussing the chromatographic separation of lipids [163,168], are -

Simple lipids - those which on saponification yield at most two types of primary hydrolysis product (fatty acids, glycerol, etc.) per mole, e.g. triacylglycerols and cholesterol esters

Complex lipids - those which on saponification yield three or more primary products per mole, e.g. phospholipids and glycolipids

The *phospholipids* and *glycolipids* can themselves be further subdivided into *glycero-* and *sphingolipids* (see below).

If a thorough analysis of a lipid sample from an organism is intended, it is necessary to separate it into simpler classes, according to the nature of the constituent parts of the molecules, and these components in turn may then have to be identified and quantified. The principal purpose of this book is to describe the analysis of lipids by means of gas chromatography, a technique which requires that the lipids be volatilised. The fatty acids and most of the simple lipids can sometimes be analysed directly by this method, but the complex lipids must be hydrolysed to simpler less-polar moieties, for example to the diacylglycerol and ceramide backbones common to glycerophospholipids and sphingolipids respectively, prior to separation. It is therefore the aliphatic part of the molecule which is of primary importance for gas chromatography, and the following text will concentrate on this aspect.

There are a number of books available that deal with lipids and their structures, and the author has found those cited to be of particular value [319,367]. Literally thousands of papers have appeared over the last 25 years detailing the structures and compositions of lipids from particular tissues and species, as determined by modern chromatographic methods, but there appears to have been very little effort to collate and critically compare these data in any systematic way, or to relate the compositions of lipids to their functions. Among other consequences of this, there remain anomalies and gaps in our knowledge. Comprehensive accounts of the lipids of the tissues of ruminant animals [162], tissue and membrane phospholipid compositions [395,970] and triacylglycerol compositions [125,553,686,824] have appeared, however, and there are miscellaneous reviews of the compositions of specific lipid classes or tissues in the literature. The author recently attempted to summarise the essential features of lipid composition in a succinct manner [168]. This cannot be repeated here, and a brief summary only of lipid structure and composition follows.

The nomenclature proposed by IUPAC-IUB commissions [415,416] has been followed throughout.

B. THE FATTY ACIDS

The fatty acids of plant, animal and microbial origin generally contain even numbers of carbon atoms in straight chains, with a carboxyl group at one extremity and with double bonds of the *cis* configuration in specific positions in relation to this. In animal tissues, the common fatty acids vary in chain-length from 14 to 22, but on occasion can span the range from 2 to 36 or even more. Individual groups of microorganisms can contain fatty acids with 80 or more carbon atoms, but higher plants usually exhibit a more limited chain-length distribution. Fatty acids from animal tissues may have one to six double bonds, those from algae may have up to five, while those of the higher plants rarely have more than three; microbial fatty acids only occasionally have more than one. Hydroxy fatty acids are synthesised in some animal tissues, but fatty acids with other functional groups, when present, have usually been taken up from the food chain. Plant and microbial fatty acids, on the other hand, can contain a wide variety of functional groups including *trans*-double bonds, acetylenic bonds, epoxyl, hydroxyl, keto and ether groups, and cyclopropene, cyclopropane and cyclopentene rings. Fatty acid structure and physical properties have been reviewed by Gunstone [318].

1. Saturated fatty acids

The most abundant saturated fatty acids in animal and plant tissues are straight-chain compounds with 14, 16 and 18 carbon atoms, but all the possible odd and even numbered homologues with 2 to 36 carbon atoms have been found in nature in esterified form. They are named systematically from the saturated hydrocarbon with the same number of carbon atoms, the final *-e* being changed to *-oic*. Thus, the fatty acid with 16 carbon atoms and the structural formula-

$$CH_3(CH_2)_{14}COOH$$

is systematically named *hexadecanoic acid*, although it is more usual to see the trivial name *palmitic* acid in the literature. It may also be termed a "C_{16}" fatty acid or with greater precision as "16:0", the number before the colon specifying the number of carbon atoms, and that after the colon, the number of double bonds. A list of the common saturated fatty acids together with their trivial names and shorthand designations is given in Table 2.1. A comprehensive list of the trivial names of fatty acids has been published elsewhere [773].

Table 2.1
Saturated fatty acids of general formula $CH_3(CH_2)_nCOOH$

Systematic name	Trivial name	Shorthand designation
ethanoic	acetic	2:0
butanoic	butyric	4:0
hexanoic	caproic	6:0
octanoic	caprylic	8:0
nonanoic	pelargonic	9:0
decanoic	capric	10:0
undecanoic	-	11:0
dodecanoic	lauric	12:0
tridecanoic	-	13:0
tetradecanoic	myristic	14:0
pentadecanoic	-	15:0
hexadecanoic	palmitic	16:0
heptadecanoic	margaric	17:0
octadecanoic	stearic	18:0
nonadecanoic	-	19:0
eicosanoic	arachidic	20:0
heneicosanoic	-	21:0
docosanoic	behenic	22:0
tetracosanoic	lignoceric	24:0

Acetic acid is not often found in association with fatty acids of higher molecular weight in esterified form in lipid molecules, but it does occur esterified to glycerol in some seed oils and in ruminant milk fats. In certain vegetable oils, it has been detected in linkage to the hydroxyl group of a hydroxy fatty acid, which is in turn esterified to glycerol.

Lipid-bound C_4 to C_{12} fatty acids are in essence only found in milk fats in animal tissues, while the medium-chain compounds occur in seed oils, such as coconut oil. *Palmitic* acid is one of the most abundant fatty acids in nature and is found in the lipids of all organisms. *Stearic* acid is also relatively common.

Odd-chain fatty acids are synthesised by many microorganisms, and are produced, but to a very limited extent, in animal tissues when the fatty acid synthetase accepts propionyl-coenzyme A as the primer molecule.

The higher saturated fatty acids are solid at room temperature and are comparatively inert chemically.

2. Monoenoic fatty acids

Straight-chain even-numbered fatty acids with 10 to more than 30 carbon atoms and containing one *cis*-double bond have been characterised from natural sources. The double bond can be in a variety of different positions, and this is specified in the systematic nomenclature in relation to the carboxyl group. Thus, the most abundant monoenoic fatty acid in tissues is probably *cis*-9-octadecenoic acid, also termed *oleic* acid, and it has the structure.

$$CH_3(CH_2)_7CH = CH(CH_2)_7COOH$$

In the shorthand nomenclature, it is designated 18:1. The position of the double bond can be denoted in the form *(n-x)*, where *n* is the chain-length of the fatty acid and *x* is the number of carbon atoms from the double bond in the terminal region of the molecule, i.e. oleic acid is 18:1(n-9). Although this contradicts the convention that the position of functional groups should be related to the carboxyl carbon, it is of great convenience to lipid biochemists. Animal and plant lipids frequently contain families of monoenoic fatty acids with similar terminal structures, but with different chain-lengths, that may arise from a common precursor either by chain-elongation or by *beta*-oxidation (Figure 2.1(a)). The *(n-x)* nomenclature helps to point out such relationships. Some obvious examples can be seen in Table 2.2.

Various positional isomers exist in nature and *cis*-6-octadecenoic acid (petroselinic acid) is found in seed oils of the Umbelliferae, for example, while *cis*-11-octadecenoic acid is the major unsaturated fatty acid in many bacterial species. Many different isomers may indeed exist in a lipid sample from a single natural source.

Monoenoic fatty acids with double bonds of the *trans*-configuration are

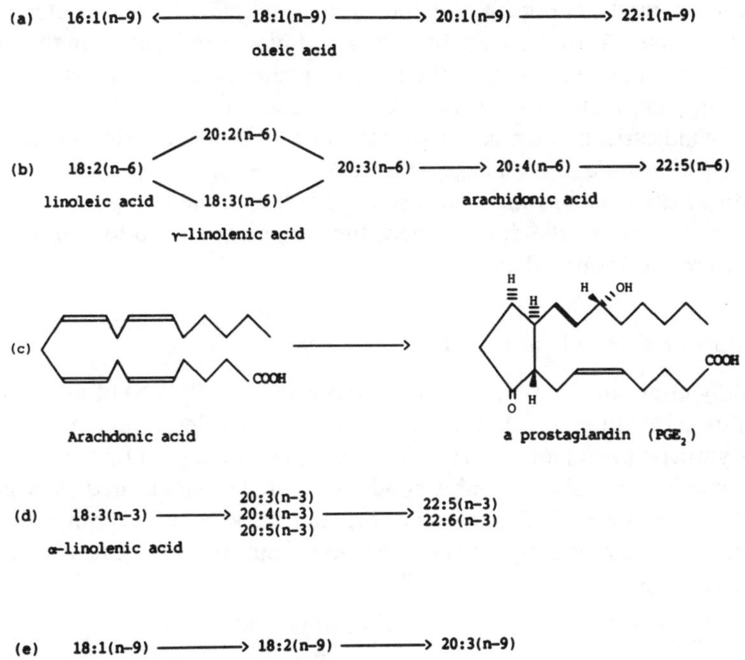

Figure 2.1 Biosynthetic relationships between unsaturated fatty acids. (a) Elongation and retroconversion of oleic acid; (b) elongation and desaturation of linoleic acid; (c) biosynthesis of prostaglandin E2 from arachidonic acid; (d) elongation and desaturation of linolenic acid; (e) elongation and desaturation of oleic acid.

Table 2.2
Monoenoic fatty acids of general formula - $CH_3(CH_2)_mCH=CH(CH_2)_nCOOH$

Systematic name	Trivial name	Shorthand designation
cis-9-tetradecenoic acid	myristoleic	14:1(n-5)
cis-9-hexadecenoic acid	palmitoleic	16:1(n-7)
trans-3-hexadecenoic acid	-	-
cis-6-octadecenoic acid	petroselinic	18:1(n-12)
cis-9-octadecenoic acid	oleic	18:1(n-9)
cis-11-octadecenoic acid	*cis*-vaccenic	18:1(n-7)
trans-9-octadecenoic acid	elaidic	-
trans-11-octadecenoic acid	vaccenic	-
cis-9-eicosenoic acid	gadoleic	20:1(n-11)
cis-11-eicosenoic acid	gondoic	20:1(n-9)
cis-13-docosenoic acid	erucic	22:1(n-9)
cis-15-tetracosenoic acid	nervonic	24:1(n)

also found in nature. For example, *trans*-3-hexadecenoic acid is always present as a substantial component of plant chloroplast lipids. *Trans*-11-octadecenoic acid *(vaccenic)* is formed as a by-product of biohydrogenation in the rumen, and thence finds its way into the tissues of ruminant animals, and via meat and dairy products into human tissues. In addition, *trans*-isomers are formed during industrial hydrogenation of fats and oils, as in margarine manufacture.

cis-Monoenoic fatty acids with 18 carbons or less melt below room temperature (*trans*-isomers have somewhat higher melting points). Because of the presence of the double bond, they are more susceptible to oxidation than are the saturated fatty acids.

3. Polyunsaturated fatty acids

Polyunsaturated fatty acids (often abbreviated to PUFA) of animal origin can be subdivided into families according to their derivation from specific biosynthetic precursors. In each instance, the families contain from two up to a maximum of six *cis*-double bonds, separated by single methylene groups (methylene-interrupted unsaturation), and they have the same terminal structure. A list of some of the more important of these acids is contained in Table 2.3

Table 2.3
Polyunsaturated fatty acids of general formula $CH_3(CH_2)_m(CH=CHCH_2)_x(CH_2)_nCOOH$

Systematic name	Trivial name	Shorthand designation
9,12-octadecadienoic*	linoleic	18:2(n-6)
6,9,12-octadecatrienoic	γ-linolenic	16:3(n-6)
8,11,14-eicosatrienoic	homo-γ-linolenic	20:3(n-6)
5,8,11,14-eicosatetraenoic	arachidonic	20:4(n-6)
4,7,10,13,16-eicosapentaenoic	-	20:5(n-6)
9,12,15-octadecatrienoic	α-linolenic	18:3(n-3)
5,8,11,14,17-eicosapentaenoic	-	20:5(n-3)
7,10,13,16,19-docosapentaenoic	-	22:5(n-3)
4,7,10,13,16,19-docosahexaenoic	-	22:6(n-3)
5,8,11-eicosatrienoic	-	20:3(n-9)

* the double bond configuration in each instance is *cis*

Linoleic acid (*cis*-9,*cis*-12-octadecadienoic acid) is the most wide-spread fatty acid of this type, and it is found in most animal and plant tissues. It is designated 18:2(n-6), using the same shorthand nomenclature as before (methylene-interrupted *cis*-double bonds are assumed). It is an essential fatty acid in animal diets, as it cannot be synthesised in animal tissues yet is required for normal growth, reproduction and healthy development. The enzymes in animals are only able to insert new double bonds between an existing double bond and the carboxyl group. Linoleic acid, therefore, serves as the precursor of a family of fatty acids that is formed by desaturation and chain elongation, in which the terminal (n-6) structure is retained (Figure 2.1(b)). Of these, *arachidonic* acid (20:4(n-6)) is particularly important as an essential component of the membrane phospholipids and as a precursor of the prostaglandins (Figure 2.1(c)). These compounds have profound pharmacological effects and are the subject of intensive study. *cis*-6,*cis*-9,*cis*-12-Octadecatrienoic acid (18:3(n-6)), an important intermediate in the biosynthesis of arachidonic acid and a constituent of certain seed oils, has been the object of considerable research in its own right.

The enzymes in plant tissues are capable of inserting a double bond in the terminal region of an existing unsaturated fatty acid, and *linolenic* acid (*cis*-9,*cis*-12,*cis*-15-octadecatrienoic acid or 18:3(n-3)) is the end-point of biosynthesis in most higher plants. When it is absorbed into animal tissues through the diet, it forms the precursor of a further family of polyunsaturated fatty acids with an (n-3) terminal structure (Figure 2.1(d)). These fatty acids are also essential dietary components, especially in fish, although the requirement in mammals is probably appreciably less than that for the (n-6) series. None the less, 20:5(n-3) and 22:6(n-3) fatty acids appear to have special functions in the phospholipids of nervous tissue and in the eye, and the former is a precursor of specific prostanoids.

Many other similar families of fatty acids exist in nature, and that derived from oleic acid (Figure 2.1(e)) tends to assume greater importance in animals suffering from essential fatty acid deficiency.

Polyunsaturated fatty acids with more than one methylene group between the double bonds, such as *cis*-5,*cis*-11- and *cis*-5,*cis*-13-eicosadienoic acids occur in marine invertebrates and some other organisms, but are rarely found in animals. Some plant species synthesise fatty acids with one or more double bonds of the *trans*-configuration (e.g. *trans*-9,*trans*-12-octadecenoic acid), with conjugated double bond systems (e.g. *cis*-9,*trans*-11,*trans*-13-octadecatrienoic or α-eleostearic acid), or with acetylenic bonds (e.g. octadec-*cis*-9-en-12-ynoic or *crepenynic* acid). The natural occurrence of such fatty acids has been reviewed [71,856].

In general, polyunsaturated fatty acids have low melting points, and they are susceptible to oxidative deterioration or autoxidation (see Section E.2 below).

4. Branched-chain and cyclopropane fatty acids

Branched-chain fatty acids occur widely in nature, but tend to be present as minor components except in bacteria, where they appear to replace unsaturated fatty acids functionally. Usually, the branch consists of a single methyl group, either on the penultimate (*iso*) or antepenultimate (*anteiso*) carbon atoms (Figure 2.2). In the biosynthesis of these fatty acids, the primer molecules for chain-elongation by the fatty acid synthetase are

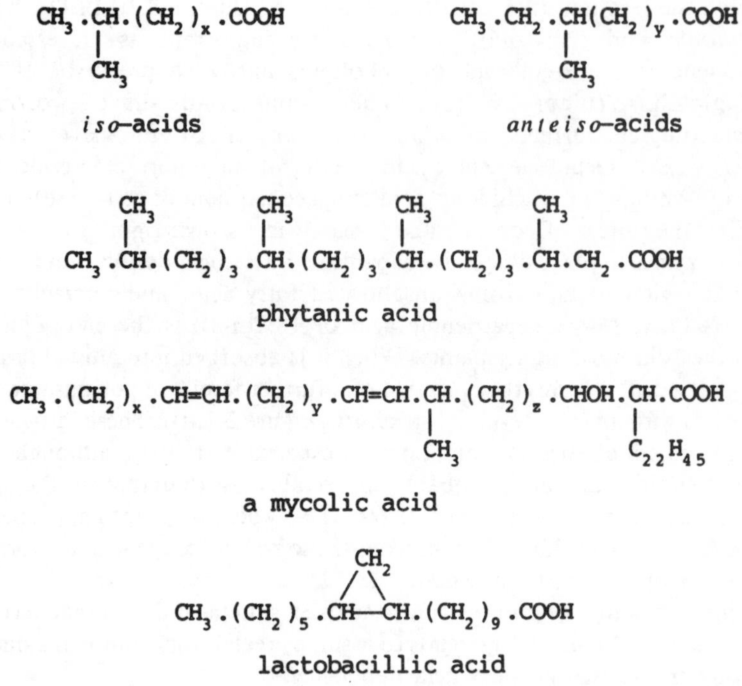

Figure 2.2 The structures of some branched-chain and cyclic fatty acids.

2-methylpropanoic and 2-methylbutanoic acids respectively. Methyl branches can be found in other positions of the chain (on even-numbered carbon atoms), if methylmalonyl-coenzyme A rather than malonyl-coenzyme A is used in for chain extension; this can occur in bacteria and in animal tissues, especially those of ruminant animals, where polymethyl-branched fatty acids even can be synthesised [275].

The commonest polymethyl-branched fatty acid is probably *phytanic* or 3,7,11,15-tetramethylhexadecanoic acid, which is a metabolite of phytol, and can be found in trace amounts in many animal tissues. It becomes a major component of the plasma lipids in Refsum's syndrome, a rare condition in which there is a deficiency in the enzymatic fatty acid -oxidation system. Lough [562] has reviewed the occurrence and biochemistry of this and other isoprenoid fatty acids. Similar fatty acids are present in the lipids of the preen gland of birds and in those of tubercle bacilli.

The Mycobacteria and certain related species contain a highly distinctive range of very long-chain α-branched β-hydroxy fatty acids, known as the *mycolic* acids, i.e. of the form -

RCH(OH)CH(R')COOH

Different species synthesise mycolic acids with quite characteristic structures and Mycobacteria, for example, produce C_{60} to C_{90} acids with C_{20} to C_{24} α -branches; the Nocardiae synthesise C_{38} to C_{60} fatty acids with C_{10} to C_{16} branches. They may also contain additional carbonyl groups, methyl branches, cyclopropane rings and isolated double bonds [367].

Fatty acids with a cyclopropane ring in the aliphatic chain, such as *lactobacillic* or 11,12-methylene-octadecanoic acid, are found in the lipids of several gram-negative and a few gram-positive bacterial families of the order Eubacteriales.

5. Oxygenated and cyclic fatty acids

In animal tissues, 2-hydroxy fatty acids are frequent components of the sphingolipids and they are also present in skin and wool wax. 4- and 5-Hydroxy fatty acids, which form lactones on hydrolysis, and keto acids are found in cow's milk. As part of the "arachidonic acid cascade", a large number of hydroperoxy, hydroxy and epoxy fatty acids (*eicosanoids*) are formed enzymatically as intermediates in the biosynthesis of prostanoids [713], e.g.

5*c*8*c*11*c*14*c*-20:4 → 11-OOH-5*c*8*c*12*t*14*c*-20:4 → 11-OH-5*c*8*c*12*t*14*c*-20:4 ("11-HETE")

This is a particularly active area of research at present, and it would not be surprising if many more novel structures and new pharmacological activities were revealed.

A large number of hydroxy fatty acids occur in seed oils [71,856], and the best known of these is probably *ricinoleic* or 12-hydroxy-*cis*-9-octadecenoic acid, which is the principle constituent of castor oil. Polyhydroxy fatty acids are present in plant cutins, while *aleuritic* or 9,10,16-trihydroxyhexadecanoic acid is one of the main components of shellac. *Vernolic or* 12,13-epoxy-*cis*-9-octadecenoic acid is one of a number of epoxy fatty acids to have been detected in seed oils.

Fatty acids containing a furanoid ring have been found in the reproductive tissues of fish, especially during starvation, in the simple lipid components, but their function is not known. They are also known to be components of at least one seed oil and of rubber latex.

Cyclopropane fatty acids were mentioned in the previous section. They sometime accompany cyclopropene fatty acids in seed oils of the Malvaceae and Bombacaceae among others. For example, *sterculic* acid is present in very small amounts in cotton seed oil, and if it is not removed during refining it can have a pharmacological effect on the consumer by inhibiting the desaturase enzyme systems. Fatty acids containing a cyclopentene ring are found in seed oils of the Flacourtiaceae, which are used in the treatment of leprosy, although there is no evidence that the acids themselves any have therapeutic value. A fatty acid with a cyclohexane ring has been found in rumen bacteria, and also occurs in the tissues of ruminants.

C. SIMPLE LIPIDS

1. Triacylglycerols and related compounds

Triacylglycerols (commonly termed "triglycerides") consist of a glycerol moiety, each hydroxyl group of which is esterified to a fatty acid. In nature, these compounds are synthesized by enzyme systems, which determine that a centre of asymmetry is created about carbon-2 of the glycerol backbone, and they exist in different enantiomeric forms, i.e. with different fatty acids in each position. A "stereospecific numbering" system has been recommended to describe these forms [415,416]. In a Fischer projection of a natural L-glycerol derivative (Figure 2.3), the secondary hydroxyl group is shown to the left of C-2; the carbon atom above this then becomes C-1 and that

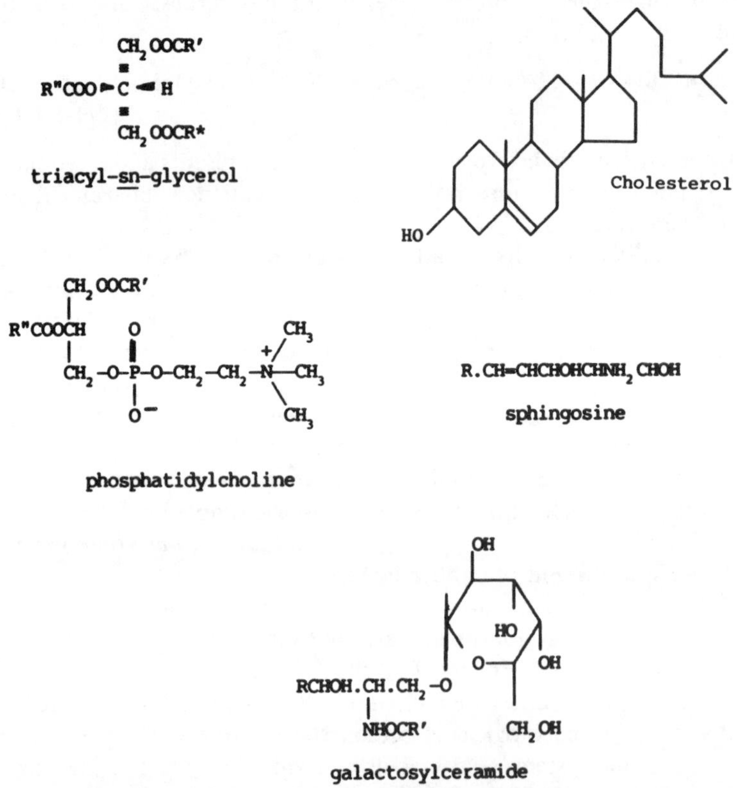

Figure 2.3 The structures of some of the more important lipid classes.

below becomes C-3. The prefix "*sn*" is placed before the stem name of the compound. If the prefix is omitted, then either the stereochemistry is unknown or the compound is racemic. Smith [857] has reviewed glyceride chirality.

Nearly all the commercially important fats and oils of animal and plant origin consist almost exclusively of this simple lipid class. The fatty acid composition can vary enormously. In seed oils, the C_{18} unsaturated fatty acids tend to predominate. In animal fats, especially those of adipose tissue origin, the fatty acid composition reflects that of the diet to some extent, but C_{16} and C_{18} fatty acids are the most abundant components. Fish triacylglycerols and those of marine mammals differ from others in that they contain a high proportion of C_{20} and C_{22} polyunsaturated fatty acids. The compositions of natural oils and fats have been reviewed recently [686].

A full analysis of a triacylglycerol requires that not only the total fatty acid composition be determined but also the distribution of fatty acids in each position. In addition, the proportions of the individual molecular species must be known and gas chromatography, among other techniques, has been used in analyses of this kind.

Some seed and fungal lipids have been found with triacylglycerol components that contain hydroxy fatty acids, the hydroxyl group of which is esterified to an additional fatty acid. These lipids are known as *estolides*.

Diacylglycerols (less accurately termed "diglycerides") and monoacylglycerols ("monoglycerides") contain two moles and one mole of fatty acids per mole of glycerol respectively, and are rarely present at greater than trace levels in fresh animal and plant tissues. Collectively, they are sometimes known as "partial glycerides". 1,2-Diacyl-*sn*-glycerols are important as intermediates in the biosynthesis of triacylglycerols and other lipids. In addition, it has become evident that they are important intra-cellular messengers, generated on hydrolysis of phosphatidylinositol and related compounds by specific enzymes of the phospholipase C type, and that they are involved in the regulation of vital processes in mammalian cells [388]. 2-Monoacyl-*sn*-glycerols are formed as intermediates or end-products of the enzymatic hydrolysis of triacylglycerols.

Acyl migration occurs rapidly with such compounds, especially on heating, in alcoholic solvents or when protonated reagents are present, so special procedures are required for their isolation or analysis if the stereochemistry is to be retained.

2. *Alkyldiacylglycerols and neutral plasmalogens*

Many glycerolipids, including simple lipids, phospholipids and glycolipids, and especially those of animal or microbial origin, contain aliphatic residues linked by an ether or a vinyl ether bond to glycerol. Their occurrence, chemistry and biochemistry have been comprehensively reviewed [586]. Alkyldiacylglycerols are lipids in which a long-chain alkyl moiety is joined via an ether linkage to position 1 of L-glycerol, positions 2 and 3 being esterified with conventional fatty acids. The alkyl groups tend to be saturated or monoenoic of chain-length 16, 18 or 20. On hydrolysis, fatty acids (2 moles) and a glycerol ether (1 mole) are the products. The trivial names *chimyl*, *batyl* and *selachyl* alcohol are used for 1-hexadecyl-, 1-octadecyl- and 1-octadec-9′-enylglycerol respectively. Alkyl ethers are found in small amounts only in most animal tissues, but they can be the major lipid class in the lipids of some marine animals.

The ether linkage in glycerol ethers is stable to both acidic and basic hydrolysis, although the ester bonds are readily hydrolysed as in all glycerolipids.

Neutral plasmalogens are related compounds in which position 1 of L-glycerol is linked by a vinyl ether bond (the double bond is of the *cis*-configuration) to an alkyl moiety. They have been detected in small amounts only in a few animal tissues. Although the vinyl ether linkage is stable to basic hydrolysis conditions, it is disrupted by acid (and by mercury salts) with the formation of a long-chain aldehyde, i.e.

$$RO\text{-}CH{=}CH\text{-}R' \xrightarrow{H_3O^+} ROH + OHC\text{-}CH_2\text{-}R'$$

The principal aldehydes usually are saturated or monoenoic compounds, 16 or 18 carbon atoms in chain-length. Their chemical and physical properties have been reviewed [576,578].

3. Cholesterol and cholesterol esters

Cholesterol is by far the most common member of a group of steroids with a tetracyclic ring system; it has a double bond in one of the rings and one free hydroxyl group (Figure 2.3). It is found both in the free state, where it has a vital role in maintaining membrane fluidity, and in esterified form, i.e. as *cholesterol esters*. The latter are hydrolysed or transesterified much more slowly than most other O-acyl lipids. (The correct generic term is indeed cholester*ol* rather than cholester*yl* esters, but the individual components are designated cholester*yl* palmitate, etc.).

Plant tissues contain related sterols, such as *beta*-sitosterol, ergosterol and stigmasterol, but trace amounts only of cholesterol, and these may also be present in esterified form. Steroid hormones and bile acids are structurally-related compounds, which differ in function from the lipids as defined above, and they are not considered further in this book.

4. Wax esters and other simple lipids

Wax esters in their most abundant form consist of fatty acids esterified to long-chain alcohols with similar aliphatic chains to the acids. They are found in animal, plant and microbial tissues and have a variety of functions, such as acting as energy stores, waterproofing and even echo-location. The fatty acids may be straight-chain saturated or monoenoic with up to 30 carbons, but branched-chain and α- and ω-hydroxy acids are present on occasion; similar features are found in the alcohol moieties.

Waxes in general can contain a wide range of different compounds, including aliphatic diols, free alcohols, hydrocarbons (especially squalene), aldehydes, ketones, hydroxy-ketones, β-diketones and sesqiterpenes. The composition and biochemistry of waxes in nature, and methods for their analysis, have been reviewed in a comprehensive monograph [491].

D. COMPLEX LIPIDS

1. Glycerophospholipids

The structures of a typical glycerophospholipid, i.e. *phosphatidylcholine*, is shown in Figure 2.3. 1,2-Diacyl-*sn*-glycerol-3-phosphorylcholine (commonly termed "lecithin") is usually the most abundant lipid in the membranes of animal tissues, and is often a major lipid component of plant membranes, and sometimes of microorganisms. Together with the other choline-containing phospholipid, sphingomyelin, it comprises much of the lipid in the external monolayer of the plasma membrane of animal cells. It shares with other glycerophospholipids a 1,2-diacyl-*sn*-glycerol backbone, and this part of the molecule can be generated by hydrolysis with phospholipase C and converted to a non-polar derivative for analysis by GLC or other techniques. (Only in the Archaebacteria do the complex lipids have the opposite 2,3-dialkyl-*sn*-glycerol structure [216]). The polar headgroup of all phospholipids prevents direct analysis by means of GLC. While this has disadvantages for certain biochemical applications, analysis via the diacylglycerol derivatives does have the merit that all phospholipids are treated in the same way, regardless of the structure of the parent compound.

In the phospholipids of animals and microorganisms, analogues containing vinyl ether and ether bonds are much more abundant than in the simple lipids. In this instance, it has been suggested that they should be termed "plasmenylcholine" and "plasmanylcholine" respectively. Phospholipid classes isolated by chromatographic means tend to be a mixture of the diacyl, alkylacyl and alkenylacyl forms. To indicate that this is so, they are sometimes termed the "diradyl" form of the appropriate phospholipid. One ether-containing phospholipid, in particular, which is presently being studied intensively because it can exert profound biological effects at minute concentrations, is 1-alkyl-2-acetyl-*sn*-glycerophosphorylcholine or "platelet-activating factor" (often abbreviated to PAF). The chemistry and biochemistry of this compound have been reviewed recently [353,866].

Lysophosphatidylcholine, which contains only one fatty acid moiety in each molecule, generally in position *sn*-1, is sometimes present in tissues also but as a minor component; it is more soluble in water than most other lipids and can be lost during extraction, unless precautions are taken.

Phosphatidic acid or 1,2-diacyl-*sn*-glycerol-3-phosphate is found naturally in trace amounts only in tissues, but it is important metabolically as a precursor of most other glycerolipids. It is strongly acidic and is usually isolated as a mixed salt. As it is somewhat water-soluble, it may be necessary to take special precautions during the extraction of tissues to ensure quantitative recovery.

Phosphatidylglycerol or 1,2-diacyl-*sn*-glycerol-3-phosphoryl-1′-*sn*-glycerol tends to be a trace constituent of tissues, although it does appear to have important functions in lung surfactant and in plant chloroplasts. Diphosphatidylglycerol (or *cardiolipin*) is related structurally to phosphatidylglycerol, and is an abundant constituent of mitochondrial lipids, especially in heart muscle; its occurrence and properties have been reviewed [410]. These lipids also are acidic.

Phosphatidylethanolamine (once trivially termed "cephalin") is frequently the second most abundant phospholipid class in animal and plant tissues, and can be the major lipid class in microorganisms. It often contains a relatively high proportion of the aliphatic moieties as the ether forms. The amine group can be methylated enzymically, as part of a vital cellular process, to yield as intermediates first phosphatidyl-*N*-monomethylethanolamine and then phosphatidyl-*N,N*-dimethylethanolamine; the eventual product is phosphatidylcholine [943]. *N*-Acyl-phosphatidylethanolamine is a minor component of some plant tissues, and is also found in animal tissues under certain conditions. Lysophosphatidylethanolamine contains only one mole of fatty acid per mole of lipid.

Phosphatidylserine is a weakly acidic lipid, so is generally isolated from tissues in salt form. It is present in most tissues of animals and plants and is also found in microorganisms. Its biochemistry has been reviewed [81]. *N*-Acylphosphatidylserine has been detected in certain animal tissues.

Phosphatidylinositol, containing the optically-inactive form of inositol - myoinositol, is a common constituent of animal, plant and microbial lipids. Often in animal tissues, it is accompanied by small amounts of phosphatidylinositol 4-phosphate and phosphatidylinositol 4,5-bisphosphate (polyphosphoinositides). These compounds have a rapid rate of metabolism in animal cells, and with their diacylglycerol metabolites have a major role in regulating vital processes. The topic has been reviewed [388].

Each phospholipid class in a tissue has a distinctive fatty acid composition, probably related to its function but in a way that is still only partly understood. In a tissue such as liver, for example, most of the glycerophospholipids contain substantial proportions of the longer-chain polyunsaturated components. Typically, phosphatidylcholine contains 50% of saturated fatty acids, while arachidonic acid constitutes 20% of the total.Phosphatidylethanolamine has a similar proportion of saturated fatty acids, but somewhat less linoleic acid and correspondingly more of the C_{20} and C_{22} polyunsaturated fatty acids. Characteristically, high proportions of stearic and arachidonic acids are present in the phosphatidylinositol, while the composition of the phosphatidylserine is similar except that the 22:6(n-3) fatty acid substitutes for part of the arachidonic acid. Diphosphatidylglycerol differs markedly from all the other glycerophospholipids in that the single fatty acid, linoleic acid, can comprise nearly 60% of the total. Similar general compositional trends are seen, although the absolute values may differ, in comparing the same lipids in other tissues and, with dietary influences superimposed, in comparing the corresponding lipids of other species.

As with the triacylglycerols, the fatty acids have distinctive positional distributions in phospholipids, with saturated components generally being concentrated in position *sn*-1 and unsaturated in position *sn*-2; they also exist in specific characteristic combinations in molecular species.

2. Glyceroglycolipids

Plant tissues especially tend to contain appreciable amounts of lipids in which 1,2-diacyl-*sn*-glycerols are joined by a glycosidic linkage at position *sn*-3 to a carbohydrate moiety. Their structures and compositions have been reviewed [366,367]. The main components are the mono- and digalactosyldiacylglycerols, but related lipids have been found containing up to four galactose units, or in which one or more of these is replaced by glucose moieties. In addition, a 6-*O*-acyl-monogalactosyldiacylglycerol is occasionally a component of plant tissues. A further unique plant glycolipid is sulphoquinovosyldiacylglycerol or the "plant sulpholipid", and contains a sulphonic acid residue linked by a carbon-sulphur bond to the carbohydrate moiety of a monoglycosyldiacylglycerol; it is found exclusively in the chloroplasts. Usually these lipids contain a high proportion of an 18:3(n-3) fatty acid, sometimes accompanied by 16:3(n-3).

Monogalactosyldiacylglycerols are known to be present in small amounts in brain and nervous tissue in some animal species, and a range of complex glyceroglycolipids have been isolated and characterized from intestinal tract and lung tissue. Such compounds would be destroyed by certain of the methods used in the isolation of glycosphingolipids, with which they frequently co-chromatograph, and may be more wide-spread than is generally thought. A complex sulpholipid, termed "seminolipid", of which the main component is 1-*O*-hexadecyl-2-*O*-hexadecanoyl-3-*O*-(3′-sulpho-β-D-galactopyranosyl-*sn*-glycerol, is the principal glycolipid in testis and sperm. The glycoglycerolipids of animal origin have also been reviewed [850].

Glycolipids, unlike phospholipids, are soluble in acetone and this property can be used in isolating them by chromatographic means.

3. Sphingolipids

Long-chain bases (sphingoids or sphingoid bases) are the characteristic structural unit of the sphingolipids, the chemistry and biochemistry of which have been thoroughly reviewed [460,974,981]. The bases are long-chain (12 to 22 carbon atoms) aliphatic amines, containing two or three hydroxyl groups, and often a distinctive trans- double bond in position 4 (see Figure 2.3). The commonest or most abundant is sphingosine ((2*S*,3*R*,4*E*)-2-amino-4-octadecen-1,3-diol). More than 60 long-chain bases have been found in animals, plants and microorganisms, and many of these may occur in a single tissue, but always as part of a complex lipid as opposed to in the free form. The aliphatic chains can be saturated, monounsaturated and diunsaturated, with double bonds of either the *cis* or *trans* configuration, and they can also have methyl substituents. In addition, saturated and monoenoic straight- and branched-chain trihydroxy bases are found. The commonest long-chain base of plant origin, for example, is phytosphingosine ((2*S*,3*S*,4*R*)-2-amino-octadecanetriol). For shorthand purposes, a nomenclature similar to that for fatty acids can be used; the chain length and number of double bonds are denoted in the same manner with the prefixes "*d*" and "*t*" to designate di- and tri-hydroxy bases respectively. Thus, sphingosine is *d*18:1 and phytosphingosine is *t*18:0.

Ceramides contain fatty acids linked to the amine group of a long-chain base by an amide bond. Generally, they are present at low levels only in tissues, but are important as intermediates in the biosynthesis of the complex sphingolipids. The acyl groups of ceramides are long-chain (up to C_{26} but occasionally longer) odd- and even-numbered saturated or monoenoic fatty acids and related 2-*D*-hydroxy fatty acids. Polyunsaturated fatty acids are rarely present at greater than trace levels. Ceramides are the basic aliphatic building blocks of the sphingolipids, and they can sometimes be generated from sphingolipids by analogous methods to those used for the diacylglycerol moiety of phospholipids. It is this part of the molecule, together with the long-chain base and fatty acid constituents, which are most relevant to a text on gas chromatography.

Sphingomyelin consists of a ceramide unit linked at position 1 to phosphorylcholine, and it is found as a major component of the complex lipids of all animal tissues, but is not present in plants or microorganisms. It resembles phosphatidylcholine in many of its physical properties, and can apparently substitute in part for this in membranes. Sphingosine is usually the most abundant long-chain base constituent, together with sphinganine and C_{20} homologues.

The most widespread glycosphingolipids are the monoglycosylceramides (or cerebrosides), and they consist of the basic ceramide unit linked at position 1 by a glycosidic bond to glucose or galactose (Figure 2.3). They were first found in brain lipids, where the principal form is a monogalactosylceramide, but they are now known to be ubiquitous constituents of animal tissues. In addition, they are found in plants (monoglucosylceramides only), where the main long-chain base is phytosphingosine.

Di-, tri- and tetraglycosylceramides (oligoglycosylceramides) are usually present also in animal tissues. The most common diglycosylceramide is lactosylceramide, and it can be accompanied by related compounds containing further galactose or galactosamine residues, for example. Tri- and tetraglycosylceramides with a terminal galactosamine moiety are sometimes termed "globosides", while glycolipids containing fucose are known as "fucolipids". Oligoglycosylceramides with more than 20 carbohydrate residues have been isolated from animal tissues, those from intestinal cells having been studied with particular intensity. They appear to form part of the immune response system. Although certain of these lipids have been found on occasion to have distinctive long-chain base and fatty acid compositions, the complex glycosyl moiety is considered to be of primary importance for their immunological function and therefore has received most attention from investigators.

Sulphate esters of galactosylceramide and lactosylceramide (often referred to as "sulphatides" or "lipid sulphates"), with the sulphate group linked to position 3 of the galactosyl moiety, are major components of brain lipids and are also found in trace amounts in other tissues; their chemistry and biochemistry have been reviewed [252].

Complex plant sphingolipids, the phytoglycosphingolipids, which contain glucosamine, glucuronic acid and mannose linked to the ceramide via phosphorylinositol, were isolated and characterised from seeds initially, but related compounds are also known to be present in other plant tissues and in fungi.

Gangliosides are highly complex oligoglycosylceramides, which contain one or more sialic acid groups (*N*-acyl, especially acetyl, derivatives of neuraminic acid, abbreviated to "NANA"), in addition to glucose, galactose and galactosamine. They were first found in the ganglion cells of the central nervous system, hence the name, but are now known to be present in most animal tissues. The nature of the long-chain base and fatty acid components of each ganglioside can vary markedly between tissues and species and is related in some way to its function.

E. EXTRACTION OF LIPIDS FROM TISSUES AND OTHER PRELIMINARIES TO ANALYSIS

1. Solvent extraction procedures

Before analysis of lipid samples can be commenced by any chromatographic procedure, it is first necessary to extract them from their tissue matrices in a relatively pure state. This should always be carried out as soon as possible after the removal of the tissue from the living organism. The author has dealt with these aspects in some detail in previous texts [163,168],so essential details only are covered here.

Lipids occur in tissues in a variety of physical forms. The simple lipids are often part of large aggregates in storage tissues, from which they are extracted with relative ease. On the other hand, the complex lipids are usually constituents of membranes, where they occur in a close association with such compounds as proteins and polysaccharides, with which they interact, and they are not extracted so readily. Generally, lipids are linked to other cellular components by weak hydrophobic or Van der Waals forces, by hydrogen bonds and by ionic bonds. For example, the hydrophobic aliphatic moieties of lipids interact with the non-polar regions of the amino acid constituents of proteins, such as valine, leucine and isoleucine, to form weak associations. Hydroxyl, carboxyl and amino groups in lipid molecules, in contrast, can interact more strongly with biopolymers via hydrogen bonds.

In order to extract lipids from tissues, it is necessary to find solvents which will not only dissolve the lipids readily but will overcome the interactions between the lipids and the tissue matrix. Various solvents or solvent combinations have been suggested as extractants for lipids, and currently some interest is being shown in isopropanol-hexane (2:3 by volume), because its toxicity is relatively low [355,743], but it does not yet appear to have been tested with a sufficiently wide range of tissues. It does not extract gangliosides quantitatively. Most lipid analysts use chloroform-methanol (2:1 by volume), with the endogenous water in the tissue as a ternary component of the system, to extract lipids from animal, plant and bacterial tissues. Usually, the tissue is homogenized in the presence of both solvents, but better results may be obtained if the tissue is first extracted with methanol alone before the chloroform is added to the mixture. With difficult samples, more than one extraction may be needed, and with lyophilised tissues, it may be necessary to rehydrate prior to carrying out the extraction. The homogenization and extraction should be performed in a Waring blender, or better in equipment in which the drive to the blades is from above, so that the solvent does not come into contact with any lubricated bearings. Generally, there is no need to heat the solvent to facilitate the extraction, although there may be times when this is necessary.

The extractability of tissues and of particular lipids is variable, and there are many instances when alternative or modified procedures must be used. Butanol saturated with water appears to be the most useful solvent mixture to disrupt the inclusion complexes of lipids in starch and gives the best recoveries of lipids from cereals [188,606,627]. This solvent combination has also been recommended for the quantitative recovery of lysophospholipids, which are more soluble in water than are many other common phospholipids [103].

Plant tissues should be pre-extracted with isopropanol to minimize artefactual degradation of lipids by tissue enzymes.

Lipid extracts from tissues, obtained in the above manner, tend to contain appreciable amounts of non-lipid contaminants, such as sugars, amino acids, urea and salts. These must be removed before the lipids are analysed. Most workers use a simple washing procedure, devised originally by Folch, Lees and Stanley [259], in which a chloroform-methanol (2:1, v/v) extract is shaken and equilibrated with one fourth its volume of a saline solution (i.e. 0.88% potassium chloride in water). The mixture partitions into two layers, of which the lower phase is composed of chloroform-methanol-water in the proportions 86:14:1 (by volume) and contains virtually all of the lipids, while the upper phase consists of the same solvents in the proportions of 3:48:47 respectively, and contains much of the non-lipid contaminants. It is important that the proportions of chloroform, methanol and water in the combined phases should be as close as possible to 8:4:3 (by volume), otherwise selective losses of lipids may occur. If a second wash of the lower phase is needed to remove any remaining contaminants, a mixture of roughly similar composition to that of the upper phase should be used, i.e. methanol-saline solution (1:1, v/v).

Any gangliosides present in the sample partition into the upper layer, together with varying amounts of oligoglycosphingolipids. They can be recovered from this layer by dialysing out most of the impurities of low molecular weight, and then lyophilising the residue [459].

Many modifications of the basic extraction procedure have been devised for use in particular circumstances, and the analyst must decide what he requires of a method. One which extracts all of the more minor lipid classes exhaustively is obviously desirable for many applications, but may be too tedious and time-consuming for routine use. On the other hand, a method which is suited to the quantitative extraction of the main lipid classes in large numbers of samples in a routine manner by relatively inexperienced staff, may not give complete recoveries of certain trace components of biological importance. The modified "Folch" procedure [967], which follows, probably falls somewhere between these extremes.

> "The tissue (1 g) is homogenized with methanol (10 ml) for 1 minute in a blender, then chloroform (20 ml) is added and the homogenization is continued for 2 minutes more. The mixture is filtered, when the solid remaining is resuspended in chloroform-methanol (2:1 by volume, 30 ml) and homogenized for 3 minutes. It is filtered again and re-washed with fresh solvent. The combined filtrates are transferred to a measuring cylinder, one fourth of the total volume of 0.88 % potassium chloride in water is added, and the mixture is shaken thoroughly before being allowed to settle. The aqueous

(upper) layer is drawn off by aspiration, one fourth the volume of the lower layer of methanol-saline solution (1:1, v/v) is added and the washing procedure is repeated. The bottom layer containing the purified lipid is filtered before the solvent is removed on a rotary film evaporator. The lipid is stored in a small volume of chloroform at -20°C, until it can be analysed."

With plant tissues as cautioned above, it is necessary to extract first with isopropanol, in order to deactivate the enzymes, as follows [664,665].

"The plant tissues are homogenized with a 100 fold excess (by weight) of isopropanol. The mixture is filtered, the residue is re-extracted with fresh isopropanol, and finally is shaken overnight with isopropanol-chloroform (1:1, v/v). The filtrates are then combined, most of the solvent is removed on a rotary evaporator, and the lipid residue is taken up in chloroform-methanol (2:1, v/v) and is given a "Folch" wash as above."

In any extraction procedure, it is important that the weight of fresh tissue extracted is recorded, together with the weight of lipid obtained from it. For some purposes, it may be desirable to determine the amount of dry matter in the tissue, so that the weight of lipid relative to that of dry matter can be calculated.

2. *Minimising autoxidation*

If they are not protected, polyunsaturated fatty acids will autoxidise very rapidly in air, and it may not be possible to obtain an accurate analysis by chromatographic means. The mechanism of autoxidation involves attack by free radicals and is exacerbated by strong light and by metal ions. Once it has been initiated, the reaction procedes autocatalytically. Linoleic acid is autoxidised twenty times as rapidly as oleic acid, and each additional double bond in a fatty acid can increase the rate of destruction by two to three fold. Natural tissue antioxidants, such as the tocopherols, afford some protection to lipid extracts, but it is usually advisable to add further synthetic antioxidants, such as BHT ("butylated hydroxy toluene" or 2,6-di-*tert*-butyl-*p*-cresol) to solvents at a level of 10 to 100 mg/litre, depending on lipid concentration. This compound need not interfere with chromatography, as it is relatively volatile and can be removed (sometimes inadvertently!) together with solvents when they are evaporated in a stream of nitrogen; it is also rather non-polar and tends to elute at the solvent front, ahead of most lipids, in many liquid chromatography systems. In contrast, excessive amounts of added antioxidants can sometimes act as prooxidants!

Wherever possible, lipids should be handled in an atmosphere of nitrogen.

On the other hand, it is rarely necessary to go to the length of constructing a special nitrogen box to contain all the equipment used in the handling of lipids. Usually it is sufficient to ensure that nitrogen lines are freely available, so that the air can be flushed out of glass containers or reaction vessels.

When it is necessary to concentrate lipid extracts, large volumes of solvents are best removed by means of a rotary film evaporator at a temperature, which in general should not exceed about 40°C. The flask containing the sample should not be too large, otherwise the lipid can spread out over a large area of glass and so be more accessible to oxygen. At the start of evaporation, it may be advisable to flush out the equipment with nitrogen, but the solvent vapours eventually will displace any air. Small volumes of solvent can be evaporated by carefully directing a stream of nitrogen onto the surface of the solvent. This should not be done too vigorously or at too high a temperature, since the more volatile fatty acid derivatives, including methyl esters of fatty acids, may also be lost by evaporation or by physical transport as an aerosol.

PLEASE NOTE!

As constant repetition is tedious, it will be assumed in all the subsequent discussion of methodology in this book that precautions will be taken at all times to minimise autoxidation.

3. Hazards

It is inevitable that lipid analysts will make use of large volumes of solvents in their work, whether in extracting lipids from tissues, in the chromatographic separation of lipid classes or in the preparation of derivatives for chromatography. Most solvents exhibit some degree of toxicity if inhaled in large amounts. Benzene, in particular, is frequently mentioned in the older literature as a solvent with valuable properties in the analysis of lipids, but it is now known to be extremely toxic and is best avoided entirely; toluene has comparable chromatographic properties and is much less hazardous. Similarly, it was once thought that chloroform was relatively safe, but it is now known that there are real hazards. Acetonitrile also has toxic properties, and indeed analysts should not view any solvent with complacency.

All solvents should be used with care in well-ventilated areas, or in fume cupboards if at all possible. Solvents should never be evaporated or distilled on an open bench. Some operations generate more vapour than others, and filtration is probably the procedure which produces most. When not in use, all solvents should be stored in well-stoppered bottles, made of dark glass, and in flame-proof cabinets. No more solvent than is required for immediate needs should be stored in the laboratory. In general, care should be taken to prevent spillages, to keep storage vessels closed, and generally to minimise any exposure of the laboratory personnel to solvent vapours.

Nor should it be forgotten that many solvents, especially low molecular weight hydrocarbons, ethers and alcohols, are highly inflammable. Ethers develop peroxides on storage, especially in bright light, and many explosions have resulted as these were concentrated when large volumes of ethers were distilled. It may seem self evident, but there should be no naked flames in laboratories in which solvents are handled. All electrical equipment, especially that in chromatographic systems, should be correctly wired and earthed to reduce the risk of sparks.

The concentrations of chloroform vapour permitted in the laboratory atmosphere have steadily been reduced in recent years as regulations have been revised. Unfortunately, there appears to be no proven substitute in many of the problems that face lipid analysts. As supplied, chloroform usually contains 0.25 to 2% of added ethanol, which acts as a stabiliser, but also has a considerable effect on the chromatographic properties of the solvent. It can be removed, if there is a need, by a simple washing procedure [163], but photochemical formation of the highly poisonous substance, phosgene, can then occur and the destabilised solvent should not be stored for any length of time. Chloroform-methanol mixtures are powerful irritants when they come in contact with the skin.

Many other reagents to be found in laboratories are known to have toxic properties, some of which may take some time to manifest themselves, and the catalogues of suppliers are often informative on the subject as should be the labels on containers. The toxicity of numerous other reagents has yet to be investigated, and it is best to err on the safe side and assume that there is some unknown hazard associated with all chemicals. They should then be handled accordingly. Similarly, the dangers associated with strong mineral acids should be well-known to analysts. They should not be stored in the same cupboard as solvents.

F. SEPARATION OF LIPID CLASSES

Gas chromatography is of most value for the separation of the aliphatic components of lipids, and some prefractionation into simpler groups or individual classes, generally by adsorption chromatography (HPLC or TLC), is often desirable. Again, the author has described these in great detail elsewhere [163,168], and the topic has been dealt with by Kuksis and Myher [507,516,517], but the brief summary which follows may be of value to some readers.

Small-scale procedures only are described as they generally give sufficient material for GC analysis.

1. Simple group separations

The complexity of natural lipid extracts is such that it is rarely possible to claim that all the lipid classes of a sample can be separated in one operation. It is, therefore, often worthwhile to be able to isolate distinct simple lipid, phospholipid or glycolipid fractions for further analysis. For example, it is frequently easier technically to isolate small amounts of pure lipid classes preparatively by means of HPLC (or other methods), after a preliminary fractionation has been carried out. Unfortunately, no procedure appears yet to have been described that is satisfactory in all respects, although some useful methods are available provided that their limitations are recognised.

The simplest small-scale procedure for isolating groups of lipids consists of making a short column of silica gel (about 1g), in a glass disposable Pasteur pipette say, and applying about 30 mg of lipid to this. Elution with chloroform or diethyl ether (10 ml) yields the simple lipids, acetone (10 ml) gives a glycolipid fraction, and methanol (10 ml) yields the phospholipids. Different brands or batches of silica gel tend to vary somewhat in their properties and some cross-contamination of fractions may be found. For example, the acetone fraction may contain some of the less polar phospholipids, especially phosphatidic acid and diphosphatidylglycerol but occasionally phosphatidylethanolamine even; if this is observed to occur, it can be minimised by adding some chloroform to the acetone prior to elution. Indeed for many purposes, there may be no need to include an elution step with acetone, as some tissues contain negligible amounts of glycolipids. On the other hand, it is possible to insert an additional elution step with methyl formate before the acetone wash, to obtain a fraction that contains most of the prostaglandins in the extract (together with some of the glycolipids) [799].

It may now be more convenient to use small proprietary pre-packed cartridges of silica gel or ion-exchange media for these small-scale group separations [101,102,453,486,777,1007]. In one application, milk fat samples (100mg), high proportions of which consist of triacylglycerols, were applied to Sep-Pak™ cartridges of silica gel (Waters Associates, Milford, U.S.A.); non-polar lipids were recovered by elution with hexane-diethyl ether (1:1, v/v; 40 ml), while the complex lipids were recovered by elution first with methanol (20 ml), and then with chloroform-methanol-water (3:5:2 by volume, 20 ml) [102]. In other work with these cartridges, chloroform (40 ml) was used to elute the simple lipids, acetone-methanol (9:1, v/v; 160 ml) gave the neutral glycosphingolipids, and chloroform-methanol (1:1, v/v; 80 ml) eluted the phospholipids [1007]. One further procedure, which appears to have much to commend it, is for the isolation of glycolipids and consists in using boric acid bound to a polymeric matrix in a column; carbohydrate moieties of glycolipids are retained much more strongly than are other lipids so facilitating separation [496].

2. High-performance liquid chromatography

Lipid analysts were relatively slow to adapt HPLC to the separation of lipid classes, largely because of limitations in the availability of a suitable detector. In spite of this, some excellent separations have now been achieved [168], and the technique is rapidly supplanting TLC in many laboratories. In comparison to the latter, it offers superior resolution, easier quantification together with a degree of automation, cleaner fractions, and a more hygienic working environment.

In discussing lipid class separations by means of HPLC, it is not possible to use a "recipe" treatment, as the approach will depend largely on the nature of the detection system available to the analyst. For example, this determines the nature of the solvents used in the mobile phase and whether gradient elution is possible. Some relevant separations are therefore described below in terms of specific detectors, as examples of what is possible.

Most lipids lack chromophores of value in spectrophotometric detection, but the absorbance of isolated double bonds (and some other functional groups) at about 205 nm in the UV range can be used successfully if care is taken in the choice of the solvents for the mobile phase. UV detection at low wavelengths has its limitations, however. For example, only a few solvents are transparent and can be used in the appropriate range (e.g. hexane, methanol, acetonitrile, isopropanol and water), and the molar extinction coefficient is so low that traces of impurities in the solvents or in the samples (e.g. hydroperoxides) can swamp the signals from the compounds of interest. It is possible to convert lipids or their component parts into derivatives with a high UV absorptivity for HPLC separation in some circumstances, and procedures of this kind are generally favoured for the separation of individual glycolipids.

One of the most popular approaches to the separation of phospholipid classes consists in the use of UV-transparent mobile phases, such as hexane-isopropanol-water or acetonitrile-methanol-water mixtures, with detection at about 200 nm. Strong acids have been used incorporated into the mobile phase in some laboratories as ion suppressants, but this can lead to degradation of plasmalogens and superior alternatives are available. Of the large number of procedures of this kind described, that of Patton and co-workers [694] appears particularly convincing, and has been adopted by a number of others. It has the additional merit of employing an isocratic elution scheme, so reducing the requirements in terms of costly equipment. Hexane-isopropanol-25 mM phosphate buffer-ethanol-acetic acid (367:490:62:100:0.6 by volume) (see the original paper for the method of mixing) is the mobile phase, at a flow-rate of 0.5 ml/min for the first hour when it is increased to 1 ml/min. The column (4.6 x 250 mm) used originally contained LiChrospher™ Si-100 silica gel, and detection was at 205 nm. Figure 2.4 illustrates the nature of the separation achieved with a rat liver extract.

Phosphatidylethanolamine eluted just after the neutral lipids, and was followed by each of the acidic lipids, i.e. phosphatidic acid, phosphatidylinositol and phosphatidylserine, then by diphosphatidylglycerol and by the individual choline-containing phospholipids. Only the phosphatidylcholine and sphingomyelin overlapped slightly. As each component was eluted, it was collected, washed to remove the buffer, and determined by phosphorus analysis. In addition, the fatty acid composition of each lipid class was obtained with relative ease, by GLC analysis after transmethylation of the fractions.

One of several bonded stationary phases to have been used in phospholipid

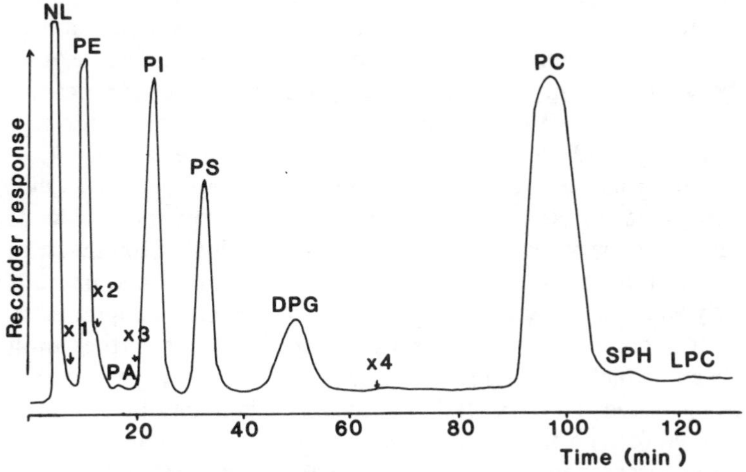

Figure 2.4 Isocratic elution of rat liver phospholipids from a column of silica gel with hexane-isopropanol-25 mM phosphate buffer-ethanol-acetic acid (367:490:62:100:0.6 by volume) as mobile phase at a flow-rate of 0.5 ml/min for the first 60 minutes then of 1 ml/min, and with spectrophotometric detection at 205 nm [694]. (Reproduced by kind permission of the authors and of the *Journal of Lipid Research*, and redrawn from the original publication). Abbreviations; NL, neutral lipids; PE, phosphatidylethanolamine; PA, phosphatidic acid; PI, phosphatidylinositol; PS, phosphatidylserine; DPG, diphosphatidylglycerol; PC, phosphatidylcholine; SPH, sphingomyelin; LPC, lyso- phosphatidylcholine; X1, X2, X3 and X4, unidentified lipids.

separations with isocratic elution had a benzene sulphonate residue as the functional group. A column (4.6 x 250 mm) of Partisil™-SCX and elution with acetonitrile-methanol-water (400:100:34 by volume) at a flow-rate of 2.5 ml/min were used to effect separation of the main ethanolamine- and choline-containing phospholipids of animal tissues [199,316]. While spectrophotometry at 203 nm was used to detect the components, phosphorus assay was preferred for quantification purposes.

Refractive index detectors also have several applications in lipid analysis. They are "universal" detectors, but lack sensitivity, require isocratic elution conditions and are sensitive to minor fluctuations in temperature. Their main value is probably in small-scale preparative applications, say with 1-2 mg of a lipid extract. For example, a refractive index detector was utilized with a column (4.6 x 250 mm) of Ultrasil™ Si (5 micron silica gel) and isocratic elution with isooctane-tetrahydrofuran-formic acid (90:10:0.5 by volume) to separate most of the common simple lipid classes encountered in animal tissue extracts, such as those of liver [304]. Cholesterol esters, triacylglycerols and cholesterol were each resolved and gave symmetrical peaks.

Transport-flame ionisation detectors probably represent the future in HPLC analysis. With such systems, the eluent is entrained on a moving belt and carried through an oven where the solvent is removed, and then in essence into a flame ionization detector, where the separated components are combusted and detected as in a gas chromatograph. A wide range of solvents can be used, and the response is highly rectilinear with respect to mass. Unfortunately, the existing commercial instruments are rather too costly for most analysts, and their reliability has still to be ascertained.

The author has been using the "mass detector", also known as the "light scattering detector" or "evaporative analyser", for some years now. It is an optical device, marketed by Applied Chromatography Systems (Macclesfield, UK). With this detector, the eluent from the column passes into a heated chimney where the solvent is evaporated in a stream of compressed air; the solute does not evaporate, if it is a lipid, and passes as a "fog" through a light beam, which is reflected and refracted. The amount of scattered light can be measured and bears a relationship to the amount of material eluting. It can therefore be termed a universal detector as it is not dependent on particular chromophores, and it can be used with gradients and a wide variety of different mobile phases. It is relatively inexpensive and rugged, but has limitations in quantitative analysis. Although the sample is lost during detection, it is possible to insert a stream-splitter between the end of the HPLC column and the detector so that a high proportion of the eluent is diverted to a fraction collector.

In order to separate and quantify the more abundant lipid classes in animal tissues, ideally on the 0.2 to 0.4 mg scale and in as short a time as could conveniently be managed, the author made use of the ACS mass detector with a ternary solvent delivery system and a short (5 x 100 mm) column, packed with Spherisorb™ silica gel (3 micron particles) [165]. In selecting a mobile phase, the choice of solvents was constrained by the need for sufficient volatility for evaporation in the detector under conditions that did not cause evaporation of the solute, and by the necessity to avoid inorganic ions, which would not evaporate. Similar restrictions apply to detectors operating on the transport-flame ionization principle. It was necessary to use a complicated ternary-gradient elution system with eight programmed steps, starting with isooctane to separate the lipids of low polarity and ending with a solvent containing water to elute the phospholipids; a solvent of medium polarity was then needed to mediate the transfer from one extreme to the other, and mixtures based on isopropanol gave satisfactory results. The three solvent mixtures selected by trial and error were isooctane-tetrahydrofuran (99:1, v/v)(A), isopropanol-chloroform (4:1, v/v)(B) and isopropanol-water (1:1, v/v)(C). In essence, a gradient of B into A was created to separate each of the simple lipids, then a gradient of C into A plus B was produced to separate each of the complex lipids; finally, a gradient in the reverse direction was generated to remove most of the bound water and to re-equilibrate the column prior to the next analysis. A relatively high flow-rate (2 ml/min) appeared to assist the separation greatly, perhaps compensating for the absence of strong acid or inorganic ions, which others have found necessary for the separation of phospholipids.

In later work [166], it was observed that much better resolution of the minor acidic components was obtained by adding small amounts of organic ions to the aqueous component of the eluent. The life time of the column was also greatly extended by this simple step. In practice, the optimum results were obtained with 0.5 to 1 mM serine buffered to pH 7.5 with triethylamine. In addition, hexane replaced isooctane in the mobile phase, in order to reduce the maximum operating pressure required.

The nature of the separation achieved with a lipid extract from rat kidney is shown in Figure 2.5 [166]. In spite of the abrupt changes in solvent composition at various points, little base-line disturbance is apparent, and each of the main simple lipid and phospholipid classes is clearly resolved in only 20 minutes. Only the highly acidic lipids, phosphatidic acid and to a lesser extent phosphatidylserine, do not give satisfactory peaks. There is no "solvent peak" at the start of the analysis, as is often seen with other detectors, and BHT added as an antioxidant evaporates with the solvent so does not interfere. After a further 10 minutes of elution to regenerate the column, the next sample can be analysed.

There are of course many other applications of HPLC in the separation of lipids that complement or otherwise assist GC analyses, and some of these are discussed in later chapters and elsewhere [168].

3. *Thin-layer chromatography*

TLC has been much used by lipid analysts over the last 30 years or so, and it has served them well. The equipment required is inexpensive and flexible, in that in can be used both analytically and preparatively with many different types of layers. While HPLC is supplanting it in several areas, it is likely to continue to be of value in many circumstances.

TLC procedures with silica gel G layers (containing calcium sulphate as binder) have been employed most frequently for lipid class separations. Commonly, the solvent elution system used is hexane-diethyl ether-formic acid (80:20:2 by volume), and this gives the separations shown in Figure 2.6.

Cholesterol esters migrate to the solvent front, and they are followed by

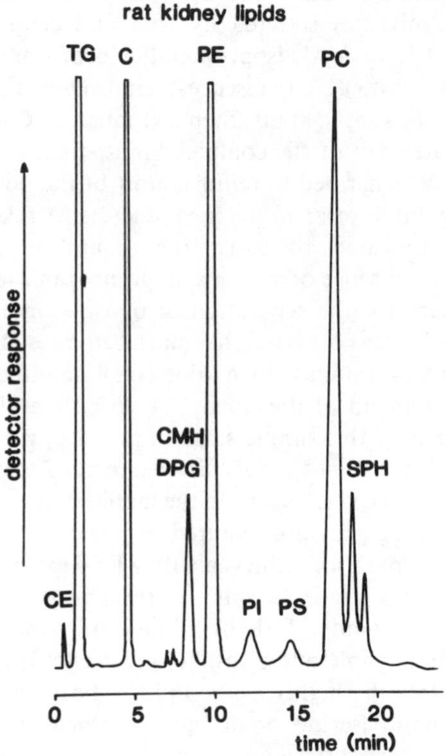

Figure 2.5 Separation of rat kidney lipids (0.35 mg) by HPLC on a column (5 x 100 mm) of Spherisorb™ silica gel (3 micron particles) with mass detection; the elution conditions are described in the text [166]. (Reproduced by kind permission of the *Journal of Chromatography*). The legend to Fig. 2.4 contains a list of abbreviations (or see Chapter 1). In addition: CE, cholesterol esters; TG, triacylglycerols; C, cholesterol; DG, diacylglycerols; MG, monoacylglycerols; FFA, free fatty acids.

triacylglycerols, free fatty acids, cholesterol, diacylglycerols, monoacylglycerols and phospholipids (with other polar lipids). For small-scale preparative purposes (2 to 20 mg), the author prefers glass plates (20 x 20 cm) coated with a layer 0.5 mm thick of silica gel G. Bands are then conveniently detected by spraying with an 0.1% (w/v) solution of 2′,7′-dichlorofluorescein in 95% methanol and viewing the dried plate under ultraviolet light; the lipids appear as yellow spots against a dark background, and they can be recovered from the adsorbent by elution with solvents for further analysis. Diethyl ether or chloroform will elute simple lipids quantitatively, while chloroform-methanol-water (5:5:1, by volume) is required for phospholipids). It is possible to add an internal standard, such as the methyl ester of an odd-chain fatty acid not present naturally in the

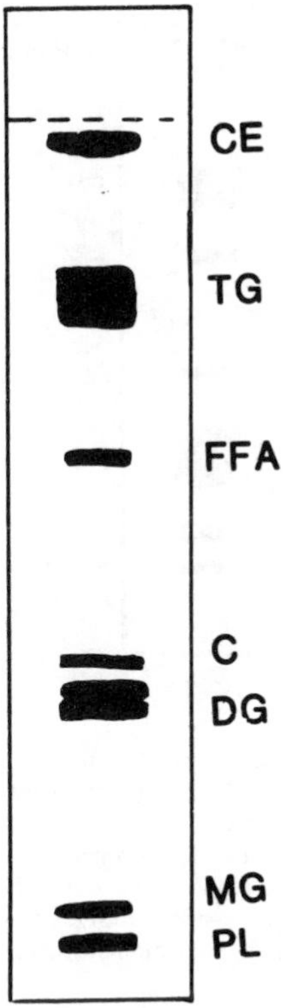

Figure 2.6 Schematic TLC separation of simple lipids on a silica gel G layer. Hexane-diethyl ether-formic acid (80:20:2 by volume) was the developing solvent. The legends to Figs 2.4 and 2.5 contain a list of abbreviations (or see Chapter 1).

sample, to transesterify, and to determine both the fatty acid composition and the amount of lipid (relative to the standard) simultaneously by GLC analysis [182]. Suitable procedures are discussed in subsequent chapters.

One-dimensional TLC procedures can be recommended for the separation of natural mixtures of phospholipids with relatively simple compositions for the most abundant components, for rapid group separations, and for small-scale preparative purposes. When acidic phospholipids are present in a sample at low levels only, the common components may be separated on layers of silica gel G, by using chloroform-methanol-water (25:10:1 by volume) as the mobile phase for development. The nature of the separation is shown in Figure 2.7 (plate A). The minor acidic lipids, such as diphosphatidylglycerol,

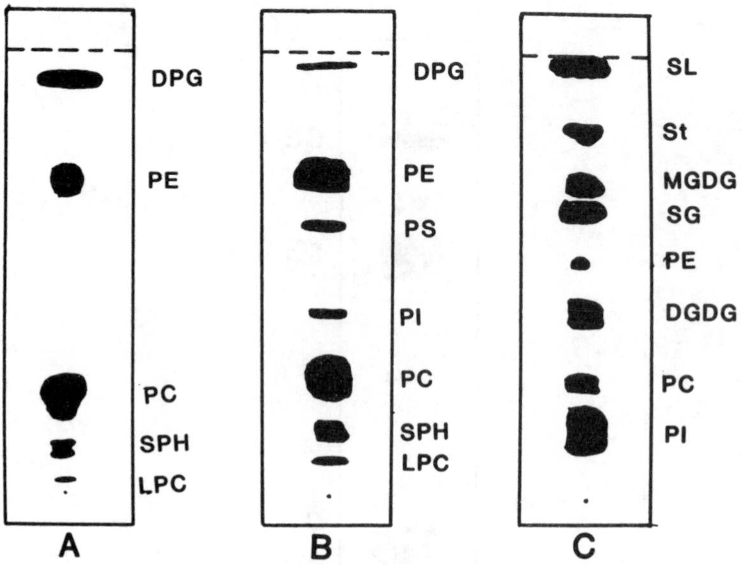

Figure 2.7 Schematic TLC separations of phospholipids. Plate A, silica gel G layer and development with chloroform-methanol-water (25:10:1 by volume); Plate B, silica gel H layer and development with chloroform-methanol-acetic acid-water (25:15:4:2 by volume) [847]; Plate C, plant complex lipids on silica gel G and development with diisobutyl ketone-acetic acid-water (40:25:3.7 by volume) [664]. The legends to Figs 2.4 and 2.5 contain a list of abbreviations (or see Chapter 1); St, sterols; SG, sterol glycosides; MGDG, monogalactosyldiacylglycerols; DGDG, digalactosyldiacylglycerols.

migrate ahead of phosphatidylethanolamine and the choline-containing phospholipids; phosphatidylserine tends to elute with phosphatidylethanolamine while phosphatidylinositol co-chromatographs with phosphatidylcholine.

In most circumstances, it is preferable to employ layers of silica gel H (without a binder) as sharper separations of most of the individual phospholipid classes, but especially of the minor acidic components, are obtained in the absence of metal ions. Unfortunately, silica gel H layers tend to be more fragile than those prepared from silica gel G, and different brands can vary greatly in their elution characteristics. The most popular separation system of this kind (especially for the lipids from animal tissues) consists of a layer of silica gel H, made in a slurry with 1 mM sodium carbonate solution to render it slightly basic, and developed in chloroform-methanol-water-acetic acid (25:15:4:2 by volume); the separation is shown in Figure 2.7 (plate B) [847]. Diphosphatidylglycerol and phosphatidic acid move to the solvent front, and phosphatidylserine and phosphatidylinositol migrate between the most abundant phospholipid constituents. If simple lipids are present in the sample, they can be run to the top of the plate by elution with acetone-hexane (1:3, v/v), prior to separation of the phospholipids [846]. Many modifications of this system have been described, often to compensate for local conditions of temperature and humidity, or for changes in the properties of particular makes of silica gel. Problems are most often manifested in the separation of phosphatidylserine and phosphatidylinositol from the other lipids, and some commercial brands of silica gel H appear to be better than others for the purpose. Other workers have reported improved separations by incorporating either ammonium sulphate [471] or EDTA [40] into the silica gel H layer.

One recently described procedure, which does appear to afford distinctive separations, makes use of silica gel layers containing boric acid, which presumably forms complexes with hydroxyl groups such as those in phosphatidylinositol, retarding the rate of migration of this lipid class; the order of elution is thus very different from those obtained with other systems, and both phosphatidylserine and phosphatidylinositol especially are separated with relative ease from the other phospholipids [256].

One-dimensional TLC systems have been used less often with plant lipid extracts, as glycolipids tend to co-chromatograph with phospholipids when many of the common elution systems are used. Nonetheless, some valuable separations have been described [627,664,722] and one is illustrated in Figure 2.7 (plate C). Mono- and digalactosyldiacylglycerols each elute as distinct bands.

In most instances, many more distinct phospholipid classes can be separated by two- than by one-dimensional TLC procedures, and this is of special value with some of the biologically-important phospholipids, which are often present in tissues at rather low levels. As innumerable examples of such separations have been published, the reader is referred to the author's previous book for further details [163].

4. Ion-exchange chromatography

Column chromatography on diethylaminoethyl(DEAE)-cellulose is a valuable but somewhat under-used method for the isolation of particular groups of complex lipids in comparatively large amounts [163,778]. The principle of the separation process is complex, involving partly ionic interactions between the packing material and the polar head groups of complex lipids, and partly adsorption effects with the polar regions of the molecules. In practice as a rough guide, about 300 mg of complex lipids can be applied to a 30 x 2.5 cm column to yield fractions with distinctive compositions and little cross-contamination. Although it has been used mainly on this scale, it is also possible to use the technique with much smaller columns and proportionately less lipid, e.g. for the isolation of plant lipids [777].

In the typical elution conditions, chloroform is employed first to obtain the simple lipids. All the choline-containing phospholipids are eluted with a chloroform-methanol mixture of relatively low polarity, while a much higher proportion of methanol is required to recover the ethanolamine-containing phospholipids; phosphatidylserine is eluted with glacial acetic acid, and a solvent of high ionic strength is required to recover the more acidic phospholipids (the salts can be removed later by a "Folch" washing step). Further separation of the individual components of particular fractions can later be achieved by means of HPLC or TLC, more easily than with the unfractionated extract. With plant lipid extracts, monogalactosyldiacylglycerols tend to elute with the chloroform fraction, but digalactosyldiacylglycerols can be recovered on their own if care is taken [666]. At the end of the analysis, it is an easy matter to regenerate the column for re-use, though it is important not to move too abruptly from solvents of high to those of low polarity.

An ion-exchange material consisting of quaternary triethylammonium (QAE) groups covalently-bound to controlled-pore glass (Glycophase^{TM}) appears to afford similar separations to those obtained with DEAE-cellulose, but the former has better packing and compressibility properties in low-pressure column chromatography applications, and will probably be more widely used in the future [76].

CHAPTER 3

GAS CHROMATOGRAPHY: THEORETICAL ASPECTS AND INSTRUMENTATION

A. INTRODUCTION

The technique of *gas chromatography* (GC) or *gas-liquid chromatography* (GLC) is a form of partition chromatography in which the mobile phase is a gas and the stationary phase is a liquid. A sample is injected into the gas phase where it is volatilised and passed onto the liquid phase, which is held in some form in a column; components spend different times in the mobile phase and the stationary phase, depending on their relative affinities for the latter, and emerge from the end of the column exhibiting peaks of concentration, ideally with a Gaussian distribution. These peaks are detected by some means which converts the concentration of the component in the gas phase into an electrical signal, which is amplified and passed to a continuous recorder, and perhaps to an integrator, so that the progress of the separation can be monitored and quantified.

Since the technique was first described in the 1950s (see Chapter 1), it has been subjected to continued development in every single aspect and indeed progress is still being made. In the second edition of the author's book "Lipid Analysis" [163], which was published as recently as 1982, only two pages were devoted to GC with capillary or open-tubular columns. This was not because the potential of these was not readily apparent or that little work had been done with them, but because they had disadvantages of high cost coupled to a limited working life; in addition, there tended to be substantial losses of polyunsaturated components on the stainless steel walls of the only columns available commercially at that time. The development of first glass and then fused silica columns has entirely removed all these disadvantages, and there would now be few lipid analysts who would consider purchasing a new gas chromatograph without facilities for capillary columns. As this new methodology has been introduced so rapidly, there must be many lipid analysts who are not fully conversant with the pleasures (and also the pitfalls) of the technique. The major emphasis in this book, must therefore be on GC with capillary columns, although this does not mean that GC with packed columns can now be neglected. Rather, it is hoped that this book will stimulate interest in the newer advances and lead to a further increase in our knowledge of lipids.

A large number of books have been published on the theory and practice of chromatography, and the author has found those cited to be of particular value [267,314,427,444,728,791]. In addition, there have been a number of review articles or books on the application of this theory to specific lipid problems, and these are listed in the appropriate Chapters below.

B. THEORY

1. Some basic considerations

Most analysts appear to manage perfectly well with only a smattering of knowledge of the theory of gas chromatography. On the other hand, some understanding of the concepts involved can be beneficial in many circumstances, if only in allowing him/her to cast a more sceptical eye over the claims of competing column manufacturers.

A basic gas chromatograph has three essential components, i.e. some form of inlet through which the sample is introduced onto the column, the column itself which contains the stationary (liquid) phase and through which the mobile (gas) phase is passed continuously, and a detector. In this Section in which the theory of the separation process is described, the column is the main component of interest, and there are three main types. The work horse of gas chromatography until relatively recent times was a packed column, consisting of a glass or metal tube, generally 2 or 4 mm in internal diameter and 1.5 to 2.5 m in length, coiled to fit the oven unit and filled with an inert solid support coated with the liquid phase. Support-coated open-tubular (SCOT) columns enjoyed a brief vogue, and are constructed of narrow bore tubing (0.5 to 1.25 mm i.d. in lengths of 10 to 15 m) and contain a finely powdered solid support coated with a liquid phase. Finally, wall-coated open-tubular (WCOT) columns consist of narrow bore tubing (0.1 to 0.3 mm i.d. and 25 or 50 m in length commonly), of glass or increasingly of fused silica, the inner wall of which is coated with the liquid phase. A conventional packed column is simple to construct and pack with the liquid phase on its particulate support, it is rugged in use and can take large sample sizes; however, resolution is limited. WCOT columns afford superb resolution, but the size of the sample and the method of its introduction are critical.

The function of the column is to allow partitioning of the constituents of the sample to be separated between the stationary and mobile phases, and this is aided by having the liquid phase as a thin film with a large surface area accessible to the flow of the gas phase. As the sample (solute) passes down the column, the molecules of each component partition between the liquid and gas phases according to a *distribution coefficient* or *constant*, K_D, i.e.

$$K_D = \frac{\text{concentration in the liquid phase}}{\text{concentration in the gas phase}} = \frac{C_L}{C_G}$$

This is a true equilibrium constant, which is specific for a given solute and liquid phase at the temperature selected. As the gas phase is moving continuously, solute molecules continue to dissolve in fresh liquid phase in relation to the equilibrium constant, while those molecules which have already dissolved overcome the various forces involved, reemerge into the gas phase and pass further down the column.

As long as a molecule is in the gas phase, it travels down the column at the same speed as the carrier gas. When a mixture of components is present in the solute, they diffuse into the liquid phase to varying degrees according to their individual equilibrium constants, and so travel down the column at different rates. In other words, the *retention times* on the column are different and the components tend to separate.

Of course, this is a rather simplified account, and the resolution can be marred by various factors, which cause band broadening. In packed columns, for example, there is a multiplicity of different flow paths around the particulate packing material, causing changes in the rate of flow of the gas in discreet areas. The particles are uneven in shape and size and the liquid phase will also consist of regions of different depth; molecules diffusing farther down into the liquid phase will emerge again much later than those which diffuse into a thin area. In a WCOT column, there is only one flow path in essence and the liquid phase is usually more uniform in thickness, so that the contribution from such factors to band broadening is minimal. Components emerge from packed columns in wide bands relative to the time spent on the column, while those emerging from a WCOT column are relatively narrow. Therefore, the efficiency of a column is dependent on the degree of band broadening which it imposes upon a solute as it passes along, in a given time. When a component emerges from the column, the shape of the concentration peak is controlled by band broadening (width) and the absolute concentration in the gas phase (height), i.e. by the equilibrium constant. Under conditions of constant temperature, the K_D does not change and the efficiency of a given column can be related to peak widths and retention times. With many practical problems, however, it is necessary to increase the column temperature during an analysis in order to bring off less volatile solutes in a reasonable time, and then the calculation of column efficiency is much more complex.

Under isothermal conditions therefore, and assuming that the sample is introduced to the column in an appropriate manner and at a suitable concentration (i.e. not overloaded) so that solutes separate in accordance with their equilibrium constants, peaks should emerge with a Gaussian distribution. Column efficiences are then defined in terms of numbers of *theoretical plates* (n), the retention time (t_r) measured from the point of injection until the peak reaches its maximum, and the width of the peak measured at half its maximum height (w_h) by using the equation:

$$n = 5.54 \times (t_r/w_h)^2$$

The same units must be used to measure t_r and w_h. These parameters are illustrated in Figure 3.1. Column efficiences may be described in terms

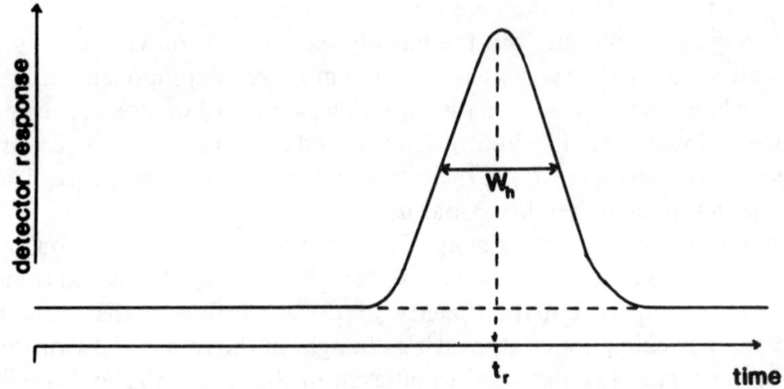

Figure 3.1 A chromatographic peak with an ideal Gaussian shape. Calculation of column efficiency.

of theoretical plates per metre, or as the column length (*plate height*) equivalent to a single theoretical plate. Ideally, retention *volumes* rather than times should be used, but the two are obviously related in a given column.

A certain volume of gas is required to carry a component which does not dissolve in the liquid phase through the column, and this is known as the *holdup volume*, although in practice, it is more convenient to measure *holdup time* (t_m). This is usually measured by injecting methane onto the column and determining the time for the leading edge of the methane peak to emerge. Every solute molecule must spend the same time (t_m) in the gas phase, the true time spent in the liquid phase is therefore an *adjusted retention time* (t^1):

$$t^1 = t_r - t_m$$

The time spent in the gas phase should not affect the quality of the separation. In WCOT columns, t_m is relatively great compared to that in packed columns, and it is necessary to take this into account in defining the true efficiency in terms of *effective theoretical plates* (N) [217,741], which are calculated as

$$N = 5.54 \times (t^1/w_h)^2$$

This figure is still mutable to some extent because it is related to the nature of the solute used as a test compound for the measurement. The factor for the solute is termed the *partition* or *capacity* ratio (k), and is the ratio of the weight (not concentration) of the solute in the liquid phase to the weight in the gas phase, and is proportional to the time spent by the solute in the liquid phase and that spent in the gas phase, i.e.

$$k = t^1/t_m$$

One of the principal differences between packed and WCOT columns lies in the relative availability of the liquid and gas phases to a solute molecule, and this is defined as the *phase ratio,* β , i.e.

$$\beta = \frac{\text{volume of gas phase}}{\text{volume of liquid phase}}$$

and in essence in a WCOT column

$$\beta = r_o/2d_f$$

where r_o is the radius of the capillary and d_f is the mean depth of the liquid film. In WCOT columns, the β values range from 100 to 300 typically (and from 5 to 20 in packed columns). As described above, the distribution constant, K_D, is based on solute concentration, but this can also be stated in terms of the weight of solute and the volume of each phase:

$$K_D = \frac{\text{weight of solute in liquid phase/volume of liquid phase}}{\text{weight of solute in gas phase/volume of gas phase}}$$

$$\frac{\text{weight of solute in liquid phase}}{\text{weight of solute in gas phase}} \times \frac{\text{volume of gas phase}}{\text{volume of liquid phase}}$$

$$= k\beta$$

where k is the partition or capacity ratio defined above.

This basic equation is of great importance in optimising all the parameters which can affect resolution in capillary GC.

2. Practical implications

The efficiency of a given column is dependent on a number of factors, including the nature and flow-rate of the carrier gas, column dimensions, liquid-phase thickness and column temperature. By optimising these, it may be possible to increase the resolution attainable quite considerably. On the other hand, this improved resolution may be bought at the expense of increased analysis time. In practice, it may be desirable to compromise and select conditions for an analysis which give adequate resolution in a reasonable time.

The nature and velocity of the carrier gas are primary considerations. At high velocities, the opportunity for band broadening through longitudinal diffusion of solute molecules (along the length of the column) are diminished, but there may then be insufficient time for them to pass into the liquid phase. When the flow-rate of the mobile phase is low, there is an increased opportunity for band broadening through longitudinal diffusion. Efficiency therefore tends to fall off both when the flow rate is too high and when it is too low. The nature of the carrier gas is also important, and hydrogen and helium but especially the former, because of their high diffusivities or low resistance to mass transfer, are greatly to be preferred to nitrogen say. It is also noteworthy that column efficiency varies much less with gas velocity over the useful working range when hydrogen is utilised, so that precise flow calibration is less critical in practice. This can be illustrated by a so-called Van Deemter plot of the variation in the height of an effective theoretical plate with carrier gas velocity for hydrogen, helium and nitrogen (Figure 3.2).

Of course if hydrogen is used as the carrier gas, it is necessary to guard against the risk of explosions, for example, through an electrical spark igniting a build up of gas caused by a leak within the instrument.

As the temperature is increased during temperature-programming, the efficiency of the column decreases, although components emerge as sharper peaks, because the vapour pressure of the solute increases as does the ratio of the concentration in the gas phase to that in the liquid phase, i.e. the capacity ratio is inversely proportional to column temperature. However, the effect can be small in relation to the advantages in terms of the sharpness of peaks and the speed of analysis with many practical problems, provided that sensible temperature-programming rates are employed. As the flow of gas through capillary columns is usually controlled by pressure regulation rather than by flow *per se*, the flow-rate through the column will actually decrease with increased temperature because of an increase in the viscosity of the gas. Therefore, a temperature-programmed analysis should ideally be started at a higher gas velocity than might be used for an isothermal run.

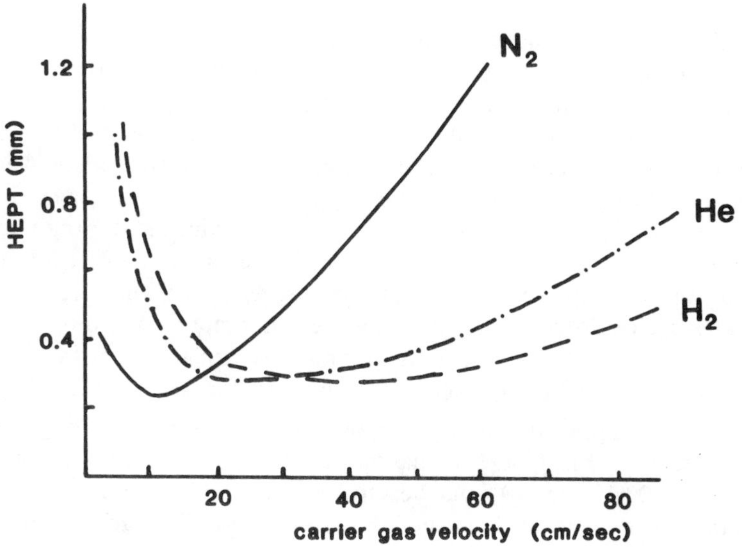

Figure 3.2 Plot of height of an effective theoretical plate (HETP) against carrier gas velocity (Van Deemter curve) for hydrogen, nitrogen and helium, obtained on a 25 m WCOT column (0.25 mm i.d.).

Again, this effect tends to be less with hydrogen than with other carrier gases.

There are three factors which are dependent on the physical characteristics of the column, i.e. column length, column internal diameter and film thickness. Of these, column length is least important as it can be established that resolution is proportional only to the square root of column length. Thus to improve by a factor of two on the resolution attainable with a 25 m column, it would be necessary to move to one 100 m in length, and this would inevitably mean that the analysis time would be increased by a factor of four.

In contrast by reference to the last equation above, it is apparent that changes in the column internal diameter or in the film thickness have a marked effect on the partition ratio. If the column diameter is reduced, this brings about a decrease in the phase ratio and a proportional increase in the capacity ratio. Retention times of solutes and the overall analysis time are increased, however. A decrease in film thickness corresponds to an increase in the phase ratio and results in a decrease in the partition ratio. In this instance, the retention time of the solute is decreased. In practice, the nature of the solute must also be considered, thick films being preferred for solutes of high volatility.

C. COLUMNS

1. Packed columns

Packed columns have been in use since the technique of gas chromatography was first developed. For many years, some analysts continued to use columns made of stainless steel, although there was some danger of rearrangement of fatty acids on the metal wall. Most prefer columns made of glass, because these are almost inert, it is easy to see how well the column has been packed and whether any gaps in the packing appear during use; deterioration of the top of the column packing at the inlet end is immediately seen. Also, glass columns are more easily emptied for reuse. It is certainly true that they are relatively fragile, but the main opportunity for breakage tends to be when columns are either mounted into or removed from the oven of the gas chromatograph; if particular care is taken when these tasks are performed, columns should last for many years.

The solid supporting materials for the liquid phases are generally diatomaceous earths, graded so that the particles are of uniform size (usually 80-100 or 100-120 mesh), and deactivated by washing with acid and by silylation in order to prolong the life of the liquid phase and to minimise any adsorptive effects on the solutes. Supports precoated with liquid phases are available commercially at prices close to those of the starting materials. If the analyst wishes to make his own, it is claimed [403] that the most uniform coating of the liquid phase on the support is obtained if a solution is filtered through a bed of the support and the whole spread out as a layer to dry; the amount of liquid phase that remains on the support must be determined by experiment and depends somewhat on the nature of the support. It is also possible to achieve satisfactory coatings by evaporating a solution of the liquid phase in the presence of the support in an indented flask (to stir the solid material) on a rotary evaporator, but great care is necessary to ensure that no damage occurs to the surface of the support.

A column is packed by adding the coated support in small amounts via a funnel, while tapping gently and applying a vacuum to the exit end. When it is filled, a glass wool plug (acid-washed and silanised) is placed on top of the packing at the inlet end to consolidate it. If columns are too tightly packed, they may block or the injection syringes may plug easily; if they are too loosely packed, the separations are poor. The column must be conditioned for up to two days at a temperature just above the maximum at which it is to be operated during routine analyses, before being used. A well-packed column (1.5 to 2m) containing a support with 10-15 % (w/w) liquid phase should have an efficiency of 3000 to 5000 theoretical plates. Such columns can have an operating life of well over a year, and this can be prolonged by periodically repacking the top 3 to 5 cm with fresh packing material.

2. *Glass WCOT columns*

It took some time for the pioneers in the use of glass WCOT columns to convince others that these were sufficiently robust for routine employment in the laboratory. There was never any doubt as to the resolution attainable or about the inertness of the surface, the latter being of particular importance to lipid analysts concerned with any potential isomerisation of polyunsaturated fatty acids. Interest was heightened when glass drawing machines for capillary production became available commercially at reasonable cost.

Before glass capillary columns can be satisfactorily coated with a thin uniform layer of a liquid phase, the surface must first be treated chemically to ensure adhesion. Otherwise, surface tension in the liquid will cause it to contract into droplets and leave the surface of the glass bare. Initially a variety of different methods of treating the surface were tried including the use of wetting agents, carbonisation and silylation, but it appears to be established that gaseous etching or aqueous leaching treatments are to be preferred. All of these methods work in essence by turning the smooth surface of the glass, at the microscopic level, to a much rougher one thereby increasing its wettability. This simply increases the contact angle between the glass and liquid phase, as pits and crevices are filled and the spreading of the film is assisted. The etching process also thoroughly cleans the glass, while chemical treatment can change the surface so that the liquid film is not deposited on silica but on salt crystals.

The mildest gaseous etching process consists in treatment with HCl or HF gas, the former only with soda glass and the latter with borosilicate glasses such as Pyrex™. With hydrogen fluoride, the silicate structure is attacked to form silicon tetrafluoride which forms crystals covering the surface. With hydrogen chloride, the reaction is extremely complex and involves a preliminary leaching of alkali metals from the surface. The surface generated in this way is then deactivated by one of several procedures before the liquid coating is applied by either dynamic or static techniques [267]. While this is indeed done by many analysts, it appears that there are some "tricks of the trade" and patented procedures are applied by commercial companies to ensure uniform reproducible products, especially with liquid phases of high polarity. As with packed columns, a new WCOT column must be conditioned for a time before use.

Glass capillary columns are much less fragile in use than might be anticipated. Again the greatest risk of breakage pertains when columns are mounted in or are removed from the oven of the gas chromatograph. One major drawback is that the column ends must be straightened before the column can be fixed in place, an operation which requires some practice and not a little manual dexterity.

3. Fused silica WCOT columns

Fused silica has proved to be an excellent medium for the manufacture of WCOT columns. They consist of an amorphous silicate material, which is free of metal oxides and is therefore very inert. Such columns are very flexible and there is no need to straighten the ends. As supplied, they are covered externally with a polymeric material which prevents brittle fractures. Coating of these columns with liquid phase does, however, appear to be a specialised task and most analysts are content to leave it to commercial manufacturers. Nowadays, the liquid phase does not merely coat the surface and adhere by physical forces, but rather it is bonded chemically to the surface by various means. In addition, the individual molecules of polymeric liquid phases are cross-linked by chemical methods, when the film is in place, to improve their stability at high temperatures. It is then possible even to remove organic impurities from the stationary phase by passing a little solvent through it. Such are the advantages of these columns indeed, that they have virtually supplanted those of glass in commerce.

D. THE LIQUID PHASE

1. Selectivity

The principal requirement of a stationary phase is that it provide the correct degree of selectivity for the separation required. At the same time, it should have reasonable chemical and thermal stability in order to prolong the working life of a column. The selectivity of a liquid phase is a product of several factors, and as it is not easy to define from first principles, it must be determined experimentally. In practice, this is accomplished by comparing the retention times of a series of standard test substances on a column containing the phase of interest against those on a non-polar phase. The Rohrschneider/McReynolds indices (named after the originators of the system) assist in identifying groups of phases with similar properties, and the choice may then depend on a factor such as temperature stability. To lipid analysts, the system has its limitations since it aids in selecting phases which separate different groups of compounds, rather than of homologuous series or closely related isomers, for example.

A major factor influencing separation according to degree of unsaturation is the polarity of the liquid phase. When packed columns only were available, this appeared to be a rather variable property, because of differences in the loading of the liquid phase, on the nature of the solid support, and on operating factors such as temperature and column age. Improvements in the manufacture of liquid phases helped a little but did not eliminate the problems. On the other hand, the thickness of the liquid film has very little effect on polarity in WCOT columns, so that the selectivity of phases can be defined with greater confidence.

In practice, lipid analysts have investigated the properties of particular stationary phases in some detail and certain have emerged as favourites for specific purposes. These are discussed in the chapters that follow in relation to each lipid class. No doubt new phases will be developed to challenge those in regular use, but changes are more likely to be made by direct comparison than by applying scientific concepts of selectivity.

2. Prolonging column life

In regular use, the stationary phase on a column can deteriorate for a number of reasons, but usually as a result of chemical attack. WCOT columns are more susceptible than packed columns, if only because they contain less stationary phase. Most polar liquid phases are very sensitive to oxygen and water, and it is strongly advised that all traces of these be removed from the carrier gas by positioning traps containing suitable molecular sieves and oxygen scrubbers (available commercially) between the gas cylinder and the column. The author once saw several columns being destroyed by a day or so of accidental exposure to an inferior grade of helium, which contained a relatively high proportion of oxygen; the traps in place could not cope. Polar solvents (e.g. chloroform, alcohols, carbon disulphide, etc.) and traces of polar impurities introduced to the columns with samples may slowly react with the liquid phase or perhaps displace it from the column, and any adverse reaction will of course be exacerbated by exposure of the column to excessive temperatures. The flow of carrier gas should never be allowed to stop as long as the column is being heated. In addition, involatile materials injected onto a column along with the components of interest may gradually build up and alter the characteristics of the liquid phase.

Most damage tends to occur in the first few coils of the column, where the liquid phase can be displaced entirely or at least suffer appreciable degeneration. When the damage is minimal, reversing the column may be all that is required. Otherwise, depending on how extensively the deterioration has progressed, anything from a few centimetres to one or two coils of the column can be broken off. Such column shortening has very little effect on resolution.

As mentioned above, packed columns can be rejuvenated by replenishing the top few centimetres of the packing material.

E. DETECTORS

1. Flame ionisation detectors

A large number of detectors operating on different principles have been developed for use in gas chromatography, but only a few of these continue to be used to a significant extent. The flame ionisation detector is now almost universally adopted as it can be used with virtually all organic compounds, and has high sensitivity and stability, a low dead volume, a fast response time and the response is linear over an extremely wide range. This detector is simple to construct and operate, and it is highly reliable in prolonged use. Only inert gases, compounds with a single carbon bound to oxygen or sulphur (e.g. CO_2, CS_2, etc.) and a few other volatile substances do not give a substantial signal.

The principle of the detector is that ions are generated by combustion of the organic compounds as they emerge from the column in a diffusion flame of hydrogen and air. The carrier gas from the column may be premixed with hydrogen, although with WCOT columns, it is usual to position the outlet of the column at the orifice of the combustion jet in a chamber through which an excess of air is passed. The collector electrode is cylindrical and is placed just above the flame, and the ion current is measured by establishing a potential between the collector and the jet tip. In order to prevent the ions from recombining, a potential is selected in the saturation region, where increasing the potential has relatively little effect on the ion current. The signal current is passed to an amplifier, which must have a linear range to match that of the detector itself, and thence is transmitted to a recorder.

The response is subject to some experimental variables, chief among which are the flow-rates of hydrogen, air and the carrier gas. The instrument manufacturer's instructions should be followed closely to optimise these, although there is usually some leeway, as it is not easy to give general guidelines for instruments of different makes.

The nature of the response in relation to specific lipids is discussed in subsequent chapters.

2. Electron-capture detectors

In the electron-capture detector, a radioactive source is used to bombard the carrier gas with *beta* particles as it passes through an ionisation chamber. Each *beta* particle can generate up to a thousand thermal electrons, which are collected by applying a voltage potential. When solutes containing electron-capturing moieties enter the cell, they interact with thermal electrons and a diminution in the background current is seen and can be measured with high sensitivity. In addition to this sensitivity, the chief virtue of this detector is its specificity, as molecules containing halogen atoms, for example, give a very marked response. Although few natural lipid molecules contain halogens, it is possible to convert lipids to halogen-containing derivatives to make use of the high sensitivity and specificity of the detector.

3. Mass spectrometry

While mass spectrometry (MS) is rarely used for detection *per se*, it has become an invaluable tool for the identification (and adventitiously for the detection) of lipids separated by gas chromatography. The principle of the technique in its simplest form is that organic molecules in the vapour phase, are bombarded with electrons and form positively charged ions, which can fragment in a number of different ways to give smaller ionised entities. These ions are propelled through a magnetic or electrostatic field and are separated according to their mass to charge (m/z) ratio; they are collected in sequence as the ratio increases, the ion current is amplified and it is then displayed by some means. The largest (or base) peak is given an arbitrary intensity value of 100, and the intensities of all the other ions are normalised to this. The ion from the parent molecule is termed the *molecular ion* (M^+). With instruments of low resolution, peaks appear at unit mass numbers, but at high resolution the masses of individual ions can be measured with sufficient accuracy for the molecular formula of each to be determined.

Molecules do not fragment in an arbitrary manner but tend to split at weaker bonds, such as those adjacent to specific functional groups, or according to certain complex rules which have been formulated empirically from studies with model compounds. Frequently, it is possible to deduce the structure of the original compound from first principles from the nature of the fragments produced. With other compounds when the results are equivocal, the spectrum can be compared with those of compounds with similar properties (nowadays with the aid of computer search facilities) until a good fit is obtained. The combination of mass spectral and GC retention data may also serve to eliminate alternative structures. In GC-MS applications, the total ion current produced from the column effluent is recorded continuously, and a trace is obtained resembling that from other detectors; spectra are also recorded continuously and can be related to specific peaks.

Of course, this description greatly oversimplifies the technique, especially as many different ionisation systems, in addition to electron-impact, are available which alter the extent and nature of the ionisation and fragmentation processes. The reader should consult one of the many authoritative texts on the subject for a more comprehensive account. There is further discussion in Chapter 7 below.

In interfacing a gas chromatograph with a mass spectrometer, a primary requirement is that the pressure must be reduced from appreciably above atmospheric in the column to about 10^{-4} torr in the ion source. The interface must be as inert as possible, there should be no cold spots where sample condensation can occur, and the extra-column volume should be as small as possible to minimise band broadening. The development of fused silica capillaries simplified the problem considerably, because a new generation of diffusion pumps had the capacity to extract helium at the same rate as it emerged from a column. It was thus possible to design the ion source so that the outlet of the capillary column passed directly in to it. The alternative is to use devices such as teflon membrane or molecular jet separators to concentrate the solute relative to the carrier gas. With this and other systems, there is inevitably some loss of the resolution achieved on the WCOT column.

F. INJECTION SYSTEMS

1. Injection technique

The choice of an injection system for packed columns is entirely straight forward. Most analysts prefer simply to inject the sample in a small volume of solvent (1 to 5 μ litres), via a syringe and through a septum, just into the top of the column packing. The sample is thus volatilised in the presence of the liquid phase so that the risk of isomerisation of double bonds or of other unwanted side reactions is minimal.

With WCOT columns, the choice of one of the many injection systems available from commercial sources may confuse the newcomer to the subject. The properties of the major types are discussed below, although it should be noted that models from different manufacturers that operate on similar principles may differ in some details. Of course, the analyst may have little choice in the matter, having to use whatever equipment is provided. It is then incumbent upon him to be aware of the potential pitfalls so that the results can be optimised. The nature of the sample is also relevant, and the topic is raised again in later chapters in relation to specific analytical problems.

During the injection process, it is extremely important that the sample should not change in composition nor should there be discrimination for or against any particular component. Ideally, thermal degradation or rearrangement should be negligible, no loss of column efficiency should be introduced, the solvent peak should not interfere with the detection of the solutes, and retention times and relative peak areas should be highly reproducible.

In addition to instrumentation, it is necessary to consider all aspects of injection technique. Problems can arise from volatilisation when the sample is in contact with the metal surface of the needle or from premature evaporation of the sample before the needle is fully inserted into the injector. In order to minimise these effects, the following method (the "hot needle" technique) is recommended:

> "The sample (0.1 to 1 μ litre) is drawn up completely into the syringe barrel, leaving the needle empty, the needle is then inserted firmly and smoothly into the injection port, and it is allowed to remain in place for about 5 seconds to allow it to warm up, before the plunger is pressed rapidly".

Some analysts prefer to use a "solvent flush" method, which is similar to the above except that a small plug of fresh solvent is drawn up ahead of the sample, and is used to push the latter into the evaporation chamber. Superficially, attention to such detail may appear trivial, but the benefits can be substantial.

2. Split injection

WCOT columns have a limited sample capacity and it is relatively easy to overload them by introducing a sample in too large a volume of solvent or at too high a concentration. Many injection systems have been developed to circumvent the problem, and split injectors are simple to use and can give excellent results in many types of analysis. A schematic diagram of an injector of this kind is shown in Figure 3.3. In this, the sample is vaporised in the carrier gas, which is divided into two streams, one of which is directed onto

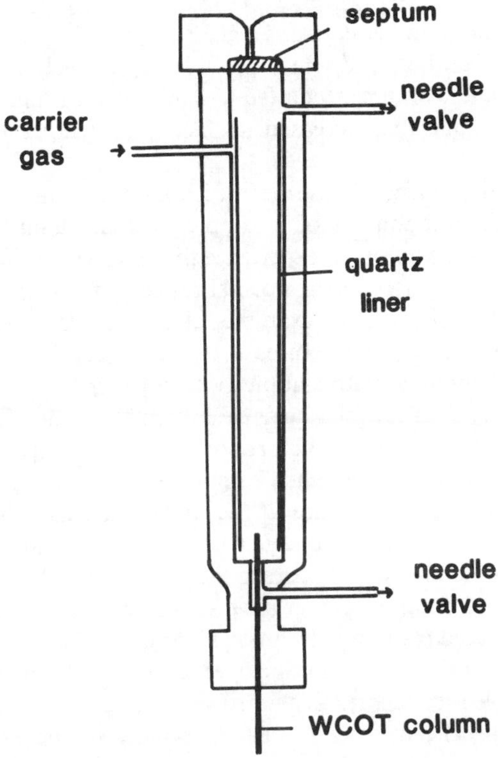

Figure 3.3 Schematic diagram of a typical split injection system for WCOT columns.

the column and the second of which is vented to the atmosphere. The flow through the latter is regulated prior to injection by a control valve to give the desired split ratio, usually from about 1:20 to 1:200. As the flow-rate through the column is commonly about 1 to 2 ml/min, the gas flow through the injector is very high (100 to 200 ml/min) and the vaporised sample is present in the injector only momentarily. This short time means that the sample is introduced onto the column as a narrow plug and is followed by the pure carrier gas.

Unfortunately, the preset split ratio differs from the true sample ratio in a complex manner, which is dependent on parameters such as sample volume, solvent, and volatility, syringe handling, and injector and column temperatures. When the sample is volatilised, it produces a pressure wave, which introduces some sample onto the column before the pressure falls back. During the second phase, little sample enters the column and most is vented. The splitter ratio in fact controls the magnitude of the pressure wave, and through this the actual sample size. Thus if the same amount of sample is injected in different volumes of solvent, sample peak areas may not be the same.

The vaporisation chamber contains a glass or quartz liner, which can easily be removed for cleaning, and which provides a relatively inert surface for vaporisation. Some workers attempted to pack these with inert materials to promote mixing, but the increased surface area tended to cause some degradation.

The principal drawback of this injector is that it may discriminate against the higher boiling components of a sample, so that quantification can be problematical. It is the author's experience that this type of injector can give excellent results in analyses of the normal range of fatty acids encountered in animal and plant tissues, for example, although it is necessary to check this with suitable calibration mixtures at regular intervals. Less satisfactory quantification is obtained with samples with a wider spread of volatilities, such as in the analysis of the fatty acids of milk fats or in the chromatography of intact triacylglycerols. As a high proportion of the sample is wasted, this injection mode is not suitable when sample size is a limitation.

For optimum results, the nature of the solvent and its volume should be kept constant, the injection technique described above should be used, the syringe needle should always penetrate to the same spot just above the column inlet and the initial column temperature should be reproduced accurately. It also helps if a "cold trapping" technique is employed during injection, i.e. the column temperature is maintained at about the boiling point of the solvent; the sample recondenses as a narrow band, and when the solvent peak is seen to emerge, the oven is heated up rapidly to the normal analysis temperature. Some commercial gas chromatographs are now designed so that cold trapping can be carried out automatically.

3. Splitless injection

The splitless sample injection method is designed to make use of the "solvent effect". In this mode, the sample is injected in a solvent with a high boiling point relative to the column temperature (but lower than the sample, of course) into a chamber through which only the carrier gas for the column is flowing. If the conditions are correct, i.e. such that the column temperature is at least 30°C below the boiling point of the solvent, the latter will recondense in the column inlet and act as a temporary thick-film stationary phase, assuming a shape in which the column phase ratio, β , decreases continuously in the direction of migration. In accord with the final equation in Section B above, the front edge of each band moving into that section must slow down faster than the rear of the band, and the sample is concentrated into a narrow volume. After a set time, roughly 1.5 times that required for the carrier gas to sweep out the injection chamber or 1 to 2 minutes, a purge valve is opened. Gas flow is directed to the bottom of the inlet, where it is divided so that one stream continues as the carrier gas while the other sweeps any residual sample from the injection chamber. A further purge stream ensures that the septum is never contaminated.

A schematic representation of a typical split/splitless injection system, which can be used in either mode, is shown in Figure 3.4. The technique has most value in the analysis of trace components.

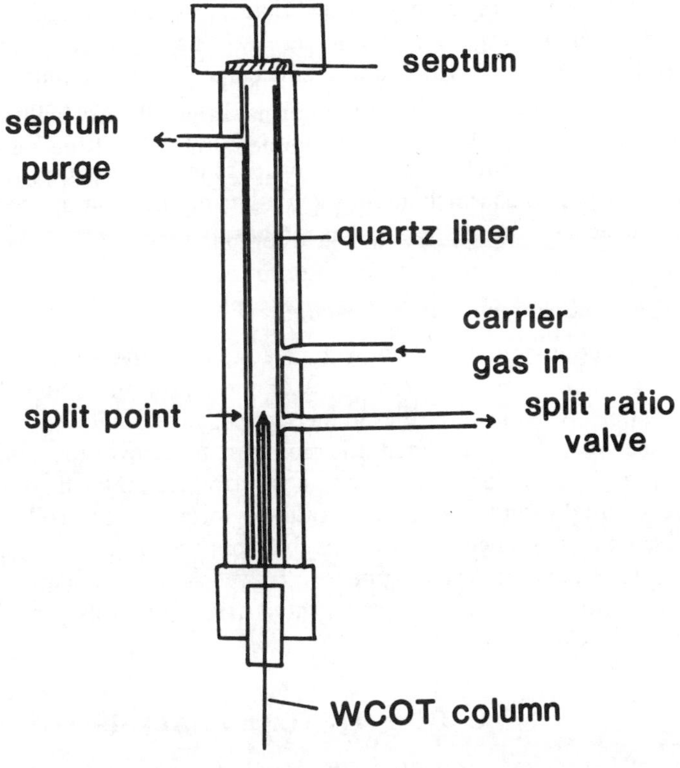

Figure 3.4 Schematic diagram of a typical split/splitless injection system for WCOT columns.

4. On-column injection

In on-column injection, the sample is injected in a solvent directly onto the column. The sample is concentrated by "cold trapping" or a "solvent effect" at the head of the column, so must be injected at a column temperature near the boiling point of the solvent. The column inlet does not have a septum but rather a "duck-bill" valve, made of a soft elastomer. This consists simply of two plastic surfaces which are pressed together by the pressure of the carrier gas in the injection port. During injection, a needle of fused silica is merely slipped between the two surfaces and is guided into the top of the WCOT column. (Some commercial systems use a more complex solenoid valve).

As no sample splitting occurs, there should be no discrimination of sample components. The sample is not vaporised instantly, so it is not stressed as much as in other procedures. In consequence, a high degree of precision can be attained in quantitative analyses

As with other injection procedures, some care must be taken to ensure that the analyst gets the most out of it. The sample dilution must be gauged correctly so that the column is not overloaded. The mechanical problems of feeding the needle into the WCOT column should be minimal with a properly designed injector, provided that the needle dimensions are correct for the column, especially when both are made of fused silica. Correct injection technique is vital, and in particular the syringe plunger should be depressed rapidly so that the sample is "sprayed" into the column. With this method, it is especially important that care is taken during the preparation of samples for chromatography, in order to ensure that they are free from involatile materials which might accumulate on the column and bring about some degeneration of the stationary phase.

5. Programmed-temperature injection

As direct on-column injection can lead to a relatively rapid deterioration of the column through interactions of the liquid phase and the injection solvent or involatile impurities in the sample, an alternative has been devised in which a specially designed injection port takes over the function of the top part of the column. The sample is introduced into this inlet, which is cooled initially, and then its temperature is raised at a controlled rate so that the sample components are selectively vaporised. Injection systems of this type are relatively new and appear to be well suited to the analysis of lipids of high molecular weight, such as intact triacylglycerols. This is discussed further in Chapter 8 below.

G. QUANTITATIVE ANALYSIS

In all aspects of chromatographic analysis, from sample preparation to the separation process itself, there are opportunities for errors to occur. Assuming everything that is possible has been done to minimise this, there can be little doubt that electronic digital integration is the most accurate and reproducible, not to mention the most rapid and convenient, means of quantifying chromatographic peaks. Yet even here, there are potential sources of error, and it is essential to ensure that the various instrumental parameters on the integrator, especially those defining the sampling rate, are appropriate to the elution volumes (peak widths) at various times during analysis. If this is not done carefully, according to the manufacturer's instruction manual, it is possible to negate the advantages of the technique. The author has noted before that it is not always easy to convince a novice analyst that a computer print-out of his results is fallible. In analyses with WCOT columns, electronic integration is the only method suitable for the quantification of complex samples.

In analyses with packed columns, electronic integration is also desirable, especially in temperature-programmed analyses. On the other hand because peak width is greater than with WCOT columns, there are some manual methods available that can give acceptable results, provided that the peaks are symmetrically shaped. One such method consists in drawing tangents to the sides of each peak to produce a triangle with the base line as the third side; the area is then proportional to the product of the height and width of the triangle. Alternatively, the height of the peak itself and its width at half the maximum height can be measured, and again the area is proportional to the product of these. Although peaks are never truly symmetrical, there have been such marked improvements in the quality of the column packing materials in recent years, that these methods are more accurate than reports in the earlier literature might indicate.

A third method, which is suited only to isothermal analyses, was once much used by the author. It consists in multiplying the retention time of a component by the peak height, this quantity also being proportional to peak area. The advantages are that two relatively large distances are measured, so errors at this stage are minimised, reasonably accurate results are obtained with incompletely resolved peaks, and it is simple and rapid. A correction factor [123] can be applied to increase the accuracy of the method and compensate for the fact that the sample is applied to the column as a finite band rather than as a point source as theory requires, i.e. the widths of the peaks at half-height are plotted against the retention times of the components and the resultant straight line intercepts the base line at a point in front of the actual point of injection. Retention times or distances are thereafter measured from this point. (In practice, the errors of extrapolation are such that the point can perhaps more easily be found by testing in the appropriate region until the results with a simple standard mixture best match the known composition).

The next step in the quantitative analysis of a sample consists in determining whether any response factors need to be applied to correct the experimental results. These must be determined by using standard mixtures, which are as similar as possible to the samples to be analysed, and comparing the analytical results under the standard chromatographic conditions with the known compositions. With most lipid samples and flame ionisation detection, these should only vary slightly from unity; if large correction factors are necessary, some stage in the chromatography has not been properly optimised. Some specific examples are described in later Chapters.

In many analyses of lipids, the results will be presented simply in terms of the relative proportions (expressed in percentage terms) of each component. There may also be instances, where the absolute amount of each component must be measured. This is best accomplished by adding a precise quantity of an internal standard, ideally a substance which resembles those to be analysed in its chromatographic properties, but does not occur naturally in the sample. If possible, this should be added when the tissue is first extracted, so that it is carried through the extraction, group separation and derivatisation steps, as well as through chromatography. This ideal situation is rarely encountered in practice, and it is usually sufficient to add the internal standard immediately prior to a derivatisation step. The areas of all the sample peaks can then be related to that of the internal standard, the absolute amount of which is known. As an example, the absolute amounts of lipid classes as well as the fatty acid compositions are frequently determined by adding the methyl ester of an odd-chain fatty acid to the sample, prior to transesterification and GC analysis.

When GC is used in routine analytical applications, it is important to set up a proper system of quality control in order to ensure that the equipment functions correctly, and that it is not subject to gradual deterioration or excessive random variation. To do this, it is necessary to establish regular checks on procedures by testing them with defined primary standards. The results should then be evaluated objectively by statistical methods. Such systematic checks may also indicate whether reagents are deteriorating with age, or whether a faulty batch of solvent say has been received. It might help in picking up unplanned changes in methodology introduced by unskilled technical support staff. Of course, these comments could equally be applied to most other aspects of lipid analysis. Quality control in the lipid laboratory has been reviewed by Naito and David [659].

PART 2

THE ANALYSIS OF FATTY ACIDS

CHAPTER 4

THE PREPARATION OF DERIVATIVES OF FATTY ACIDS

A. INTRODUCTION

Before the fatty acid components of lipids can be analysed by GLC, it is necessary to convert them to low molecular weight non-polar derivatives, such as methyl esters. In addition, it may be advisable to mask other polar functional groups in a similar manner, or to prepare specific derivatives as an aid to identification. Peak shape and resolution are greatly improved at the same time. It is only possible to identify fatty acids tentatively by GLC retention times alone, but GLC used in combination with derivatisation and chemical degradative or spectroscopic procedures, especially mass spectrometry, can be an extremely powerful means of characterisation. Of course, such derivatives are also of value in isolating specific fatty acids by other chromatographic methods. Therefore, before any chromatographic procedure is undertaken, it is necessary to consider what derivatives should be prepared and what method should be employed for the purpose.

Because of the high sensitivity of chromatographic analysis procedures, small amounts of material (usually less than 1 mg, and certainly less than 10 mg) may be all that is required, and most of the procedures described below are on this scale. Any conventional Pyrex glassware with ground-glass joints can be used for the reactions, but for many the author has found it convenient to use test-tubes of about 15 ml capacity with a standard ground-glass joint and stopper. Condensers and other equipment can be connected to these when required. Organic and aqueous layers can be separated efficiently in such tubes with the aid of Pasteur pipettes, perhaps after brief centrifugation to ensure a clean separation of the layers. Usually it is necessary to dry the solvents prior to evaporation by allowing them to stand over anhydrous sodium sulphate, but if hexane-containing layers are separated with care, this step can frequently be omitted. If an internal standard is to be added, it should be introduced into the sample at the earliest possible stage. Precautions should be taken at all times to prevent autoxidation of lipids (see Chapter 2). Methods of preparing derivatives in general have been reviewed [111,230].

B. HYDROLYSIS (SAPONIFICATION) OF LIPIDS

Lipids can be hydrolysed by heating them under reflux with an excess of dilute aqueous ethanolic alkali and the fatty acids, diethyl ether-soluble non-saponifiable materials and any water-soluble hydrolysis products recovered for further analysis. When the water-soluble components (such as glycerol, glycerophosphorylcholine, etc.) are required, special procedures must be used and most of these are outwith the scope of this book. The free fatty acids and the non-polar non-saponifiable components are separately recovered in the following procedure:

> "The lipid sample (10 mg) is refluxed with a 1M solution of potassium hydroxide in 95 % ethanol (2 ml) for 1 hour; alternatively, reaction at room temperature overnight is equally effective. The solution is cooled, water (5 ml) is added and the mixture is extracted thoroughly with hexane-diethyl ether (1:1, v/v; 3 x 5 ml). It may be necessary to centrifuge to break any emulsions that form. The solvent extract is washed with water, dried over anhydrous sodium sulphate and the non-saponifiable materials are recovered on removal of the solvent in a rotary evaporator. The water washings are added to the aqueous layer, which is acidified with 6 M hydrochloric acid and extracted with diethyl ether-hexane (1:1, v/v; 3 x 5 ml). The free fatty acids are recovered after washing the extract with water, drying it over anhydrous sodium sulphate and removing the solvent by evaporation."

The non-saponifiable layer will contain any hydrocarbons, long-chain alcohols and sterols originally present in the lipid sample in the free or esterified form. If the sample contained any glycerol ethers or plasmalogens, the deacylated residues will also be in this layer. With single lipid classes, isolated by chromatographic means, the preliminary extraction of the alkaline medium may not be required, and this step can be omitted from the procedure.

When short-chain fatty acids (C_{12} or less) are present in lipids, it is necessary to extract the acidified solution much more exhaustively, and even then it may be almost impossible to recover the fatty acids of shortest chain-length, such as butyric, quantitatively. Epoxyl groups and cyclopropene rings in fatty acids are normally disrupted by acid, but with care they will survive the above procedure if the exposure to the acidic conditions is short.

Cholesterol esters are hydrolysed very slowly by most reagents, and they may not react completely under the conditions above. Therefore, when they are major components of the lipid sample, longer reflux times are necessary.

Similarly, *N*-acyl derivatives of long-chain bases are saponified only slowly by alkali, but hydrolysis is virtually complete when sphingomyelin, for example, is refluxed for 10 hours in 1M KOH in methanol-water (9:1, v/v) [463]. An alternative procedure has been described in which acidic hydrolysis conditions are employed, i.e. the sphingolipids are hydrolysed with 0.5 M hydrochloric acid in acetonitrile-water (9:1, v/v) for 45 minutes at 100°C or for 4 hours at 70°C [65]. Long-chain bases can also be isolated by procedures of this kind, although some precautions are necessary to prevent any degradation occurring (see Chapter 10).

The non-saponifiable materials and the free fatty acids can be obtained by separating the total acidified extract by adsorption chromatography techniques (TLC or HPLC), as described in Chapter 3 and elsewhere [163,168], eliminating the step in which the alkaline solution is extracted. The free fatty acids are easily separated from the other products of hydrolysis, which can be individually isolated and identified. As an alternative, acidic and neutral materials can be separated by ion-exchange chromatography using the following procedure [1014].

> "DEAE-Sephadex™ (1 g; type A-25; capacity 3.2 m-equiv./g; Pharmacia, Sweden) on a Buchner funnel is washed successively with small amounts of 1M hydrochloric acid, water, 1M potassium hydroxide and water again (the procedure is repeated three times), the last wash until neutral. It is then washed twice with methanol (25 ml), and with 25 ml of diethyl ether-methanol-water (89:10:1 by volume). It is slurried in the latter solvent mixture, left to equilibrate overnight and packed into a small column. The sample, containing up to 100 mg of fatty acids, is washed through with 25 ml of the solvent mixture; the neutral materials are eluted while the acids remain on the column. The latter can be recovered with a mobile phase of diethyl ether-methanol (9:1, v/v) saturated with carbon dioxide."

Polyunsaturated fatty acids are not altered by the mild hydrolysis conditions described above. On the other hand, if the reaction time is prolonged unduly or if too strong an alkaline solution is used, some isomerisation of double bonds can occur.

C. THE PREPARATION OF METHYL AND OTHER ESTERS OF FATTY ACIDS

The preparation of the methyl ester derivatives of fatty acids must be by far the commonest chemical reaction performed by lipid analysts, yet it is often poorly understood; the topic has been comprehensively reviewed [160,205,839]. There is no need to hydrolyse lipids to obtain the free fatty acids before preparing the esters as most lipids can be transesterified directly.

No single reagent will suffice, however, and one must be chosen that best fits the circumstances. Esters prepared by any of the following methods can be purified if necessary by adsorption chromatography (see below). Care should be taken in the evaporation of solvents as appreciable amounts of esters up to C_{14} can be lost if this step is performed carelessly. In particular, an over vigorous use of nitrogen to blow off solvents must be avoided. Esters other than methyl may be required from time to time for specific purposes.

1. Acid-catalysed esterification and transesterification

Free fatty acids are esterified and *O*-acyl lipids transesterified by heating them with a large excess of anhydrous methanol in the presence of an acidic catalyst. If water is present, it may prevent

$$RCOOR' + CH_3OH \overset{H^+}{\rightleftharpoons} RCOOCH_3 + R'OH$$

$$RCOOH + CH_3OH \overset{H^+}{\rightleftharpoons} RCOOCH_3 + H_2O$$

the reaction going to completion. The commonest and mildest reagent is 5 % (w/v) anhydrous hydrogen chloride in methanol. It is most often prepared by bubbling hydrogen chloride gas (which is commercially available in cylinders or can be prepared by dropping concentrated sulphuric acid slowly on to fused ammonium chloride or into concentrated hydrochloric acid) into dry methanol. A simpler procedure is to add acetyl chloride (5 ml) slowly to cooled dry methanol (50 ml). Methyl acetate is formed as a by-product, but it does not interfere with methylations at this concentration. It is usual to heat the lipid sample in the reagent under reflux for about 2 hours, but they may also be heated together in a sealed tube at higher temperatures for a shorter period. Alternatively, equally effective esterification is obtained if the reaction mixture is heated in a stoppered tube at 50°C overnight (also incidentally reducing the glassware requirements).

A solution of 1-2 % (v/v) concentrated sulphuric acid in methanol transesterifies lipids in the same manner and at much the same rate as methanolic hydrogen chloride. It is very easy to prepare, and it is thus the author's preferred reagent for esterification of free fatty acids, but utilised at a temperature below reflux. If the reagent is used carelessly, some decomposition of polyunsaturated fatty acids may occur.

Boron trifluoride in methanol (12-14 % w/v) has also been used as a transesterification catalyst and in particular as a rapid means of esterifying free fatty acids. The reagent has a limited shelf-life, even when refrigerated, and the use of old or too concentrated solutions often results in the production of artefacts and the loss of appreciable amounts of polyunsaturated fatty acids. In view of the large amount of acid catalyst used in comparison with other reagents and the many known side reactions, it is the author's opinion that boron trifluoride in methanol has been greatly over-rated, and that it is best avoided.

Boron trichloride in methanol does not appear to have been much used for transesterification of lipids, but is almost as effective as boron trifluoride-methanol and does not appear to bring about the same unwanted side reactions [129,485].

Non-polar lipids, such as cholesterol esters or triacylglycerols, are not soluble in reagents composed predominantly of methanol, and will not react in a reasonable time, unless a further solvent is added to effect solution. Benzene was once employed regularly to this end, but because of its great toxicity, it is advisable to use some other solvent such as toluene or tetrahydrofuran.

Methanolic hydrogen chloride (5 %) or sulphuric acid (1 %) are then probably the best general purpose esterifying agents. They methylate free fatty acids very rapidly and can be employed to transesterify other O-acyl lipids efficiently; they are generally used as follows:

> "The lipid sample (up to 50 mg) is dissolved in toluene (1 ml) in a test tube fitted with a condenser, and 1 % sulphuric acid in methanol (2 ml) is added. The mixture is left overnight in a stoppered tube at 50°C (or is refluxed for 2 hours), then water (5 ml) containing sodium chloride (5 per cent) is added and the required esters are extracted with hexane (2 x 5 ml), using Pasteur pipettes to separate the layers. The hexane layer is washed with water (4 ml) containing potassium bicarbonate (2 %) and dried over anhydrous sodium sulphate. The solution is filtered and the solvent removed under reduced pressure in a rotary film evaporator or in a stream of nitrogen."

No solvent other than methanol is necessary if free fatty acids alone are to be methylated (also only 20 minutes at reflux, or 2 hours at 50°C, is required), or if polar lipids such as phospholipids are to be transesterified. The reaction can be scaled up considerably; for example, 50 g of lipid in 100 ml of toluene can be transesterified with 200 ml of methanol containing 4 ml of concentrated sulphuric acid. *N*-acyl lipids are transesterified very slowly with these reagents (see below). If acidic reagents are permitted to super-heat in air, some artefact formation is possible.

The same method is used to prepare dimethylacetals from aliphatic aldehydes or plasmalogens (see Chapter 10 below), and when it is used on lipid samples containing such compounds, acetals are formed which may contaminate the methyl esters. The two classes of compound can be separated by saponification or better by means of preparative TLC with toluene [582] or dichloroethane [980] as the solvent for development (esters migrate ahead of acetals). Acetals are not formed during base-catalysed transesterification.

With lipid samples from animal tissues, it is sometimes necessary to purify methyl esters after transesterification has been carried out in order to eliminate cholesterol, which can be troublesome when the esters are subjected to gas chromatography. This can be accomplished by adsorption chromatography with a short column (approx. 2 cm) of silica gel or Florisil™ in a Pasteur pipette plugged with glass wool, and eluted with hexane-diethyl ether (95:5, v/v; 10 ml). The cholesterol and other polar impurities remain on the column. Commercial pre-packed columns (Bond Elut™ or Sep-Pak™) can be used in a similar way. Methyl esters can also be purified by preparative TLC, with hexane-diethyl ether (9:1, v/v) as the mobile phase.

2. Base-catalysed transesterification

O-Acyl lipids are transesterified very rapidly in anhydrous methanol in the presence of a basic catalyst. Free fatty acids are **not** normally esterified, however, and care must be taken to exclude water from the reaction medium to prevent their formation as a result of hydrolysis of lipids. 0.5 M Sodium methoxide in anhydrous methanol, prepared simply by dissolving fresh clean sodium in dry methanol, is the most popular reagent, but potassium methoxide or hydroxide have also been used as catalysts. The reagent is stable for some months at room temperature, especially if oxygen-free methanol is used in its preparation. The reaction is very rapid; phosphoglycerides, for example, are completely transesterified in a few minutes at room temperature. It is commonly performed as follows:

> "The lipid sample (up to 50 mg) is dissolved in dry toluene (1 ml) in a test-tube, 0.5 M sodium methoxide in anhydrous methanol (2 ml) is added, and the solution is maintained at 50°C for 10 min. Glacial acetic acid (0.1 ml) is then added, followed by water (5 ml). The required esters are extracted into hexane (2 x 5 ml), using a Pasteur pipette to separate the layers. The hexane layer is dried over anhydrous sodium sulphate and filtered, before the solvent is removed under reduced pressure on a rotary film evaporator."

As with acid-catalysed transesterification procedures, an additional solvent, such as toluene or tetrahydrofuran, is necessary to solubilise non-polar lipids such as cholesterol esters or triacylglycerols, but is not required if they are not present in the sample. Chloroform should not be used in this way, because it contains ethanol as a stabiliser, and because dichlorocarbene, which can react with double bonds, is generated by reaction with sodium methoxide. Again, cholesterol esters are transesterified very slowly and may require twice as long a reaction time as that quoted. The quantities of lipid used can be scaled up considerably; for example, 50 g of lipid is transesterified in toluene (50 ml) and methanol (100 ml) containing fresh sodium (0.5 g) in 10 minutes at reflux, and a related procedure has been used to transesterify litre quantities of oils [658]. Under the conditions described above, no isomerisation of double bonds in polyunsaturated fatty acids occurs, though prolonged or careless use of basic reagents can cause alterations to fatty acids.

The author has made extensive use of the following convenient micro-scale procedure, in which methyl acetate is added to the medium to suppress the competing hydrolysis reaction [164].

> "The lipids (up to 2 mg) are dissolved in sodium-dried diethyl ether (0.5 ml) and methyl acetate (20 μl.). 1 M Sodium methoxide in dry methanol (20 μl.) is added, and the solution is agitated briefly to ensure thorough mixing. The solution immediately becomes cloudy as sodium-glycerol derivatives are precipitated. After 5 min. at room temperature, the reaction is stopped by the addition of acetic acid (2 μl.), the solvent is evaporated in a stream of nitrogen (taking care not to blow out the solid precipitate), hexane (1 ml) is added and the mixture is centrifuged at about 1500 *g* for 2 min. The supernatant layer is decanted, and an aliquot is taken directly for GLC analysis."

Amide-bound fatty acids, as in sphingolipids, are not affected by alkaline transesterification reagents under such mild conditions, and this fact is sometimes used in the purification of such lipids. Also, aldehydes are not liberated from plasmalogens with basic reagents, in contrast to when acidic conditions are employed.

Although free fatty acids are not esterified under the basic conditions described above, methyl esters can be prepared by exchange with N,N-dimethylformamide dimethyl acetal in the presence of pyridine [913]. Similarly, methyl iodide reacts with sodium or potassium salts of fatty acids in the presence of a polar aprotic solvent such as dimethylacetamide to form methyl esters [41,186].

Quaternary ammonium salts of fatty acids are converted to methyl ester derivatives pyrolytically in the injection port of a gas chromatograph. Of a number of reagents which have been described for the purpose, it appears that trimethylsulphonium hydroxide is the most powerful and exhibits fewer side reactions; it can be used for the simultaneous transesterification of lipids and esterification of free acids [144].

3. Diazomethane

Diazomethane reacts rapidly with unesterified fatty acids in the presence of a little methanol, which catalyses the reaction, to form methyl esters. The reagent is generally prepared as a solution in diethyl ether by the action of alkali on a nitrosamide, e.g. *N*-methyl-*N*-nitroso-*p*-toluene-sulphonamide (Diazald™, Aldrich Chemical Co., Milwaukee, U.S.A.). Solutions of diazomethane are stable for short periods if stored refrigerated in the dark over potassium hydroxide pellets. If they are kept too long, polymeric by-products form which may interfere with the subsequent GLC analysis.

Diazomethane is highly toxic, carcinogenic and potentially explosive, so great care must be exercised in its preparation; in particular, strong light and apparatus with ground glass joints must be avoided, and the reagent should only be prepared in an efficient fume cupboard. In addition, the intermediate nitrosamines are among the most potent carcinogens known. Accordingly, diazomethane should only be used when no other reagent is suitable.

The procedure of Schlenk and Gellerman [800] is particularly convenient for the preparation of small quantities of diazomethane for immediate use. In this instance, there is very little by-product formation and, if sensible precautions are taken, the risk to health is minimal.

> "A simple apparatus is required that can be quickly assembled by a glassblower. It consists of three tubes with side arms that are bent downwards and arranged so that the arm of each projects into and is near the bottom of the next tube. A stream of nitrogen is saturated with diethyl ether in the first tube and carries diazomethane, generated in the second tube, into the third tube where it esterifies the acids. The flow of nitrogen through diethyl ether in tube 1 is adjusted to 6 ml per min. Tube 2 contains 2-(2-ethoxyethoxy)ethanol (0.7 ml), diethyl ether (0.7 ml) and 1 ml of an aqueous solution of potassium hydroxide (600 g/l). The fatty acids (5-30 mg) are dissolved in diethyl ether-methanol (2 ml; 9:1 by vol) in tube 3. About 2 mmole of *N*-methyl-*N*-nitroso-*p*-toluene-sulphonamide per mmole of fatty acid in ether (1 ml) is added to tube 2 and the diazomethane which is formed is passed into tube 3, until the yellow colour persists. Excess reagent is then removed in a stream of nitrogen."

An alternative small-scale procedure has recently been described [963].

4. Special cases

(i) Short-chain fatty acids

Short-chain acids are completely esterified in all of the procedures described above, but quantitative recovery of the esters from the reaction medium can be very difficult because of their high volatility and partial solubility in water. As short-chain acids are major components of such commercially-important fats and oils as milk fats or coconut oil, a great deal of attention has been given to the problem. Diazomethane can be used to esterify free fatty acids quantitatively in ethereal solution, and a portion of the reaction medium may then be injected directly onto the GLC column so that there are no losses. On the other hand, if the free acids have to be obtained by hydrolysis of lipids, it is not easy to ensure that there are no losses at this stage. The best methods are those in which there are no aqueous extraction, or solvent removal steps and in which the reagents are not heated. The alkaline transesterification procedure of Christopherson and Glass [185], on which the following method is based, meets these criteria better than most.

> "The oil (20 mg) is dissolved in hexane (2.5 ml) in a stoppered test-tube, and 0.5M sodium methoxide in methanol (0.1 ml) is added. The mixture is shaken gently for 5 min at room temperature then acetic acid (5 μ l.) is added followed by powdered anhydrous calcium chloride (about 1 g). After 1 hour, the mixture is centrifuged at 700 *g* for 2 to 3 minutes to precipitate the drying agent. An aliquot of the supernatant liquid is taken for GLC analysis."

The method described above or variations upon it are widely used, and can give excellent results if care is also exercised during the chromatography stage [72,78]. Others have argued that more reproducible gas chromatographic analyses are obtained by preparing butyl ester derivatives [419,420].

If the sample contains both *O*-acyl bound and unesterified fatty acids, the latter can be esterified with diazomethane first, before the former are transesterified. Alternatively, for safety reasons, the procedures of Thenot *et al.* [913] or of Martinez-Castro *et al.* [599] might be used.

(ii) Unusual fatty acids

The methods described above can be used to esterify all fatty acids of animal origin without causing any alteration to them. Many fatty acids from plant sources and certain of bacterial origin are more susceptible to chemical attack. For example, cyclopropene, cyclopropane and epoxyl groups in fatty acids are disrupted by acidic conditions, and lipid samples containing such acids are best transesterified with basic reagents; the free fatty acids can be methylated safely with diazomethane.

Conjugated polyenoic fatty acids such as α-eleostearic acid undergo *cis-trans*-isomerisation and double bond migration when esterified with methanolic hydrogen chloride, and all acidic reagents can cause addition of methanol to conjugated double bond systems. Similar reactions occur under acidic conditions with fatty acids containing a hydroxyl group immediately adjacent to a conjugated double bond system (e.g. dimorphecolic acid, 9-hydroxy,10-*trans*,12-*trans*-octadecadienoic acid), and dehydration and other unwanted side reactions may also take place (reviewed elsewhere [160,163]). However, no side effects occur when basic transesterification is used.

Diazomethane can add to double bonds and keto groups in 2,3-unsaturated and 2-keto acids respectively [89].

(iii) Amide-bound fatty acids

Sphingolipids, which contain fatty acids linked by *N*-acyl bonds, are not easily transesterified under acidic or basic conditions. If the fatty acids alone are required for analysis, the lipids may be refluxed with methanol containing concentrated hydrochloric acid (5:1 v/v) for 5 hours or by maintaining the reagents at 50°C for 24 hours, and the products worked up as described above for the anhydrous reagent [160,895]. Unfortunately, a small proportion of free fatty acid is also formed. As an alternative, the specific hydrolysis methods, described above (Section B), can be used to generate the free acids quantitatively, and these can then be methylated by an appropriate procedure. Traces of degradation products of the bases that might interfere with subsequent analyses can be removed by adsorption chromatography (see Section C.1 above). If the long-chain bases are also required for analysis, suitable procedures are available (see Chapter 10) [168]. With *N*-acylphosphatidylserine and related lipids, the *O*-acyl bound fatty acids can be released by mild alkaline methanolysis and so distinguished from the *N*-acyl components which require much more vigorous hydrolytic conditions.

(iv) Esterification on TLC adsorbents

After lipids have been separated by TLC, the conventional procedure is to elute them from the adsorbent before transesterifying for GLC analysis. A number of methods have been described for transesterifying lipids on silica gel without prior elution, with the objective of simplifying the methodology and of reducing the opportunities for contamination. Regrettably, it has been the author's experience that poor recoveries of esters are obtained when basic transesterification reagents are used, probably because water bound to the silica gel causes some hydrolysis. Acid-catalysed procedures give better results, but when the ratio of silica gel to lipid is very high (>4000:1), poor recoveries are again the rule. In practice, such high ratios may not often be obtained and satisfactory methylation is achieved by direct transesterification.

The favoured technique is to scrape the band of adsorbent containing the lipid into a test-tube, then to add the reagent (e.g. 2% methanolic sulphuric acid) with efficient mixing, and to carry out the reaction as if no adsorbent were present. On working up the aqueous mixture obtained when the reaction is stopped, it is necessary to centrifuge to precipitate all the silica gel and to extract with a more polar solvent than hexane, for example diethyl ether, to ensure quantitative recovery of the methyl esters. Unfortunately, cholesterol esters are not transesterified readily in the presence of silica gel, and it is still necessary to elute these from the adsorbent prior to reaction.

(v) Side reactions

Methyl esters are the derivatives of choice for gas chromatography but in choosing an appropriate reagent, it is necessary to consider its effect on lipid components other than fatty acids and on gas chromatography stationary phases since artefacts may be produced which interfere with subsequent analyses (see also Section C.4.ii above). For example, BHT may be partially methylated by boron trifluoride-methanol reagent giving rise to an extraneous peak which tends to emerge from a GLC column together with or just in front of the C_{16} fatty acid derivatives [380].

If cholesterol esters are esterified directly and the free cholesterol is not removed prior to GLC analysis, it may dehydrate to form cholestadiene on the column and this may obscure some of the C_{22} components [499]. Similarly, cholestadiene and cholesterol methyl ether are generated to some extent when most acidic reagents are used for transesterification, and analogous by-products are formed from plant sterols [376,494,626,835]. This does not occur with base-catalysed transesterification. Other hydrolysis products of low molecular weight from lipids, such as phytol and aldehydes, can be troublesome in some circumstances. If need be, the methyl ester derivatives can be purified by adsorption chromatography as discussed in Section C.1 above.

If the methyl ester derivatives are not cleaned up properly, traces of the transesterification reagents injected onto the column will bring about degradation of the stationary phase, and cause spurious peaks to emerge.

5. Preparation of esters other than methyl

(i) Other alkyl and aromatic esters

Esters other than methyl may be required for a variety of reasons, for example to diminish the volatility of short-chain fatty acids or to introduce aromatic groupings into molecules so that the UV-detection systems can be used in HPLC. The latter aspect has been reviewed elsewhere [168]. Many of the methods described above can be adapted to the preparation of alternative alkyl esters simply by substituting the appropriate alcohol for methanol; for example, either 1 M sodium butoxide or 2 % sulphuric acid in butanol may be utilised for the preparation of butyl esters. Where the alcohol has a high boiling point or is comparatively expensive, the acid chloride or anhydride can be prepared and reacted with a slight excess of the appropriate alcohol in the presence of a base such as pyridine [160,205]. Section (iii) below contains an example of this. Analogous methods have been used for the preparation of fluorinated esters for GLC with electron-capture detection [869].

Trimethylsilylesters of unesterified fatty acids have been prepared for GC analysis by the same methods used to prepare the trimethylsilylether derivatives of hydroxyl groups (see Section D.3 below) [526]. Similarly, *t*-butyldimethylsilyl esters have proved of value for the analysis of saturated (including deuterated) fatty acids by mass spectrometry [692,709,1004].

(ii) Pyrrolidides

Amide derivatives of fatty acids can be prepared by reaction of acid chlorides or anhydrides with an amine. *N*-Acylpyrrolidines (Figure 4.1(a))

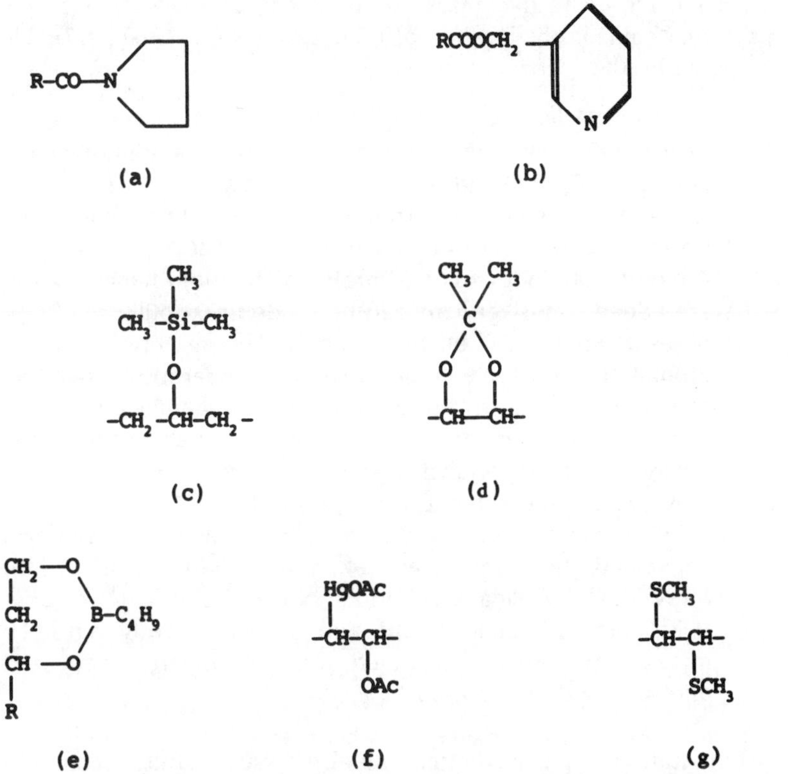

Figure 4.1 Derivatives of fatty acids. (a) pyrrolidide; (b) picolinyl ester; (c) trimethylsilylether; (d) isopropylidene derivative; (e) butylboronate derivative; (f) mercuric acetate adduct; (g) dimethyldisulphide adduct.

are prepared for mass spectrometric analysis by reaction of methyl esters with pyrrolidine and acetic acid (see Chapter 7) [51]. The reaction is carried out as follows:

> "The fatty acid methyl ester (10 mg) is dissolved in freshly distilled pyrrolidine (1 ml), acetic acid (0.1 ml) is added, and the mixture is heated at 100°C for 1 hour. On cooling, the amide is taken up in dichloromethane (8 ml) and is washed repeatedly with 2 M hydrochloric acid then water (4 ml portions). After drying over anhydrous sodium sulphate, the required product is obtained on evaporation of the solvent."

(iii) Picolinyl esters

It is increasingly being recognised that picolinyl ester derivatives (Figure 4.1(b)) offer distinctive mass spectrometric fragmentation patterns of particular value in the location of double bonds (see Chapter 7). They are prepared by the following procedure:

> "Unesterified fatty acids (5 mg) are dissolved in trifluoroacetic anhydride (0.5 ml) and are left for 30 min at 50°C. The excess reagent is blown off in a stream of nitrogen. As soon as it has gone, 3-hydroxymethylpyridine (20 mg) and 4-dimethylaminopyridine (4 mg) in dichloromethane (0.2 ml) are added. (This reagent can be made up in bulk; for 1 mg or less, simply add half the volume). The mixture is left in a stoppered tube for 3 hours at room temperature, then the solvent is removed in a stream of nitrogen, and hexane (8 ml) and water (4 ml) are added. After thorough shaking and vortex mixing, the hexane layer is washed twice more in the same way, before the solvent is evaporated.
>
> To remove any residual free fatty acids, with small samples especially, the products are taken up in diethyl ether (1 ml), a few mg of a silica-NH_2 bonded phase (Bond Elut™ or Sep-Pak™) are added and the mixture is shaken briefly. After 10 min, the mixture is centrifuged, then the solvent is decanted carefully and evaporated."

Purification of picolinyl esters can also be successfully accomplished by elution from a small column of Florisil™ with hexane-diethyl ether (1:4, v/v).

D. DERIVATIVES OF HYDROXYL GROUPS

Many fatty acids with hydroxyl substituents exist in nature and the polar functional moieties must be masked by derivatisation prior to analysis by GLC. In addition, the free hydroxyl groups of long-chain alcohols, glycerol ethers, and mono- and diacylglycerols are frequently converted to the same non-polar derivatives for chromatographic analysis. Acyl migration of fatty acids in partial glycerides is prevented, and more symmetrical peaks are obtained on gas chromatography than could be obtained with the native compounds. As essentially the same procedures are used for these classes of lipids, they are all considered for convenience here. The choice of derivative will depend on the nature of the compound and the separation to be attempted, and on occasion it may be necessary to prepare several types of derivative to confirm identifications. With hydroxy fatty acids, it is normal practice to derivatise the carboxyl group first. The following types of derivative are among the more useful.

1. Acetylation

Acetyl chloride and pyridine at room temperature, or prolonged heating with acetic anhydride, can be used to acetylate lipids, but the mildest reagent is probably acetic anhydride in pyridine (5:1 v/v), which is used as follows [763]:

> "The lipid (up to 50 mg) is dissolved in acetic anhydride in pyridine (2 ml, 5:1. v/v), and is left at room temperature overnight. The reagents are then removed in a stream of nitrogen with gentle warming and the acetylated lipid is purified, if necessary, by preparative TLC on silica gel layers, generally with hexane-diethyl ether (80:20, v/v) as the mobile phase."

Free amino groups are also acetylated with this reagent. *N*-acetylation without simultaneous *O*-acetylation (e.g. of long-chain bases) can be accomplished by reaction with acetic anhydride in methanol (1:4, v/v) at room temperature overnight (see Chapter 10 for details) [280]. Acetylmethanesulphonate in microcolumns of celite acetylates alcohols very rapidly and is claimed to be suited to the routine analysis of large number of samples [823]. It is also possible to acetylate lipids with [1-^{14}C]-acetic anhydride in order to make use of the high sensitivity and precision of liquid scintillation counting for quantification following chromatographic separation [80].

2. Trifluoroacetates

Trifluoroacetate derivatives of hydroxy acids, monoacylglycerols and glycerol ethers are comparatively volatile and are sufficiently temperature-stable to be subjected to gas chromatographic analysis. They are prepared simply by dissolving the hydroxy compound in excess of trifluoroacetic anhydride, leaving for 30 minutes and then removing most of the excess reagent on a rotary film evaporator. Diacylglycerol trifluoroacetates are not sufficiently stable at high temperatures for gas chromatography, however [501]. Trifluoroacetates hydrolyse very rapidly, even in inert solvents such as hexane, and it is necessary to store them and to inject them onto the gas chromatographic column in a solution containing some trifluoroacetic anhydride. Column packings must be conditioned by repeatedly injecting trifluoroacetic anhydride into them before being used, but they are rendered acidic and may no longer be suitable for other analyses. Such a procedure is not recommended with capillary columns. Trifluoroacetic acid also appears to attack the methylene group between double bonds in some circumstances, causing losses of polyunsaturated components [999].

3. Trimethylsilyl ether and related derivatives

Trimethylsilyl (often abbreviated to TMS) ether derivatives (Figure 4.1(c)) are a useful alternative to acetates for gas chromatographic analysis. They are much more volatile than the latter, but are not as stable, particularly to acidic conditions, and will hydrolyse slowly on TLC adsorbents. The preparation and properties of TMS and related derivatives have been reviewed [230,727,729]. Probably, the most popular reagent for the preparation of TMS ethers in the earlier literature consists of a mixture of hexamethyldisilazane, trimethylchlorosilane and pyridine (3:1:10 by volume), and it is used as follows.

> "To the hydroxy-compound (up to 10 mg) is added pyridine (0.5 ml), hexamethyldisilazane (0.15 ml) and trimethylchlorosilane (0.05 ml). The mixture is shaken for 30 seconds and then is allowed to stand for 5 min. An aliquot can then be injected directly into a gas chromatography column; alternatively, the reaction mixture can be taken to dryness on a rotary evaporator, the products extracted with hexane (5 ml), the hexane layer washed with water (1 ml) and dried over anhydrous sodium sulphate, before the solvent is removed. The derivatives are stored in fresh hexane and are stable at -20°C for long periods".

More recently, many simpler more-powerful silylating reagents have become available, and that used most frequently in lipid analysis is probably bis(trimethylsilyl)acetamide ("BSA"). Reaction is carried out simply by dissolving the lipid in a solvent such as acetone or acetonitrile and adding a ten fold excess by weight of BSA; with the normal range of unhindered alcohols likely to be encountered, the reaction is complete in 10 minutes at room temperature. While some analysts inject an aliquot of the reaction mixture directly onto the GLC column, the author prefers to clean the sample up as described above, immediately prior to analysis, in order to prolong column life.

t-Butyldimethylsilyl (*t*-BDMS) etherification has been proved to be of particular value for the preparation of lipid derivatives of higher molecular weight. They are approximately 10^4 times more stable than the corresponding TMS ethers and only hydrolyse at an appreciable rate under strongly acidic conditions. The following method of preparation is recommended [195].

> "The silylation reagent consists of *t*-butyldimethylsilyl chloride (1 mmole) and imidazole (2 mmole) in *NN*-dimethylformamide (10 ml). The reagent (0.5 ml) is added to the lipid (up to 10 mg) and heated at 60°C for 30 min. After rapid cooling, hexane (5 ml) is added, and the mixture is washed with water (3 x 1 ml). The solvent is dried over anhydrous sodium sulphate, filtered or decanted and evaporated in a stream of nitrogen to yield the required derivative."

In contrast to TMS derivatives, it is possible to purify *t*-BDMS derivatives by preparative TLC.

Recently, nicotinates [96,364,950] and certain hybrid derivatives, i.e. picolinyldimethylsilyl ethers [362], have been shown to be of value for the identification of alcohols by means of GC/mass spectrometry (see Chapter 10).

4. Isopropylidene compounds

Isopropylidene derivatives of vicinal diols, for example glycerol ethers, 1-monoacylglycerols or dihydroxy acids, are prepared by reacting the diol with acetone in the presence of a small amount of an acidic catalyst (Figure 4.1(d)). Anhydrous copper sulphate is probably the mildest catalyst and is used as follows:

> "The esters (10 mg) are dissolved in dry acetone (3 ml) and anhydrous copper sulphate (50 mg) is added. After 24 hr at room temperature (or 3 hr at 50°C), the solution is filtered, the copper salts are washed with dry ether and the combined solutions are evaporated *in vacuo*."

The use of perchloric acid as a catalyst permits a more rapid reaction, but is potentially much more hazardous [985]. Isopropylidene compounds are stable under basic conditions, but are hydrolysed by aqueous acid; the original diol compound can be regenerated by shaking with a 1 M solution of concentrated hydrochloric acid in 90 % methanol.

5. n-Butylboronate derivatives

Alkyl-boronic acids, such as n-butylboronic acid, react with 1,2- or 1,3-diols or with α- or β-hydroxy acids to form 5- or 6-membered ring non-polar boronate derivatives (Figure 4.1(e)). They are prepared simply by adding n-butylboronic acid to a solution of the hydroxy-compound in dimethylformamide. The reaction is complete in 10-20 minutes at room temperature and the reaction mixture can be injected directly into a gas chromatographic column for analysis [55,133]. As alternatives, cyclic di-*tert*-butylsilylene derivatives have been shown to be of value in the analysis of diols and hydroxy acids [132]. The preparation and use of cyclic derivatives for the analysis of bifunctional compounds have been reviewed [730].

E. DERIVATIVES OF DOUBLE BONDS

The double bonds of unsaturated fatty acids may be reacted to form various addition compounds, as an aid to the isolation of individual fatty acids or as part of a method for establishing the configuration or location of the double bonds in the aliphatic chain.

1. Mercuric acetate derivatives

Mercuric acetate in methanol solution reacts with the double bonds of unsaturated fatty acids to form polar derivatives (Figure 4.1(f)), which are more easily separated by adsorption or partition chromatography, according to degree of unsaturation, than are the parent fatty acids. They can be prepared by the following procedure.

> "The lipid sample is refluxed with a 20 % excess of the theoretical amount of mercuric acetate in methanol (2 ml per g of mercuric acetate) for 60 min. After cooling to room temperature, a volume of diethyl ether 2.6-fold that of the methanol is added to precipitate inorganic mercury salts, and the solution is filtered. The mercuric acetate adducts are obtained on evaporation of the solvent."

When fatty acids with five or six double bonds are present, it has been recommended that the reaction mixture should be heated at 100°C for two hours in a sealed tube [827,830]. Although silver ion chromatography has superceded the use of mercuric acetate derivatives for most analytical purposes, the latter may still be useful for the bulk preparation of concentrates of single fatty acids, for example.

The original double bond is regenerated by reaction with aqueous acid with no double bond migration or *cis-trans* isomerisation if the following method is used.

> "10 g of the adduct in methanol (20 ml) is mixed with concentrated hydrochloric acid (50 ml) and a stream of hydrogen chloride gas is bubbled into the solution for a few minutes. The solution is extracted twice with hexane (50 ml portions), and the extracts are combined and washed with water. If a test of the organic layer with a solution of diphenylcarbazone in methanol reveals that mercury is still present, a single repeat of the process will remove it entirely."

Acetylenic groups react with 2 mols of mercuric acetate to form adducts, but the original bonds cannot be regenerated and a keto derivative is formed on acidic hydrolysis. This reaction has been turned to advantage to locate the position of the original triple bond in a fatty acid derivative by mass spectrometry [140,480].

Mercuric acetate derivatives can be further reacted with inorganic halides yielding methoxyhalogenomercuri-derivatives, which are less polar and more volatile than the original mercury compounds and are more easily separated by chromatography [972,973] (see Chapter 6). The reaction is accomplished simply by adding a 10 % excess of sodium bromide in methanol to the mercuration mixture whenever the adduct formation is complete. After about 2 minutes, the new derivative is formed and the mixture is worked up as before.

Mercury adducts can also be reacted with sodium borohydride and converted to methoxy compounds (oxymercuration-demercuration) [321,613], or reacted with halogens in methanol to form methoxyhalogen compounds [611]. Such derivatives have been utilised for locating double bonds in fatty acids by mass spectrometry.

2. *Hydroxylation*

Double bonds can be oxidised to vicinal diols by a variety of reagents of which the most useful are alkaline potassium permanganate and osmium tetroxide. With both reagents, *cis*-addition occurs to yield diols of the *erythro* configuration from *cis*-double bonds, and diols of the *threo* configuration from *trans*-double bonds. The following procedure utilises alkaline permanganate and is particularly suited to reaction with monoenes and dienes [671].

"The fatty acid (0.1-100 μ mole) is dissolved in 0.25M sodium hydroxide solution (0.2 ml) and diluted with ice water (1 ml). 0.05 M Potassium permanganate (0.2 ml) is added and after 5 min, the solution is decolorised by bubbling sulphur dioxide into it. The fatty acid derivatives are extracted with chloroform-methanol (7 ml, 2:1 v/v), and the lower layer is collected and dried over anhydrous sodium sulphate before the solvent is evaporated."

Osmium tetroxide gives higher yields of multiple diols from polyunsaturated fatty acids [671].

"The fatty acid (0.1-100 μ moles) is dissolved in dioxane-pyridine (1 ml, 8:1 v/v), and a 5 % solution of osmium tetroxide in dioxane (0.1 ml) is added. After 1 hour at room temperature, methanol (2.5 ml) and 16 % sodium sulphite in water (8.5 ml) are added, and the mixture is allowed to stand for 1 hour more. After centrifuging to precipitate and compact the sodium sulphite, the supernatant solution is diluted with 4 volumes of methanol and filtered. The filtrate is evaporated to dryness and suspended in methanol (2 ml). Chloroform (4 ml) is added, the suspension is filtered once more and the solvent evaporated."

Such hydroxylated fatty acids in the form of less polar derivatives, for example isopropylidene, TMS ether or methoxy compounds, have been used in conjunction with mass spectrometry to locate the positions of the original double bonds (see Chapter 7). Because the isopropylidene derivatives of *threo* and *erythro* compounds are separable by GLC, the configuration of the original double bonds can also be established [572,985].

3. Epoxidation

Epoxides are formed from olefins by the action of certain per-acids. *Cis*-addition occurs, so *cis*-epoxides are formed from *cis*-olefins and *trans*-epoxides from *trans*-olefins. The following procedure is based on that of Gunstone and Jacobsberg [324].

"The monoenoic ester (20 mg) is reacted with *m*-chloroperbenzoic acid (16 mg) in chloroform (2 ml) at room temperature for 4 hr. Potassium bicarbonate solution (5 %, 4 ml) is added, and the product is extracted thoroughly with diethyl ether. After drying the organic layer over anhydrous sodium sulphate and removing the solvent, the required epoxy ester is obtained by preparative TLC on silica gel G layers with hexane-diethyl ether (4:1 v/v) as the mobile phase."

Mass spectrometry of epoxy derivatives has been used to locate the position of double bonds in fatty acids, and since *cis*- and *trans*-isomers can be separated by GLC, as a means of estimating fatty acids with *trans*-double bonds (see Chapter 5) [243,244].

4. Dimethyldisulphide addition

One of the most convenient methods for the location of double bonds by mass spectrometry involves the addition of dimethyldisulphide across the double bond, a reaction catalysed by iodine (Figure 4.1(g)) [261]. It is carried out as follows:

> "The monoenes (1 mg) are dissolved in dimethyldisulphide (0.2 ml) and a solution (0.05 ml) of iodine in diethyl ether (60 mg/ml) is added. The mixture is stirred for 24 hours, then hexane (5 ml) is added, and the mixture is washed with dilute sodium thiosulphate solution, dried over sodium sulphate and evaporated to dryness. The product is taken up in fresh hexane for injection directly onto the GC column."

Some residual starting material may remain, but it elutes substantially ahead of the product when this is subjected to GC analysis. By using a higher temperature, the reaction can be taken to completion but some by-product formation may occur (see Chapter 7). The author (unpublished) observed excessive by-product formation when the dimethyldisulphide was evaporated off before the washing step.

5. Hydrogenation

Hydrogenation of lipids is undertaken prior to confirming the chain-lengths of aliphatic moieties, or to protect lipids (and simultaneously simplify the chromatograms) during high temperature GC analysis of intact lipids. Many of the published hydrogenation procedures are needlessly complex and the following method is adequate for most purposes.

> "The unsaturated ester (1-2 mg) in a test-tube is dissolved in methanol (1 ml) and Adams' catalyst (platinum oxide; 1 mg) is added. The tube is connected via a two-way tap to a reservoir of hydrogen (e.g. in a balloon or football bladder) at or just above atmospheric pressure and to a vacuum pump. The tube is alternatively evacuated and flushed with hydrogen several times to remove any air, then it is shaken vigorously

while an atomosphere of hydrogen at a slight positive pressure is maintained for 2 hr. At the end of this time, the hydrogen supply is disconnected, the tube is flushed with nitrogen and the solution is filtered to remove the catalyst. The solvent is evaporated under reduced pressure, and the required saturated ester is taken up in hexane or diethyl ether for GLC analysis."

Hexane may be used as the solvent for the hydrogenation reaction if the fatty acid is still esterified to glycerol as in a triacylglycerol or diacylglycerol acetate, but the hydrogenated compounds must later be recovered from the catalyst with a more polar solvent such as chloroform.

6. Deuteration

Deuterium can be added across double bonds to assist in their location by mass spectrometry, by means of deuterohydrazine reduction [221]. Oxygen is required to generate d_2-diimine for reaction to occur, so the procedure must be carried out in air.

"The unsaturated ester (0.5 mmole) is dissolved in anhydrous dioxane (5 ml) at room temperature, and deuterohydrazine (5 mmole) in twice its volume of deuterium oxide is added. The mixture is stirred at 55-60°C in air, but in the absence of atmospheric moisture, for 8 hr when the reagents are evaporated *in vacuo*. The product is purified by preparative silver ion chromatography (see Chapter 6)".

Others recommend a procedure in which diimine is generated by the reaction of acetic acid with potassium azodicarboxylate [472,484,809].

"$^2H_4$-acetic acid (0.3 ml) is added in small aliquots to a constantly-stirred slurry of potassium azodicarboxylate (500 mg) and the fatty acid (up to 10 mg) in ^2H-methanol (5 ml) over a period of 3 to 5 hours. The progress of the reaction can be checked by gas chromatography. The methanol is evaporated in a stream of nitrogen, and the reaction is repeated twice more. The product is purified as above."

Potassium azodicarboxylate is not available commercially, but can be prepared by reaction of 40% aqueous potassium hydroxide on azodicarbonamide [94]. Neither procedure is entirely satisfactory, and occasionally can refuse to work for no apparent reason. A further approach has still to be tried with fatty acids *per se*, but does appear to give excellent results with intact lipids. It involves catalytic deuteration with deuterium gas and Wilkinson's catalyst (tris(triphenylphosphine)rhodium (I) chloride) [219].

CHAPTER 5

GAS CHROMATOGRAPHIC ANALYSIS OF FATTY ACID DERIVATIVES

A. INTRODUCTION

The advent of gas-liquid chromatography revolutionised the analysis of the fatty acid components of lipids, and it is undoubtedly the technique that would be chosen in most circumstances for the purpose. It is now possible to obtain a complete quantitative analysis of the fatty acid composition of a sample in a very short time. Individual fatty acids can usually be identified by GC with reasonable certainty from their relative retention times, especially if the analysis is carried out with a variety of stationary phases, and taking into account the large body of knowledge that now exists on the compositions of specific tissues or organisms. On the other hand, there are many circumstances when it must be recognised that GC analysis permits a tentative identification only. In the first analysis of any new sample, for example, confirmation of fatty acid structures may have to be obtained by unequivocal chemical degradative and spectroscopic procedures, although some of these may also benefit from an involvement of GC.

The only technique to compare with GC for the analysis of fatty acid derivatives is HPLC in the reversed-phase mode with UV-absorbing or fluorescent derivatives (reviewed elsewhere [168] and briefly in the next Chapter). Both the capital and running costs of this technique are appreciably higher than for GC, and identification of components emerging from HPLC columns is rarely easy, because of the complex nature of the separation process. On the other hand, HPLC certainly has advantages for the isolation of specific components on a small scale for structural analysis or for radioactivity measurements, and for the analysis of fatty acids with labile functional moieties such as hydroperoxy or cyclopropene groups.

GC procedures only for the analysis of fatty acid derivatives are considered in this Chapter, and the topic is discussed in terms of both packed and WCOT columns. The superb resolution attainable with WCOT columns can present the analyst with the problem of identifying large numbers of minor components, many of which have little metabolic or nutritional relevance. Analysis with packed columns gives simplified chromatograms, but often with much of the essential information. Several review articles have appeared recently on the subject [14,163,428,430,508,546,922]. Alternative or complementary methods for identification of fatty acids are described in Chapters 6 and 7.

Precautions should be taken at all times to prevent or minimise the effects of autoxidation (see Chapter 2).

B. COLUMN AND INSTRUMENTAL CONSIDERATIONS

1. Liquid phases

The liquid phase used in a GC column is the principal factor determining the nature of the separations that can be achieved. Non-polar silicone liquid phases, such as SE-3O™, OV-1™, JXR™ or QF-1™, permit the separation of fatty acid esters mainly on the basis of their molecular weights, when in packed columns. However, there can be separation of unsaturated fatty acids of the same chain-length with WCOT columns (c.f. Figure 5.7 below) or when the amount of the stationary phase on the support (in packed columns) is low (1-3 %). Non-polar phases are of value in the analysis of fatty acid derivatives of higher than normal molecular weight, especially with WCOT columns.

Liquid phases consisting of high molecular weight hydrocarbons, such as the Apiezon™ greases (of which the most popular is Apiezon L™) also separate saturated and unsaturated components of the same chain-length, unsaturated esters eluting before the related saturated compounds, but in packed columns there is very little separation of esters of a given chain-length differing in the number of double bonds in the molecule; these phases are now less used than formerly having apparently been superceded by methylsilicone polymers.

Polar polyester liquid phases are much more suited to fatty acid analysis as they allow clear separations of esters of the same chain-length, but with zero up to six double bonds, unsaturated components eluting after the related saturated ones. These phases can be subdivided into four main classes: group a, the highest polarity phases, e.g. alkylpolysiloxanes containing various polar substituents including nitrile groups, marketed under trade designations such as Silar 10C™, Silar 9CP™, SP 2340™ and OV-275™; group b, highly polar phases, e.g polyethyleneglycol succinate (EGS), polydiethyleneglycol succinate (DEGS), EGSS-X™ (a copolymer of EGS with a methyl silicone), CP-Sil 84™ and CP-Sil 88™; group c, medium polarity phases, e.g. polyethyleneglycol adipate (PEGA), polybutanediol succinate (BDS) and EGSS-Y™ (a copolymer of EGS with a higher proportion of the methysilicone than in EGSS-X™); group d, low polarity phases, e.g. polyneopentylglycol succinate (NPGS), EGSP-Z™ (a copolymer of EGS and a phenyl silicone), Carbowax 20M™ and Silar 5CP™. In work with packed columns, EGSS-X™, EGSS-Y™ and newer phases of equivalent polarity are widely accepted as the most useful representatives of the second and third groups respectively because of their relatively high thermal stability, particularly in packed columns. Low polarity phases are utilised principally in WCOT and SCOT columns, because saturated and monoenoic components of the same chain-length are poorly separated when these phases are used in packed columns. In packed column work, the phases from the group of highest polarity are stable at temperatures above those possible with EGSS-X™ (providing that oxygen is rigidly excluded), for example, and they afford excellent separations of polyunsaturated fatty acids, complementing those obtained with the more common stationary phases; they are of particular value in the separation of *cis*- and *trans*-isomers. Examples of the properties of many of these phases in specific applications are given below.

It is occasionally advantageous to be able to subject fatty acids in the unesterified form to GLC analysis. In this instance, acidic liquid phases such as DEGS containing 3 % phosphoric acid, FFAP™ ("free fatty acid phase"), Carbowax 20M™-terephthalic acid or the structurally-related phase, SP-1000™, are used (see Section F.8 below).

The quality and to some extent the nature of the separations achieved is influenced by the amount of liquid phase applied to the inert support in packed column GC. Low levels of polyesters (1-3 % by weight relative to that of the support) are occasionally suggested for use with highly inert supports, as methyl esters of long-chain polyunsaturated fatty acids will elute from them at comparatively low temperatures. On the other hand, greater amounts of polyester (10-15 % by weight) offer more protection to polyunsaturated esters at high temperatures and this is generally preferred. The retention times of fatty acid esters relative to a chosen standard ester (usually 16:0 or 18:0) tend to decrease as the amount of liquid phase on the support is decreased and therefore, for reproducible work, it is advisable to determine the optimum amount of liquid phase necessary for a given separation and to standardise the chromatographic conditions accordingly. Unfortunately, columns age with use as the stationary phase polymerises further or bleeds from the column, and some changes in the retention characteristics of esters inevitably occur.

When the technique was first introduced, there were a number of reports of losses of esters of polyunsaturated fatty acids on GC columns. These were attributed partly to the use of too active support materials and partly to transesterifications of methyl esters with the polyester liquid phase. The latter effect was caused by residues of the catalyst required for the preparation of the polyester, so that those components remaining longest on the columns suffered the greatest losses. Now, polyester liquid phases are made without the aid of catalysts and such losses should not be significant.

As long as packed columns continue to be employed for fatty acid analysis, it would aid interlaboratory comparisons, if more general use were made of a select number of phases; the author has found that three columns containing

15 % EGSS-Y™, EGSS-X™ and Silar 10C™ on 100-120 mesh acid-washed and silanised supports cover a sufficient range of polarity for the analysis of most polyunsaturated fatty acids. Fatty acids which co-chromatograph on a given column will often be separable on one of the others.

With the advent of WCOT columns of fused silica, the number of different stationary phases in use for the analysis of fatty acid derivatives has appeared to diminish, although phases covering the extremes of the polarity spectrum do have specific uses. Polyglycol phases based on Carbowax 20M™ seem to have been used in a high proportion of recent published papers. Many of these are sold under trade designations so that their chemical derivation is not immediately obvious, but Carbowax-20M™, FFAP™, Supelcowax-10™, and SP-1000™, for example, are very similar. Ackman [14] has shown that these phases, variously in glass, fused silica or stainless steel WCOT columns, have very similar effects on the relative retention times of a wide range of fatty acid derivatives. Similarly, cross-linking or bonding to the support did not appear to affect relative retention times greatly, although others found that small but significant improvements in the resolution of certain critical pairs could be obtained with a cross-linked and bonded phase [493]. In polarity, these stationary phases fall into the last of the groups described above, but the inherent resolution of WCOT columns is such that satisfactory resolution is achieved for most practical purposes. Indeed, they can have distinct advantages in many applications (see below).

Ackman [14] has proposed that phases of the Carbowax 20M™ type should be utilised in the "'standard' reference WCOT column for interlaboratory studies as well as for application in its own right". For the moment and until any new phase with demonstrable advantages is introduced, this suggestion seems eminently sensible. The author has made considerable use of fused silica columns coated with Silar 5CP™, a phase which is slightly more polar than Carbowax 20M™.

2. Carrier gases

It was demonstrated in Chapter 3 that hydrogen had undoubted advantages as a carrier gas in WCOT columns in terms of efficiency. On the other hand, the potential consequences of a leak of hydrogen within the oven of a gas chromatograph should be obvious to all. In modern instruments with solid state electronics, the risk is perhaps less than it was formerly, but it is still advisable to install leak detection equipment as a precaution. Helium is much safer, if rather costly outwith the U.S.A., and still gives excellent results. Both gases should be passed through oxygen and moisture traps in advance of the column to give greater base-line stability at high sensitivity and to prolong column life.

With packed columns, nitrogen (containing less than 5 ppm of oxygen) affords adequate resolution at the standard flow-rates. Argon is more expensive than nitrogen, but it is generally obtainable with a very low oxygen content so has advantages in the analysis of sensitive compounds.

3. The oven of the gas chromatograph

It should go without saying that the oven in a gas chromatograph should have a highly-sensitive means of controlling the temperature, and that the temperature should be uniform in all regions. Unfortunately, when WCOT columns of fused silica were introduced, it became apparent that higher standards were needed in some commercial instruments. The reasons for this are still not entirely clear, but it may be related to the thinner walls or the rate of transmission of heat through fused silica rendering such columns more sensitive to minor temperature fluctuations or uneven heat distributions within the oven. In practice, the effects were manifested by poor peak shapes, often exhibiting spiking. Ackman [14] solved the problem in his laboratory by enclosing the column in a simple chamber made from disposable aluminium pie plates. The author merely wrapped aluminium foil round his columns to resolve the difficulty (at the suggestion of the instrument manufacturer). New equipment ought not to suffer from this problem.

C. PROVISIONAL IDENTIFICATION USING STANDARDS OR RETENTION TIME RELATIONSHIPS

Lipid analysts soon acquire an intuitive understanding of the relationship between the retention times of peaks on a GC trace and their identity. For example, a typical fingerprint of the fatty acids from animal tissue phospholipids would have the 16:0 component standing in relative isolation, followed by the three peaks for the C_{18} components (18:0, 18:1 and 18:2), then a gap to the next substantial peak for 20:4(n-6), followed by a further gap to the C_{22} components, the last of which is 22:6(n-3). Many of the minor peaks can be identified tentatively according to their proximity to these major components.

Please note that in this section and in the remainder of the Chapter, it will be assumed that fatty acids are being subjected to GC in the form of the methyl ester derivatives, unless it is stated otherwise.

When the fatty acids of a simple material such as maize (corn) oil is analysed in a laboratory for the first time, there should be no problems of identification, because its composition has been so well documented in the literature. This may also be true of some more complex lipid samples, such as extracts of rat liver. However, problems of identification can arise whenever any new or unknown sample is analysed, or when trace components are seen in otherwise familiar samples, now a relatively common situation with WCOT columns. If the interpretation of metabolic events in tissues hinges on the recognition of a particular fatty acid, intuitive labelling will not suffice. Components can be identified with much more certainty from a systematic study of retention time relationships on particular liquid phases, as described in the remainder of this section. Ultimately, it may be necessary to use the methods described in the next two Chapters for complete certainty.

Standard mixtures containing accurately known amounts of methyl esters of saturated, monoenoic and polyenoic fatty acids are available commercially from a number of reputable biochemical suppliers. These are invaluable for checking the quantification procedures used (see Section G below), and also for the provisional identification of fatty acids by direct comparison of the retention times of their methyl esters with those of the unknown esters on the same columns under identical conditions. Comparisons should be made on at least two columns with phases of different polarity.

The lipids in animal tissues usually contain a much wider spectrum of fatty acids than is available commercially. It is then helpful to obtain a secondary external reference standard consisting of a natural fatty acid mixture of known composition. This can be a common natural product that has been well characterised or a mixture of natural esters, the composition of which has been accurately established by the procedures described in Chapters 6 and 7. Ideally, it should be similar to the samples under investigation; for example, Ackman and Burgher [15] used cod liver oil in this way to identify the fatty acids of other marine animals, and Holman *et al.* [392,393] have used the fatty acids of bovine and porcine testes in the same manner in analyses of animal tissues. Rat liver fatty acids are frequently used for the latter purpose. In work with packed columns, the author used a simple mixture made up of the fatty acids of pig liver lipids and cod liver oil together with a little linseed oil, as this contains significant amounts of all the major fatty acid classes (saturated and mono-, di-, tri-, tetra-, penta- and hexaenoic components of both the (n-3) and (n-6) families), including most of the fatty acids likely to be encountered in animal tissues; the ingredients are readily obtainable from local shops.

Figure 5.1 illustrates separations of such a standard mixture, on 15 per cent EGSS-X™ and EGSS-Y™ in packed columns. Both phases give excellent separations of fatty acid esters of a given chain-length that differ in degree of unsaturation. Separation of esters which differ only in the positions or configurations of the double bonds, where these are approximately central, is not easily achieved with monoenoic acids, but is possible with esters of polyunsaturated fatty acids. For example, on both of these columns, 18:3(n-3) and 18:3(n-6) are separated, as are three isomers of 20:3, two isomers of 20:4 and two isomers of 22:5 as is apparent in Figure 5.1. With the methyl esters of the more common families of polyunsaturated fatty acids, the shorter the distance between the last double bond and the end of the molecule, the longer the retention time of the isomer.

The principle disadvantage of these columns is that there is some overlap

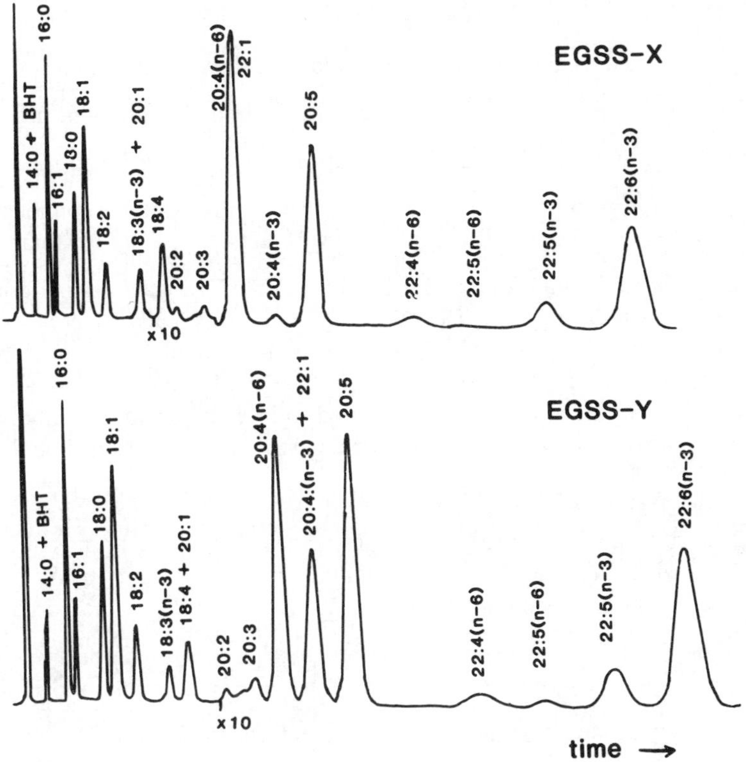

Figure 5.1 GC analysis of a complex mixture of natural fatty acids (as the methyl ester derivatives) on packed columns with EGSS-X™ and EGSS-Y™ as stationary phases (see footnote to Table 5.1 for further chromatographic details).

of fatty acids of different chain lengths. For example on EGSS-X™, 18:3(n-3) and 20:1(n-9) coincide, as do 20:4(n-6) and 22:1(n-9). On EGSS-Y™, these pairs can be separated, but 20:0 and 18:3(n-3), 18:4(n-3) and 20:1(n-9) or 22:1(n-9) and 20:4(n-3) are not separable. If Silar 10C™ had been selected as a stationary phase rather than EGSS-Y™, a different range of separations again would be seen, complementing those on EGSS-X™. Again, by a judicious use of two or more columns differing in polarity in this way, most of the fatty acids of metabolic importance can be separated and estimated.

With WCOT columns, the inherent resolution is such that there tend to be fewer problems of overlap of major components. On the other hand, a multiplicity of peaks may be revealed, so compounding the identification problems. It is possible to eliminate the difficulties with overlapping components of different chain-lengths by using low polarity polyester liquid phases such as those of group d above (but not with packed columns in which the resolution is markedly inferior to that obtained with more polar liquid phases).

As cautioned earlier, it should be noted that the retention times of esters and the separations achieved are all dependent on the precise column conditions used and may vary with such factors as the temperature or the age of a column and the amount of stationary phase on the support (with packed columns).

The absolute retention time of an ester on any GC column has very little meaning as a measure of its elution characteristics, because slight changes in the operating conditions or in the character of the packing material (in its origin or on ageing) can affect this parameter drastically. In contrast, the retention time of a fatty acid ester relative to that of a chosen standard commonly-occurring component (usually 16:0 or 18:0) has a greater absolute significance and is a quantity more suited to inter-laboratory comparisons, i.e. the *relative retention time* ($r_{18:0}$) of an ester is its retention time divided by that of 18:0.

Theoretically, retention times should be measured from the time of emergence of an unretained sample on the gas chromatographic column to the time when the peak is at its maximum, but as very large peaks may be skewed, it has been suggested that the distances should be measured from the point at which the solvent first emerges to that where the tangent drawn to the leading edge of the peak intercepts the base line [7]. Also, as the retention times of esters are influenced to some extent by components eluting immediately adjacent to them, relative retention times should be measured on pure compounds or on simpler fractions isolated by silver ion chromatography (see Chapter 6) say, whenever this is possible. The retention times relative to that of 18:0 ($r_{18:0}$) of the component esters of the external reference standard, separated on packed columns of EGSS-X™ and EGSS-Y™ (as illustrated in Figure 5.1), are listed in Table 5.1, as are similar values for a Silar 10C™ column. An interesting feature of the Silar 10C™ column is that the retention time of 22:6(n-3), the last component to emerge in most analyses of animal lipids, relative to that of 18:0 is much lower than that obtained with the other polyester columns. In practice, this can mean that shorter analysis times are possible while maintaining a good spread of peaks in the chromatogram in isothermal analyses especially. Relative retention times vary somewhat with the conditions of the column packing materials and with other operating parameters, such as temperature or flow-rate, but these variations are comparatively small and are in the same direction for all components.

Kovats' retention indices are more generally accepted as a standard means of recording GC retention data, but have been little used for the esters of fatty acids. Analogous parameters known as *equivalent chain-lengths* (abbreviated to ECLs) [615] or *carbon numbers* [1003] (the latter term is now little used because of its ambiguity) have considerable utility, however. ECL

TABLE 5.1

Equivalent chain lengths and relative retention times of some unsaturated esters on packed columns with EGSS-X, EGSS-Y and SILAR 10C as stationary phases.

Methyl ester	ECLs EGSS-X	ECLs EGSS-Y	ECLs Silar 10C	Relative retention times* EGSS-X	Relative retention times* EGSS-Y	Relative retention times* Silar 10C
16:0	16.00	16.00	16.00	0.58	0.57	0.67
16:1 (n-9)	16.57	16.62	17.08	0.69	0.68	0.84
16:2 (n-6)	17.65	17.45	—	0.90	0.85	—
18:0	18.00	18.00	18.00	1.00	1.00	1.00
18:1 (n-9)	18.53	18.52	19.00	1.19	1.16	1.23
18:2 (n-6)	19.42	19.20	20.07	1.52	1.41	1.56
18:3 (n-6)	20.00	19.67	21.10	1.80	1.60	1.75
18:3 (n-3)	20.40	20.02	21.28	2.02	1.78	2.01
18:4 (n-3)	21.05	20.52	22.12	2.40	2.04	2.37
20:0	20.00	20.00	20.00	1.77	1.76	1.53
20:1 (n-9)	20.50	20.45	20.90	2.08	2.01	1.84
20:2 (n-6)	21.40	21.15	—	2.67	2.44	—
20:3 (n-9)	21.63	21.33	22.22	2.87	2.57	2.22
20:3 (n-6)	21.77	21.53	22.72	2.99	2.72	2.42
20:3 (n-3)	21.95	21.60	23.10	3.13	2.78	2.64
20:4 (n-6)	22.43	22.00	23.53	3.59	3.10	3.14
20:4 (n-3)	23.00	22.47	24.13	4.18	3.54	3.55
20:5 (n-3)	23.50	22.80	24.77	4.85	3.91	4.06
22:0	22.00	22.00	22.00	3.14	3.12	2.31
22:1 (n-9)	22.43	22.35	22.80	3.59	3.44	2.71
22:4 (n-6)	24.45	24.00	25.70	6.34	5.48	4.94
22:5 (n-6)	24.57	23.85	25.96	6.55	5.27	5.14
22:5 (n-3)	25.53	24.80	26.83	8.60	6.92	6.20
22:6 (n-3)	26.18	25.50	27.40	9.10	7.74	6.89
24:0	24.00	24.00	24.00	5.59	5.50	3.50

* Relative to 18:0

EGSS-X and EGSS-Y: Data obtained with 2m x 4mm i.d. glass columns packed with 15% (w/w) stationary phase on Chromasorb W (100-120) mesh, acid washed and silanised). Carrier gas, Nitrogen at 50ml/min; column temperatures, 178°C (EGSS-X) and 194° (EGSS-Y).

Silar 10C: Data obtained with 3m x 4mm i.d. glass column packed with 10% Silar 10C on Gaschrom Q (100-120 mesh). Carrier gas, nitrogen at 11.5ml/min; column temperature, 210°C

values can be calculated from an equation similar to that for Kovats′ indices [436], but are usually found by reference to the straight line obtained by plotting the logarithms of the retention times of a homologous series of straight-chain saturated fatty acid methyl esters against the number of carbon atoms in the aliphatic chain of each acid (Figure 5.2). (Semilogarithm paper is particularly convenient for the purpose). The retention times of the unknown acids are measured under identical ***isothermal*** operating conditions and the ECL values are read directly from the graph. The ECL values of the esters of the component acids separated as illustrated in Figure 5.1 are listed in Table 5.1, and were obtained in this way. Such values have more obvious physical meaning than relative retention times and are more easily remembered. It should be noted, however, that outwith the normal range of chain-lengths (C_{14} to C_{22}), a straight line relationship between log retention time and number of carbon atoms may no longer hold [436].

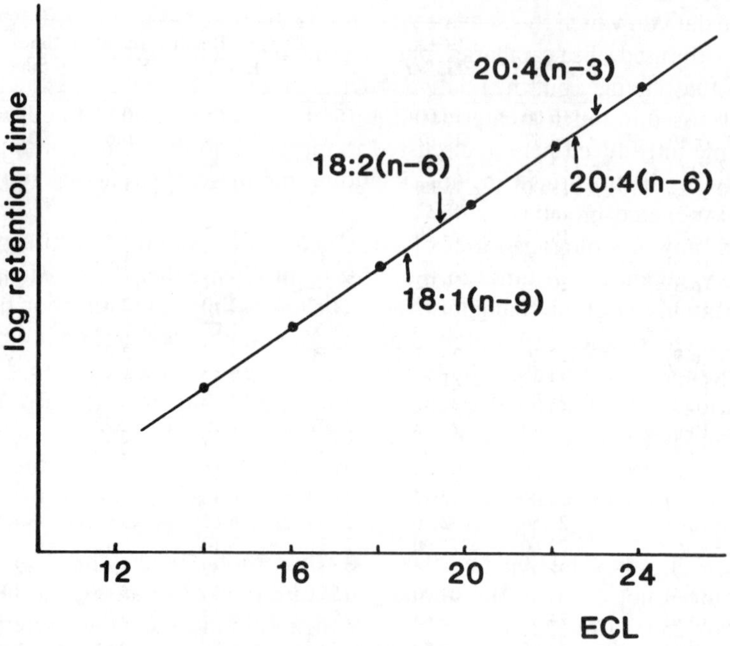

Figure 5.2 Plot of the logarithms of the retention times of the methyl ester derivatives of a homologous series of saturated fatty acids against the number of carbons in the aliphatic chains (*Equivalent Chain-Lengths* (ECL)), on a packed column of EGSS-X™ (see footnote to Table 5.1 for further chromatographic details). The elution times of some unsaturated fatty acids are indicated.

The increment in ECL value of a given ester over that of the saturated ester of the same chain-length, sometimes known as the *fractional chain-length* (or FCL) value, is dependent on the structure of the compound, and it is influenced by the number of double bonds in the aliphatic chain and the distance of the double bonds from the carboxyl and terminal ends of the molecule. From studies with synthetic fatty acids, such as the complete series of C_{18} monoenoic (2-18:1 to 17-18:1) [85,323], methylene-interrupted dienoic (2,5-18:2 to 14,17-18:2) [158] isomers and many others [326,387], a picture began to emerge and was well documented by Jamieson (but with special emphasis on packed columns) [435] and more recently by Ackman [13]. A single double bond in the centre of a long aliphatic chain gives an FCL value, when compared with that of the corresponding saturated compound, of about 0.4 to 0.5 on a DEGS column, and as the double bond nears the carboxyl end of the molecule, the FCL value tends to increase slightly; as the double bond nears the terminal end of the molecule, the FCL value increases somewhat more rapidly. Proximity of a functional group of any kind to the terminal end of the molecule appears to have a greater effect on FCL values than when it is a similar distance from the carboxyl end.

The data on which this scheme was based may now be superceded by new work with fused silica capillaries, but the principle still holds true; the absolute values may change, but the relative order of elution does not. These matters are discussed in much greater detail in the following section. Although the principle introduced here is discussed in terms of double bonds, it applies equally to all substituent groups, including methyl-branches, ring systems and oxygenated moieties.

With any homologous series of fatty acids that contain a substituent in the alkyl chain, the distance of the substituent from either the proximal or the terminal end of the chain must vary. The logarithms of the retention times of the esters of such series plotted against the numbers of carbon atoms in the chain do not therefore lie in straight lines unless the series are short. Deviations are greatest for short-chain esters, but with longer-chain compounds, the FCL values obtained for esters with similar groups of double bonds and terminal structures are approximately constant. Using the data in Table 5.1 for the EGSS-X™ column, 20:4(n-6) has an ECL value of 22.43 (i.e. 20.00 + 2.43) and that of 22:4(n-6) is 24.45 (i.e. 22.00 + 2.45); 20:4(n-3) has an ECL value of 23.00 (i.e. 20.00 + 3.00) so we can predict that 22:4(n-3), which does not occur in the standard mixture, will have an ECL value of close to 25.00 (i.e. 22.00 + 3.00). With a suitable secondary reference standard and packed columns, the ECL values of the esters of most of the fatty acids likely to be encountered in animal tissues can be measured or predicted. With WCOT columns, a simple arithmetic procedure of this kind will certainly indicate the general area in which a specific component can be expected, but may not be sufficiently precise in all regions of a chromatogram for positive identification unless particular care is taken in making the measurements.

Marine oils, which may contain more families of polyunsaturated fatty acids, present more complicated identification problems and there are difficulties in applying FCL factors to esters of shorter chain-length. Ackman [7] has proposed the use of an alternative series of systematic separation factors, which can be calculated from the relative retention times of esters, to assist in overcoming these problems, and these are discussed below.

Again it must be emphasised the ECL values may vary a little with column conditions (e.g. temperature and carrier gas flow-rates), with the nature of the support, with the amount of liquid phase and with the age of columns. ECL values obtained in allegedly similar circumstances in other laboratories may be taken as a guide, but should be applied with caution. A method of overcoming some of the difficulties of interlaboratory comparisons has been developed in Jamieson's laboratory [436,440]. Polyester liquid phases are considered as a single class of substances varying continuously in polarity, and the ECL value of methyl linolenate (18:3(n-3)) on a given column is a function of this polarity. If the ECL value of 18:3(n-3) is known, those of other fatty esters can be obtained from tabulated values, or from a simple equation and computer-derived constants.

It is usually stressed that ECL values must be obtained under isothermal conditions. However, it has been shown that the relationship between ECL values and temperature is a rectilinear one, and that in linear temperature-programmed analyses with gradients of 0.5 to 3.5°C/min, this can be expressed by -

$$\text{ECL} = A + B\text{T}$$

- where A and B are constants depending on the nature of the solute and the chromatographic system [497]. This finding may be useful when isothermal analysis is impracticable.

Of the many ancillary techniques that can help with provisional identification of fatty acids, silver ion chromatography is probably the most useful or cost-effective (see Chapter 6). Mass spectrometry of course has enormous advantages, when it is available, for unequivocal identifications (Chapter 7).

D. POSITIONAL AND CONFIGURATIONAL ISOMERS OF UNSATURATED FATTY ACIDS

1. Cis- and trans-monoenoic fatty acids.

Most lipids of animal origin contain a wide range of isomeric fatty acids in which the positions of the double bonds differ. In addition, isomeric fatty acids are generated during commercial processing of fats and oils, especially during the partial hydrogenation step which is employed to raise the melting point of fish and vegetable oils in margarine manufacture. Many different isomeric fatty acids thus enter the food chain and appear in human tissues. The magnitude of the analytical task thus varies with the nature of the sample, and may not always be soluble by GC methods alone. Before discussing practical examples, it may be instructive to examine some data obtained with pure compounds. It should perhaps be reiterated at the outset of this discussion that ECL values from a particular laboratory under specified conditions can rarely be reproduced in another, a problem compounded by the changes in the nature of column materials in recent years. Nevertheless, while the specific numerical values have no absolute significance, the order of elution of particular components does have considerable relevance. Ackman [13] has compiled an exhaustive list of retention data for unsaturated fatty acids. A relatively few examples only are listed here.

A complete series of C_{18} *cis*-monoenoic fatty acids (2-18:1 to 17-18:1) was synthesised first by Gunstone and colleagues in St Andrews, and they obtained retention (ECL) data for the methyl ester derivatives on several different columns [85,323]. To illustrate the pattern, some of this data is depicted graphically in Figure 5.3. While the absolute values differ for each stationary

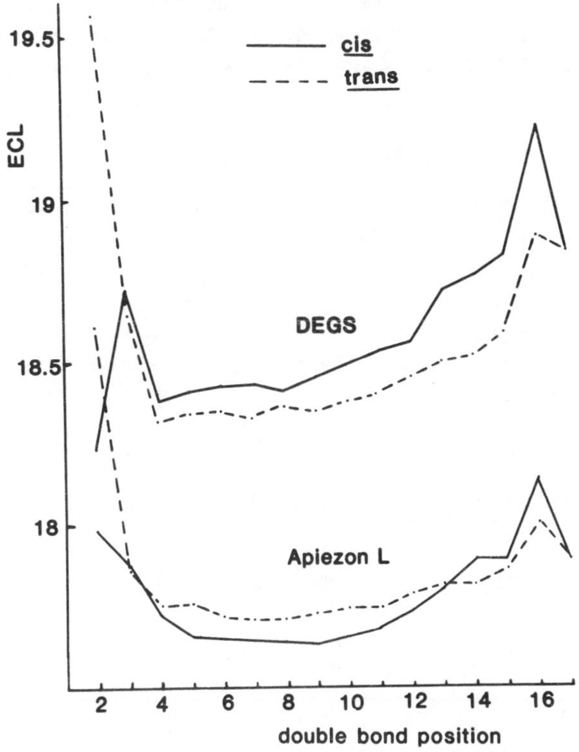

Figure 5.3 Variation of ECL values for the methyl ester derivatives of the isomeric octadecenoic acids (*cis* and *trans*) with double bond position on WCOT columns coated with Apiezon L^{TM} and DEGS [85].

phase, increasing with the polarity of the phase, the elution patterns are broadly similar. Thus the ECL values are lowest when the double bonds are approximately central, i.e. in positions 8 or 9. They increase relatively rapidly as the double bonds near the terminal (methyl) end of the molecule, reaching a maximum with 16-18:1, before falling slightly for the 17-isomer (the geometrical configuration of which cannot be designated). Similarly, the ECL values increase, although rather more slowly, as the double bonds near the carboxyl group and reach a second lower maximum at 3-18:1, before dropping once more for the 2-isomer. In addition, the ECL value of the 6-isomer is sometimes sightly out of line, being higher than those of adjacent isomers. The phenomenon has been explained theoretically in terms of the shapes of molecules and the opportunities for interaction between the double bonds and the walls of a WCOT column [92]. Analogous data for these and many other monoenoic fatty acids have been published [13,22,565, 815,828,845]. The author recently had the opportunity to re-determine ECL data for the C_{18} *cis*-monoenes on WCOT columns of fused silica and coated with one non-polar (a 5% phenylmethyl silicone) and three polar stationary phases (Carbowax 20M™, Silar 5CP™ and CP-Sil 84™), and this is listed in Table 5.2 [170].

Table 5.2
Equivalent Chain-Lengths of the methyl ester derivatives of isomeric C_{18} mono- and dienoic fatty acids [170].

Fatty acid	Stationary phase			
	Silicone	Carbowax	Silar 5CP	CP-Sil 84
3-18:1	17.91	18.44	18.47	18.64
4-18:1	17.80	18.19	18.21	18.30
5-18:1	17.72	18.09	18.17	18.29
6-18:1	17.75	18.18	18.25	18.43
7-18:1	17.72	18.14	18.24	18.40
8-18:1	17.72	18.14	18.29	18.43
9-18:1	17.73	18.16	18.30	18.47
10-18:1	17.75	18.19	18.32	18.49
11-18:1	17.78	18.23	18.36	18.54
12-18:1	17.83	18.30	18.46	18.62
13-18:1	17.89	18.37	18.52	18.67
14-18:1	17.95	18.46	18.57	18.76
15-18:1	18.00	18.56	18.62	18.83
16-18:1	18.19	18.84	18.91	19.14
17-18:1	17.94	18.54	18.61	18.82
3,6-18:2	17.73	18.69	18.74	19.14
4,7-18:2	17.58	18.43	18.48	18.82
5,8-18:2	17.53	18.36	18.48	18.85
6,9-18:2	17.59	18.47	18.63	19.04
7,10-18:2	17.57	18.45	18.70	19.02
8,11-18:2	17.61	18.50	18.75	19.12
9,12-18:2	17.65	18.58	18.80	19.20
10,13-18:2	17.73	18.67	18.88	19.30
11,14-18:2	17.81	18.79	18.98	19.41
12,15-18:2	17.89	18.91	19.08	19.51
13,16-18:2	18.12	19.28	19.41	19.90
14,17-18:2	17.90	18.98	19.06	19.52
5,12-18:2	17.56	18.40	18.60	18.93
6,12-18:2	17.60	18.48	18.69	19.06
7,12-18:2	17.56	18.40	18.64	18.99
8,12-18:2	17.63	18.49	18.70	19.07
6,10-18:2	17.58	18.42	18.61	18.94
6,11-18:2	17.52	18.38	18.62	18.93

As isomers differing in ECL value by about 0.04 should be separable on most WCOT column, it would be expected that those fatty acids with central double bonds (about 4-18:1 to 9-18:1) will not be easily resolved; petroselinic (6-18:1) and oleic acids occur together in some seed oils and are not readily resolved by GC (although this is possible with reversed-phase HPLC [168]). In the monoenoic fatty acids from animal tissues, there tend to be isomers in which the double bond positions are two carbon atoms apart, because they are formed biosynthetically from homologous fatty acids by chain-elongation or by beta-oxidation, in each instance the difference being two carbon atoms (see Chapter 2). Thus 16:1(n-9) and 16:1(n-7), 18:1(n-9) and 18:1(n-7), and 20:1(n-11), 20:1(n-9) and 20:1(n-7) are frequently found together and they are usually separable. A good example of this is seen in the next Chapter (Figure 6.4(B)). As the separation improves with increasing distance of the double bonds from the carboxyl group, resolution of isomers from particular biosynthetic families tends to improve with increasing chain-length. Monoenoic acids with double bonds in even numbered positions are not usual constituents of animal tissues, and the presence of 12- and 14-16:1 (separated on a WCOT column) in rat hepatoma was seen as evidence of an error of metabolism, for example [993]. In hydrogenated fats, double bonds are found in the even positions as often as in the odd, and this is discussed further below.

ECL data for the complete series of C_{18} monoenes with double bonds of the *trans*-configuration have also been published [85,323], and they are likewise depicted schematically for phases of medium and low polarity in Figure 5.3. The pattern tends to resemble that for the monoenes in that ECL values are lowest when the double bond is in the centre of the chain, and increase as it nears either end of the molecule. In this instance, the 2-isomer often has the highest ECL value. For a given double bond position, the *trans*-isomer nearly always elutes a little before the corresponding *cis* compound on polar phases, the difference tending to increase with increasing polarity [845,882]. On non-polar phases, *cis*-isomers elute before *trans* [323,870].

In practice then, *cis*- and *trans*-isomers of mono-unsaturated fatty acids are not easily separated on the more widely-used polyester stationary phases, but some excellent resolutions have been recorded with the newer highly-polar cyanoalkylpolysiloxane phases such as Silar 10C™, SP-2340™ or OV-275™. For example, near base-line separations of methyl oleate and methyl elaidate have been achieved on longer than usual packed columns (6 to 7 m x 0.2 mm i.d.) containing these phases [193,222,297,682,706,949]. Obviously, much better analyses of the pure compounds can now be obtained with the same phases in WCOT columns. In addition, excellent resolution of the *cis*- and *trans*-isomers of 11- and 13-22:1 fatty acids was achieved on a glass WCOT column (44 m x 0.3 mm) coated with the non-polar phase OV-73™ [870]. Acceptable separations of isomers have been achieved on a liquid-crystal phase [679].

It should be noted, however, that many of the published separations have been accomplished with simple model mixtures. With many lipids of potential commercial interest, such as partially-hydrogenated vegetable or fish oils, there is a wide range of positional as well as configurational isomers. Ruminant fats also contain many different positional isomers, as by-products of biohydrogenation of dietary fatty acids by microorganisms in the rumen [162]. The problems involved in this type of analysis have been reviewed [190,813,882]. It appears that acceptable results can be attained by GC with the more polar phases, especially with hydrogenated vegetable oils, as opposed to fish oils; the former are used almost exclusively in the U.S.A, while the latter are employed extensively in much of the rest of the World for margarine manufacture. For example, in a large collaborative trial [297], a method described originally by Perkins *et al.* [706] gave consistently better results than the more traditional infrared spectrophotometric procedure. A glass packed column (6.1 m x 2 mm i.d.) with 15% OV-275™ stationary phase on 100-120 mesh Chromosorb P™ (acid-washed and silanised), helium or nitrogen as the carrier gas and an isothermal column temperature of about 220°C were recommended. The nature of the separation is illustrated in Figure 5.4A. Although individual isomers of the *cis*- and *trans*-monoenes are not separated, this is probably not necessary in many circumstances.

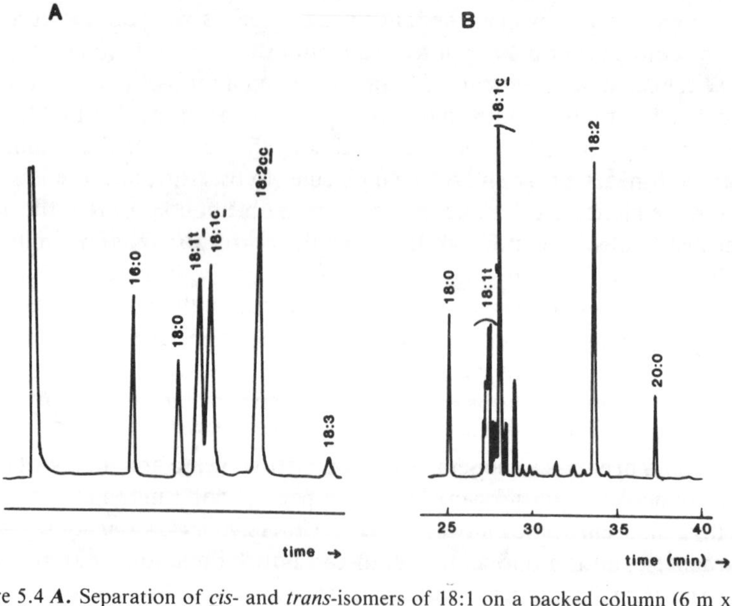

Figure 5.4 ***A***. Separation of *cis*- and *trans*-isomers of 18:1 on a packed column (6 m x 2 mm i.d.) containing 15% OV-275™ on Chromosorb P AW-DCMS™ (100-120 mesh). Helium was the carrier gas and the operating temperature was 220°C [297]. (Reproduced by kind permission of the authors and of the *Journal of Official Analytical Chemists*, and redrawn from the original paper).
B. As A, but with a WCOT column (100 m x 0.25 mm) coated with SP-2560™, with hydrogen as the carrier gas and at a temperature of 175°C (redrawn from the original figure) [891].

Many published separations testify to the resolving power of WCOT columns in analyses of hydrogenated vegetable fats and of ruminant fats [115,535,551,683,785,851,891], and one such is illustrated in Figure 5.4B; *cis*- and *trans*-isomers elute as series of partially resolved peaks [891]. Regrettably, it is not easy to state exactly where the division between the two groups of peaks lies as *trans*-components with double bonds close to the terminal part of the molecule may overlap with isomers with *cis*-double bonds nearer to the centre. Van Vleet and Quinn [949] have argued that many workers have exaggerated the quality of the separation that can be achieved, since *trans*-11- and *cis*-9-18:1 are not readily resolved on a WCOT column coated with SP-2340™; they contend that the best results are obtained with lengthy packed columns of OV-275™. Similar conclusions were reached by Strocchi *et al.* [885]. Hydrogenated fish oils contain a wider spread of isomers and of chain-lengths, and determination of the *cis*- and *trans*-content of these cannot yet be achieved by GC methods alone, in spite of occasional claims to the contrary [16,680,829,845,870,894].

The author believes that even those claims for packed columns may prove to be illusory on rigorous examination, and suspects that other procedures, especially silver ion HPLC, will ultimately prove to be more reliable. On the other hand, it is probably true to say that there is no method currently available that can be recommended with complete confidence. The best GC method for the determination of the relative proportions of the total *cis*- and *trans*-isomers of monoenoic esters involves epoxidation of the double bond followed by analysis of the products, *trans*- eluting ahead of the corresponding *cis*-isomers [243,244,514]. (A suitable procedure for the reaction is described in Chapter 4). Individual positional isomers are not resolved, but near base-line resolution of the configurational isomers is possible, especially with WCOT columns.

2. Dienoic fatty acids

By far the most abundant dienoic fatty acid in nature is linoleic acid (18:2(n-6)), but other methylene-interrupted dienes are found in tissues, and dienes with more than one methylene group between the double bonds occur in some organisms. ECL data for the complete series of synthetic methylene-interrupted C_{18} dienes (2,5- to 14,17-18:2) have been published for packed and capillary columns [158]. In this instance too, the author has been able to re-determine the data for modern columns of fused silica, at the same time as for the C_{18} monoenes, and this is also listed in Table 5.2 [170]. With each of the phases examined, the ECL values tended to increase with the distance of the double bonds from the carboxyl group, though there are discontinuities for the 3,6- and 13,16-isomers, where the ECL values are higher than those of adjacent compounds (c.f. the data for the 3- and 16-monoenoic isomers). If the FCL values for the complete series of monoenes [323] are used to predict the ECL values for these dienes, the calculated values are somewhat lower than those actually found. Thus in the earlier work, there was found to be a mean difference between the actual and predicted values of 0.13 on the Apiezon L™ column and 0.16 to 0.18 on more polar phases. This probably means that there is some interaction (possibly homo-conjugation) between the double bonds or with the diallyl methylene group, that increases the dipole moment of the unsaturated system. Comparable results were obtained in other studies, and the same principle held whether the double bonds were of the *cis*- or the *trans*-configuration [22,815,828]. Using the newer data of Table 5.2, the discrepancy between the actual and found values tends to vary with the position of the double bonds as well as with the stationary phase, and is highest for double bonds in positions 5 to 11 (0.10 to 0.15) and diminishes towards either end of the molecule. The FCL values for monoenes together with factors for the interaction with the appropriate methylene groups (the difference between the actual and predicted results for the dienes) have been used by Ackman and coworkers especially for the prediction of ECL values; some examples are given below.

Isomeric forms of linoleate with *trans*-double bonds are frequently found in hydrogenated fats. The four possible isomers, i.e. 9-*trans*,12-*trans*-, 9-*trans*,12-*cis*-, 9-*cis*-,12-*trans*- and 9-*cis*,12-*cis*-18:2, have been separated in the order stated on WCOT columns coated with polar cyano-polysiloxane phases, as illustrated in Figure 5.5 [489,535,683].

ECL data have also been obtained for a number of synthetic octadecadienoates with more than one methylene group between the double bonds [22,326,532,533,545,815,816,828], and some newly-acquired information is listed in Table 5.2. Fatty acids of this type occur at trace levels in a variety of natural sources, and can be major components of the lipids of marine invertebrates. Although data for *cis,cis*-, *trans,trans*- and some *cis,trans*-dienes is available, only the first will concern most analysts. Once more, if FCL values from the monoene data are used to calculate ECL values for these components, the difference between the actual and predicted results was found to be small, i.e. 0.07, when there are two methylene groups between the double bonds; it becomes negligible, i.e. 0.00 to 0.02, when there are more than two methylene groups [326]. Again, this is confirmed by the newer data in Table 5.2, although there appear to be some differences according to the polarity of the stationary phase. Similar results were reported by others [21,22,828].

The presence of conjugated double bond systems in the alkyl chain increases the retention time of an ester considerably over that of a similar compound with methylene-interrupted double bonds. Methyl 9-*cis*,11-*trans*-octadecadienoate, for example, is a common minor constituent of ruminant and other tissues and has ECL values of 20.48 on EGSS-X™ and 20.24 on EGSS-Y™, i.e. appreciably greater than the corresponding values for methyl linoleate [161]. In addition, the configuration of the double bonds has a much more pronounced effect and some separation of geometrical

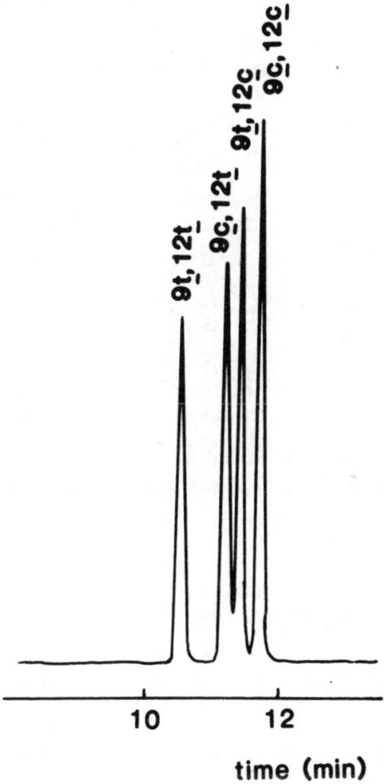

Figure 5.5 Separation of the four geometric isomers of linoleic acid on a glass WCOT column (30 m x 0.3 mm i.d.) coated with SS-4^{TM} [489]. Nitrogen was the carrier gas and the operating temperature was 190°C. (Reproduced by kind permission of the author and of the *Journal of Chromatography*, and redrawn from the original figure).

isomers of conjugated esters may be possible on columns containing the conventional polyester phases as well as on the cyanoalkylpolysiloxanes. ECL data for further conjugated dienoic isomers have been published [19,226,326].

Again, the epoxidation procedure described for monoenoic acids in the previous section has been employed in the GC analysis of configurational isomers of dienes and trienes. However, the author has observed (unpublished) that the method does not appear to work with fatty acids having conjugated double bond systems, and such compounds are formed during commercial hydrogenation reactions.

3. Polyenoic fatty acids

C_{18} and C_{20} fatty acids of the (n-9), (n-6) and (n-3) families are the most abundant trienoic fatty acids in animal tissues; they elute in this order on polar stationary phases and some ECL data are contained in Table 5.3. All

Table 5.3
Equivalent chain-lengths of the methyl ester derivatives of some natural fatty acids [170].

No	Fatty acid	Stationary phase			
		Silicone	Carbowax	Silar 5CP	CP-Sil 84
1	14:0	14.00	14.00	14.00	14.00
2	14-isobr	14.64	14.52	14.52	14.51
3	14-anteiso	14.71	14.68	14.68	14.70
4	14:1 (n-5)	13.88	14.37	14.49	14.72
5	15:0	15.00	15.00	15.00	15.00
6	16:0	16.00	16.00	16.00	16.00
7	16-isobr	16.65	16.51	16.51	16.50
8	16-anteiso	16.73	16.68	16.68	16.69
9	16:1 (n-9)	15.76	16.18	16.30	16.48
10	16:1 (n-7)	15.83	16.25	16.38	16.60
11	16:1 (n-5)	15.92	16.37	16.48	16.70
12	16:2 (n-4)	15.83	16.78	16.98	17.47
13	16:3 (n-3)	15.69	17.09	17.31	18.06
14	16:4 (n-3)	15.64	17.62	17.77	18.82
15	17:0	17.00	17.00	17.00	17.00
16	17:1 (n-9)	16.76	17.20	17.33	17.50
17	17:1 (n-8)	16.75	17.19	17.33	17.51
18	18:0	18.00	18.00	18.00	18.00
19	18:1 (n-11)	17.72	18.14	18.24	18.40
20	18:1 (n-9)	17.73	18.16	18.30	18.47
21	18:1 (n-7)	17.78	18.23	18.36	18.54
22	18:2 (n-6)	17.65	18.58	18.80	19.20
23	18:2 (n-4)	17.81	18.79	18.98	19.41
24	18:3 (n-6)	17.49	18.85	19.30	19.72
25	18:3 (n-3)	17.72	19.18	19.41	20.07
26	18:4 (n-3)	17.55	19.45	19.68	20.59
27	19:1 (n-8)	18.74	19.18	19.32	19.47
28	20:1 (n-11)	19.67	20.08	20.22	20.35
29	20:1 (n-9)	19.71	20.14	20.27	20.41
30	20:1 (n-7)	19.77	20.22	20.36	20.50
31	20:2 (n-9)	19.51	20.38	20.59	20.92
32	20:2 (n-6)	19.64	20.56	20.78	21.12
33	20:3 (n-9)	19.24	20.66	20.92	21.43
34	20:3 (n-6)	19.43	20.78	21.05	21.61
35	20:3 (n-3)	19.71	20.95	21.22	21.97
36	20:4 (n-6)	19.23	20.96	21.19	21.94
37	20:4 (n-3)	19.47	21.37	21.64	22.45
38	20:5 (n-3)	19.27	21.55	21.80	22.80
39	22:1 (n-11)	21.61	22.04	22.16	22.30
40	22:1 (n-9)	21.66	22.11	22.23	22.36
41	22:3 (n-9)	21.20	22.52	22.78	23.25
42	22:3 (n-6)	21.40	22.71	22.99	23.47
43	22:4 (n-6)	21.14	22.90	23.21	23.90
44	22:5 (n-6)	20.99	23.15	23.35	24.19
45	22:5 (n-3)	21.18	23.50	23.92	24.75
46	22:6 (n-3)	21.04	23.74	24.07	25.07

of the eight possible geometrical isomers of 9,12,15-octadecatrienoic acid (α-linolenic acid) have been prepared by nitrous oxide-catalysed elaidinisation [21-23,815] and by total synthesis [747]. Ackman and Hooper [21-23] were able to predict their ECL values, from data for the appropriate monoenes obtained on a WCOT column coated with Silar 5CP™, by applying diethylenic coupling constants for each pair of double bonds. Some of these isomers were detected in deodorised vegetable oils [24]. On the other hand, any dubiety about the order of elution was removed by having compounds of defined structure, and Rakoff and Emken [747] were able to demonstrate resolution into six components, eluting in the order *ttt* (ECL = 19.93), *ctt* and *tct* (20.11 and 20.12), *cct* and *ttc* (20.20 and 20.23), *tcc* (20.38), *ctc* (20.39), and *ccc* (20.52), on a WCOT column of Silar 10C™ as shown in Figure 5.6. ECL data for some C_{20} trienes, prepared by partial reduction of 20:5(n-3) with hydrazine, have been published [828].

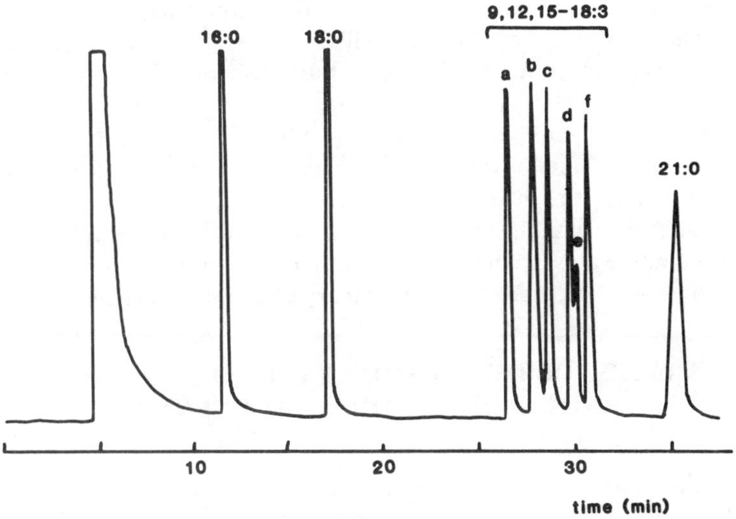

Figure 5.6 Separation of the geometrical isomers of 9,12,15-octadecatrienoic acid on a glass Quadrex™ WCOT column (50 m x 0.25 mm i.d.) coated with Silar 10C™ [747]. The carrier gas was helium and the operating temperature was 170°C. (Reproduced by kind permission of the authors and of *Chemistry and Physics of Lipids*, and redrawn from the original figure).

In some early work, conjugated trienes were reported to undergo *cis-trans* isomerisation and double bond migration on packed columns containing polyester stationary phases [609,619]. On the other hand, when an "all-glass" WCOT system was used with the non-polar stationary phase, OV-1™, it proved possible to separate a number of geometrical isomers, eluting in the order *ctc*-9,11,13- (ECL = 18.95), *ctt*-9,11,13- (18.99), *ttc*-8,10,12- (19.09), *ttc*-9,11,13- (19.10), *ttt*-9,11,13- (19.36) and *ttt*-8,10,12-18:3 (19.39), without any losses [900]. Comparable results have been obtained by others [226,282,394]. Indeed, somewhat better resolution was obtained with a WCOT column coated with Carbowax 20M™ (ECLs = 21.50 to 22.19), from which components eluted without loss in the same order as on an OV-1™ column [282].

Methylene-interrupted tetra-, penta- and hexaenoic fatty acids of the (n-6) and (n-3) families, eluting on GC in this order, are ubiquitous components of animal tissues and present no problems in analysis provided that glass or fused silica WCOT columns are employed. Some ECL data for the more frequently encountered components of this type, and obtained on modern WCOT columns of fused silica, are given in Table 5.3, and additional information can be found in most of the review articles cited above (see also Table 5.1). Even with compounds of this kind, it is possible to predict ECL values from the FCL values of the appropriate monoenes by applying correction factors for interactions with the methylene groups, and 18:5(n-3) was identified as a component of a marine alga in this way, for example [25].

The data in Table 5.2 can be used to illustrate the principles involved [170]. Thus, an ECL value for an 18:4(n-3) fatty acid on Carbowax 20M™ is equal to 18 + the FCL values (0.18 + 0.16 + 0.30 + 0.56 = 1.20) + the methylene group factors (0.13 + 0.12 + 0.05 = 0.30), i.e. ECL = 19.50; the actual value found by direct measurement is 19.45 (Table 5.3). With the silicone phase for this acid, the calculated and actual values were only 0.02 units apart, while with the more polar phases, the difference was 0.07 in each case. It therefore appears that this approach to the identification of unknowns is of some value, provided that the primary data are of sufficient accuracy.

E. SEPARATION OF THE COMMON FATTY ACIDS OF PLANT AND ANIMAL ORIGIN ON WCOT COLUMNS

It could be argued that almost any polar stationary phase may be used for the analysis of the methyl ester derivatives of the fatty acids of the common seed oils of commerce, which tend to contain a limited range of C_{16} and C_{18} components. Yet even these will be found to have a number of trace constituents, when the attentuation of the gas chromatograph is turned up somewhat, and it may sometimes be necessary to utilise similar techniques for identification as are applied to the more complex samples of animal origin. If the analyst wishes to look hard enough, all tissues will be found to contain an enormous range of different fatty acids; for example, 437 different fatty acids, including positional and geometrical isomers, have been found in cow's milk fat [697]. No single GC system could hope to resolve more than 15 to 20% of these, and a barrage of combinations of complementary techniques, of the kind described in Chapter 6, were applied to make the identifications. In this section, it is assumed that the goal of most analysts is to identify only those naturally occurring components present at a level of 0.1 % or more in a sample, relying mainly on relative retention times. To illustrate this aspect, a relatively few "typical" examples of applications to real samples have been selected. It is assumed that appropriate methylation procedures will be used to eliminate artefact formation (see previous chapter).

The "standard" WCOT column in general use nowadays is made of fused silica, and is probably 25 m in length and 0.2 to 0.25 mm in internal diameter. Narrower bore columns (0.1 mm i.d.) are now becoming available, and they hold promise of improved resolution. The author is not convinced of the value of wide-bore (0.5 mm i.d.) WCOT columns. As discussed in Chapter 3, a substantial increase in the length of a column is necessary before an appreciable improvement in resolution is achieved; there may be occasions when this is of value, but as a corollary there will be a substantial increase in the time required for an analysis. An alternative philosophy has been to make use of shorter columns, losing something of the resolution but gaining by reducing analysis times. For example, in a comparison of 100 m, 10 m and 2 m glass columns coated with SP-2340™ for the analysis of animal and hydrogenated vegetable fats, the 100 m column gave superb resolution as expected, but with analysis times of 2 hours or more, the 10 m column gave adequate resolution for the major components, with analysis times of under 30 minutes, while the resolution with the 2 m column was inadequate although analyses were completed in less than 4 minutes [536]. A 15 m fused silica column coated with Carbowax 20M™ gave acceptable resolution of the fatty acids of a rape seed oil with a high erucic acid content in about 3 minutes [542].

Polar phases are used almost universally for fatty acid analysis, although the inherent resolution of WCOT columns is such that some remarkable separations can be achieved even with non-polar silicone phases, which are more stable at elevated temperatures. Such columns are easier to manufacture than are polar ones. For example, the separation of a hydrogenated fish oil is a horrendous problem for any stationary phase, yet a published chromatogram obtained with a 44 m column coated with the non-polar OV-73™ (175,000 theoretical plates) is probably as good as any in which polar phases have been used [870]. A set of retention data has been published for cod liver oil fatty acids on a 50 m fused silica column coated with SP-2100™; a large number of isomeric branched-chain monoenoic and polyenoic components were clearly resolved [298]. A short column of this type was used for the analysis of plasma fatty acids [925].

The nature of the separation attained on non-polar columns is rather different from that with polar columns as can be seen from Figure 5.7, in which a separation of pig testis fatty acids on a 12 m fused silica column coated with BP5™ (a 5 % phenylmethylsiloxane polymer) is illustrated. Unsaturated components emerge ahead of saturated fatty acids of the same chain-length. Isomeric fatty acids differing in the positions of double bonds

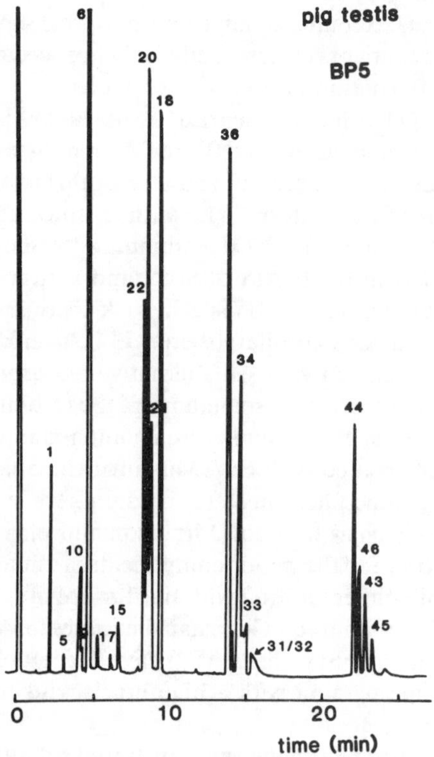

Figure 5.7 GC analysis of the fatty acids (methyl esters) of pig testis on a fused silica column (12 m x 0.22 mm i.d.) coated with BP5 (a 5 % phenylmethylsiloxane) (SGE Ltd, Milton Keynes, U.K.). Helium was the carrier gas at a flow-rate of 0.96 ml/min. The oven of the Packard Model 428 gas chromatograph, fitted with split injection (ratio 100:1), was maintained at 190°C for 3 min, then was temperature-programmed to 220°C at 1°/min. Peaks can be identified by reference to the numbers in Table 5.3.

are usually clearly resolved, thus 18:1(n-9) and 18:1(n-7) are separated as are many of the polyenes. Indeed, the C_{22} fatty acids are possibly almost as well separated as on a polar column of the same length. Unfortunately, there are substantial overlaps among the C_{18} fatty acids, and 18:2(n-6) is not fully resolved from 18:1(n-9); 18:2(n-6) and 18:3(n-3) merge completely, and this is also true of the corresponding C_{20} and C_{22} compounds [715]. As linoleate and linolenate are essential fatty acids with major nutritional importance, the deficiency in this aspect of the separation is likely to mitigate against a wider use of non-polar columns. The order of elution of the C_{22} components is not that which might expected intuitively, i.e. it is 22:5(n-6), 22:6(n-3), 22:4(n-6) and 22:5(n-3). In this instance, it appears that the position of the double bonds has a greater effect on retention time than does the number of double bonds. Non-polar phases do have advantages in specific applications, e.g. with fatty acids of high molecular weight or containing thermally-labile functional groups, and in GC-mass spectrometry, where their stability at high temperatures, their considerable degree of inertness and their low rate of bleed are virtues.

Nearly all analysts then are going to make use of polar stationary phases for the major proportion of their work. Any attempt at a comprehensive account of what has been achieved with columns of this type would be only slightly more readable than a telephone directory. I have therefore elected to show what can be accomplished with a "standard" 25 m WCOT column of fused silica coated with Carbowax 20M™, and with a similar column coated with a highly-polar phase CP-Sil 84™. As the first substrate for analysis, pig testis (obtained from an abattoir) lipids were selected, as the fatty acids have been well-characterised and have been used as an external standard in the Hormel Institute for some years [392,393]; they contain a wide range of fatty acids of the (n-6) series, encountered typically in animal tissues. The second substrate is cod liver oil (obtainable from any pharmacy), which has also been well-characterised and is used as an external standard in the analysis of lipids of marine origin [15]; it contains many different fatty acids, and especially those of the (n-3) family. Both of these materials were used by the author in studies of the efficacy of picolinyl ester derivatives of fatty acids for identification by GC-mass spectrometry (see Chapter 7), so all the main components have been identified unequivocally [173,184].

The separation of the methyl esters of the fatty acids of pig testis lipids on the Carbowax 20M™ column is illustrated in Figure 5.8. Each of the main chain-length groups is reasonably well resolved. For example, three 16:1 isomers are seen and they are distinct from the C_{17} fatty acids. Similarly the important C_{18} components are clearly separated from a minor 19:1 fatty acid, and this is in a different region from each of the C_{20} unsaturated constituents. With the last, the only serious overlap problem is with 20:3(n-3), which co-chromatographs with 20:4(n-6); these are, however, just separable on a slightly more polar Silar 5CP™ column. Finally, all the biologically important C_{22} fatty acids are cleanly resolved.

An analogous separation of the fatty acids of cod liver oil is illustrated in Figure 5.9. Here the C_{16} region contains a wider range of fatty acids than does the previous sample, including branched-chain and polyunsaturated fatty acids, but all the main isomers are resolved. Similarly, the C_{18} polyenoic fatty acids are eluted before the C_{20} fatty acids begin to emerge, and 20:5(n-3) comes well before the first of the C_{22} fatty acids.

With the more polar CP-Sil 84™ column, there is again excellent resolution of the pig testis fatty acids (Figure 5.10). Individual unsaturated esters are particularly well resolved, and for example there is near base-line separation of 18:1(n-9) and 18:1(n-7). On the other hand, 18:3(n-6) emerges after the minor C_{19} fatty acid. The C_{20} group are all well separated from

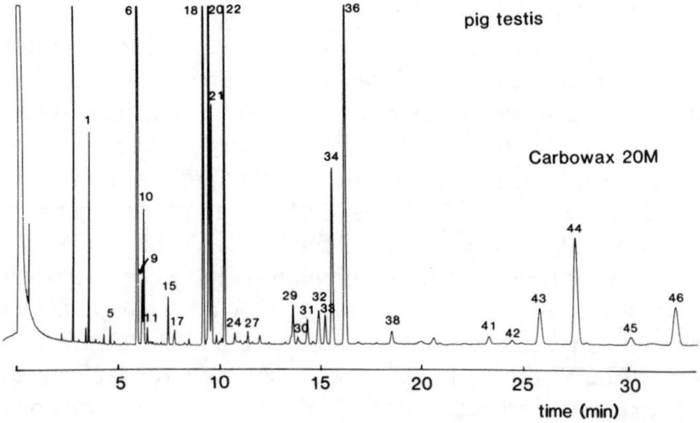

Figure 5.8 Separation of the fatty acids (methyl esters) of pig testis on a fused silicia column (0.25 m x 0.22 mm) coated with Carbowax 20M^{TM} (Chrompack UK). Hydrogen was the carrier gas at a flow-rate of 1 ml/min; the temperature was maintained at 165°C for 3 min, then was raised at 4°C/min to 195°C, where it was maintained for a further 23 min. A Carlo Erba Model 4130 gas chromatograph fitted with a split/splitless injection system was used with the split ratio set to 100:1. Peaks can be identified by reference to the numbers in Table 5.3.

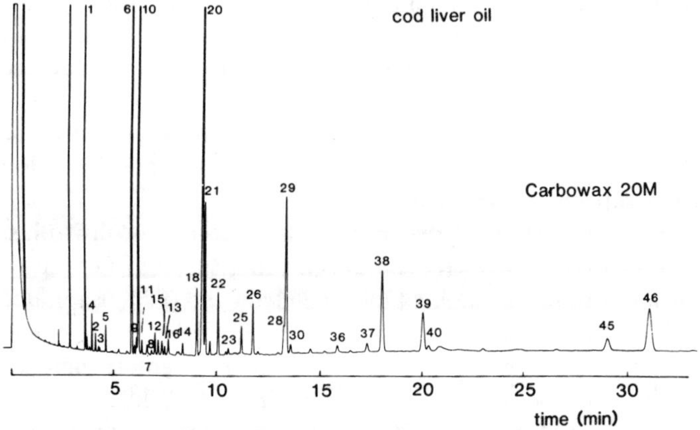

Figure 5.9 Separation of fatty acids (methyl esters) of codliver oil on a fused silica column coated with Carbowax 20M^{TM}. Conditions and peak identification are as in the legend to Figure 5.8.

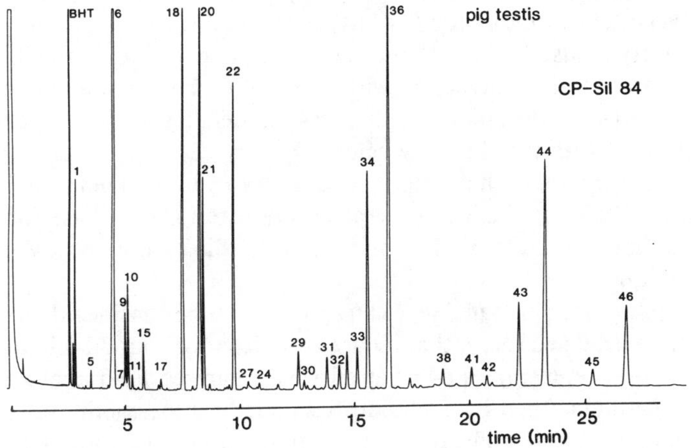

Figure 5.10 Pig testis fatty acids (methyl esters) separated on a fused silica WCOT column (25 m x 0.22 mm) coated with CP-Sil 84[TM] (Chrompak UK). The oven was maintained at 150°C for 5 min, then was raised at 2°C/min to 180°C, where it remained for a further 10 min. Other conditions and peak identifications are as in the legend to Figure 5.8.

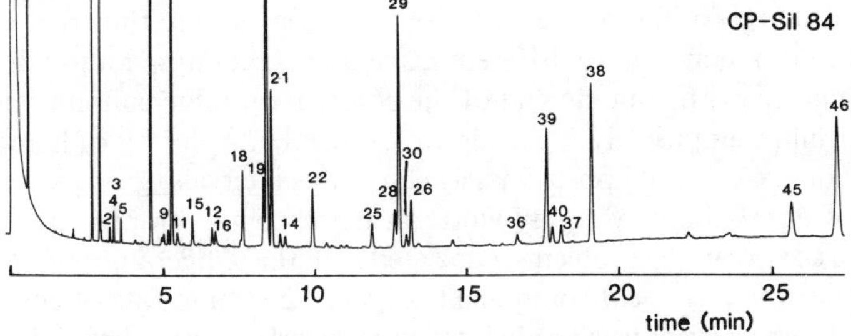

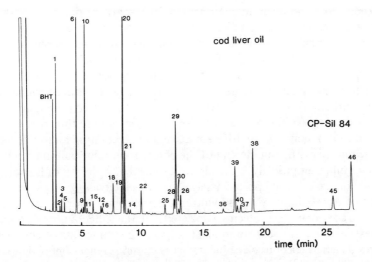

Figure 5.11 Cod liver oil fatty acids (methyl esters) separated as in the legend to Figure 5.10.

each other but are beginning to run into an area occupied by C_{22} fatty acids. This last effect can be more troublesome with fish oils, which contain appreciable amounts of 22:1 isomers as in the cod liver oil sample (Figure 5.11). The latter compounds emerge only just before 20:4(n-3). Similarly the C_{16} polyenes elute among the C_{18} fatty acids, and the C_{18} polyenes run into the C_{20} fatty acids. The C_{16} branched and monoenoic constituents tend to co-chromatograph. Nonetheless, with tissue lipids from plants and terrestrial animals especially, the polar column gives excellent results provided that care is taken in identifying the fatty acids.

These chromatographic traces lend support to Ackman's view that Carbowax 20M™ is the best general-purpose stationary phase for WCOT columns, and it is certainly recommended by this author to newcomers to the technique.

The separations illustrated here can by no means be considered the ultimate that can be achieved (Note that all are complete in about 30 minutes only). Rather, they are examples of what all well organised lipid analysts should be able to attain in their own laboratories under routine conditions. By using longer columns, extending the analysis time and taking particular care to optimise the separation conditions, greatly improved resolution is possible. It is invidious to have to select examples from the literature to illustrate the "state of the art", but analyses of natural mixtures that fall into this category include plasma lipids [115,429], baby food formulae [851], and sea urchin lipids (Takagi and Itabashi, reproduced by Ackman [14]).

F. SOME APPLICATIONS TO LESS COMMON FATTY ACIDS

1. *Short-chain fatty acids*

When short and medium chain-length fatty acids are present in a sample in addition to the common range of longer chain-length components, it is possible to obtain an analysis by isothermal gas chromatography at two or more different temperatures, if a peak common to each chromatogram is used as a reference point for quantification purposes. It is generally much more convenient and accurate, however, to use temperature-programming in such circumstances, i.e. to start the analysis at a low column temperature and raise it at a fixed reproducible rate to an appropriate final temperature. The optimum temperature differentials and programming rates will depend on the nature of the sample and of the chromatographic column, and must be determined empirically in each instance. Ideally, members of homologous series should emerge at approximately equal time intervals, giving symmetrical peaks of approximately equal width on the chromatographic trace. Some valuable advice on the problems associated with the temperature-programmed analysis of the fatty acid components of some common fats of commercial interest, such as coconut oil or butter fat, is contained in a brief review article [418].

Because of its commercial importance, milk fat has provided many analysts with a challenge. As mentioned earlier, 437 different fatty acids have been reported as being present, but fortunately only a few of these have any nutritional significance and only 20 to 30 are likely to concern most analysts. The separation itself is a small part of the problem, and quantification presents more difficulties; the various steps in the process have been studied by several research groups [72,78,420]. First, it is essential that transesterification (to methyl or butyl esters) should be carried out by a method that ensures neglible losses, and that described by Christopherson and Glass [185], or appropriate modifications of it, should be used for the purpose (see Chapter 4 for details). Next, there should be no discrimination during the introduction of the sample onto the gas chromatograph; a non-vaporising on-column injection technique is strongly recommended, especially with WCOT columns. As cautioned above, the rate of temperature-programming during the chromatographic run should be such that the major components emerge at approximately equal time intervals. A typical chromatogram is illustrated in Figure 5.12. There is a larger than usual solvent

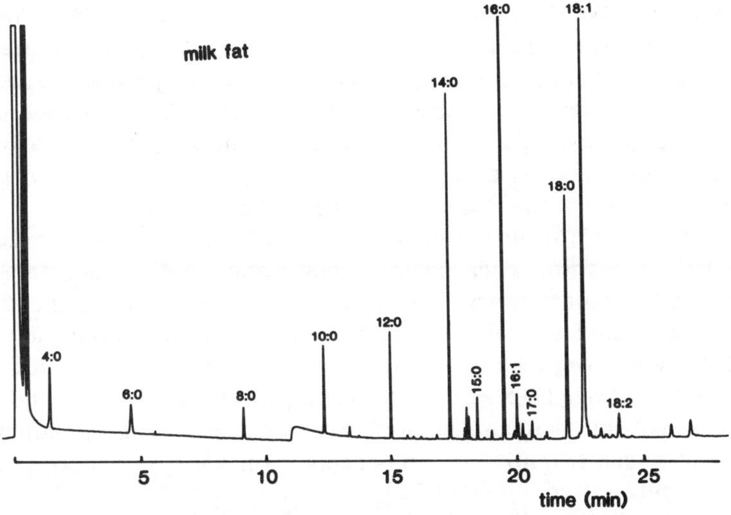

Figure 5.12 Milk fatty acids (methyl esters) separated on a fused silica column coated with CP-Sil 84^{TM}. The oven was held at 30°C for 3 min, then was raised at 8°C per min to 160°C and was held at this point for a further 10 min. Other conditions are as in the legend to Figure 5.8.

peak, because of the low initial temperature and the presence of methanol, and the base-line is almost steady before butyrate emerges followed by each of the remaining fatty acids in turn. Perhaps the single most comprehensive analysis is that by Strocchi and Holman [884]. Many different stationary phases have been used for milk fat analyses, and there is no clear favourite.

Some workers who have studied the problem systematically indicate that small response factors only, and ideally the theoretical factors for flame ionisation detectors, calculated by Ackman and Sipos [27,28], are all that need be applied [78,197]. On the other hand, others claim that the theoretical response factors cannot be applied to short chain esters [72,73]. It is certainly advisable to determine what response factors for 4:0 to 8:0 fatty acids are necessary under the analyst's own conditions. This is discussed in greater detail in Section G below.

2. Fatty acids of longer than usual chain-length

When chromatographing the fatty acids of animal tissues, most analysts have tended to assume that they can terminate the analysis after the 22:6(n-3) fatty acid has emerged from the column, but it is increasingly being realised that components of even longer chain-length will often be present. Their detection has been facilitated by the inertness of fused silica WCOT columns and the high temperature stability of modern stationary phases, which also reduce the times necessary for elution. For example, Grogan [315] identified 26:4(n-6), 26:5(n-6), 28:5(n-6) and 30:5(n-6) fatty acids from rat testis on a fused silica WCOT column coated with the highly polar SP-2340^{TM} and temperature-programmed to 260°C. The ECL values obtained under isothermal conditions were consistent with the structures proposed. Fatty acids up to C_{26} were found in plasma by GC-MS [62], but more spectacular discoveries of C_{26} to C_{38} fatty acids with 4, 5 and 6 double bonds of both the (n-6) and (n-3) series have been found in the brain of patients with Zellweger's syndrome (a peroxisome deficiency) [733,838], in ram spermatozoa [732] and in vertebrate retina [64,66]. A non-polar methylsilicone (BP-1^{TM}) in a fused silica column, temperature-programmed to 320°C, resolved the brain fatty acids sufficiently well for GC-MS identification, as shown in Figure 5.13 [838]. In contrast, polar phases in packed and WCOT columns were used for the unusual fatty acids from retina. Fatty acids of this type have long been known as constituents of marine lipids, and in one typical study 30:4(n-6) and 30:5(n-3) were identified on packed columns of Silar 10C^{TM} and OV-101^{TM} as components of the lipids of the sponge *Cliona celata* [559].

Bacterial lipids may contain fatty acids with very long chain-lengths, and the methyl ester derivatives of mycolic acids, silylated to reduce the polarity of the free hydroxyl group, from *Corynebacterium diphtheriae* were found to consist of more than 20 components with up to 36 carbon atoms and 4 double bonds [272]; these were resolved on a fused silica column coated with a non-polar phase (OV-1^{TM}) and temperature-programmed to 330°C. The record must, however, go to methyl ester/silylether derivatives of C_{66} to C_{84} mycolic acids from *Mycobacterium* species, which were partially resolved on a short packed column (0.3 to 0.4 m x 3 mm) of 1% OV-101^{TM}, operated isothermally at temperatures of 320 to 340°C [457].

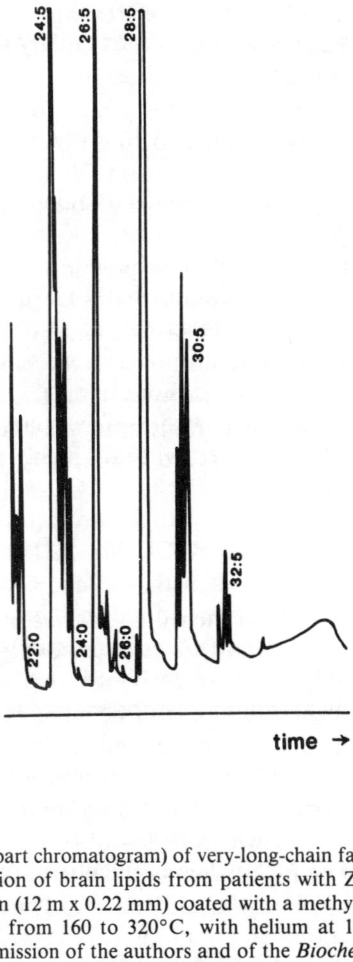

Figure 5.13 Separation (part chromatogram) of very-long-chain fatty acids (methyl esters) from the cholesterol ester fraction of brain lipids from patients with Zellweger syndrome [838]. A fused silica WCOT column (12 m x 0.22 mm) coated with a methylsilicone phase (BP-1™) was temperature-programmed from 160 to 320°C, with helium at 1 ml/min as the carrier gas. (Reproduced by kind permission of the authors and of the *Biochemical Journal*, and redrawn from the original papers.

Mono-, di- and trimeric fatty acids, formed by thermal polymerisation, were resolved on a short packed column, containing 3 % JXR™ (a methylsilicone) and temperature-programmed to 350°C [1013].

The analysis of fatty acids of high molecular weight is one of the areas where HPLC with reversed-phase columns may have advantages in comparison to GC [168], and this technique was indeed also used to complement the GC analyses in some of the studies just described.

3. Acetylenic fatty acids

An isolated triple bond has a similar effect on the retention characteristics of a fatty acid as three methylene-interrupted double bonds, and methyl stearolate (methyl octadec-9-ynoate) is eluted with or slightly after methyl linolenate on columns packed with DEGS [333] or PEGA [620]. Retention data on several different stationary phases have been published for the complete series of methyl octadecynoates (i.e. in positions 2 to 17) [85,323]. The ECL values tend to be parallel to those of the corresponding *trans* isomers, but differing by 0.6 to 0.7 units on DEGS and 0.1 to 0.2 units on Apiezon L™. In addition, ECL data are available for a number of synthetic C_{18} diynoates [326,533,545].

In certain species of moss, there are polyunsaturated fatty acids with acetylenic bonds separated from double bonds by single methylene groups. It proved possible to use separation factors and FCL values to predict the retention times of these coumpounds as with the more common range of unsaturated fatty acids as described above [441]. Distinctive separations of acetylenic fatty acids are obtained on the cyanoalkylpolysiloxane phases, in that FCL values for triple bonds are appreciably lower than on more conventional polar phases [437].

4. Branched-chain fatty acids

While distinctive long-chain branched fatty acids occur in bacteria, fatty acids with simple methyl branches are encountered most often in microorganisms and in animal tissues. Generally, only a single methyl branch is present, but multibranched fatty acids (including isoprenoids) are found in ruminant and certain other tissues, but especially in the preen glands of birds. Normally the acyl chain is saturated, but in some bacteria there may also be a single double bond. The separation of branched-chain fatty acids by GC has been reviewed elsewhere [6,114,854].

Retention data, including ECL values, were reported for the complete series of isomeric methyl-branched octadecanoates, and all are eluted before methyl nonadecanoate on both polar and non-polar stationary phases [6,10,11]. In addition, analogous data for fatty acids of different chain-lengths have been published [668]. Elution patterns similar to those seen with the unsaturated series discussed above are found, in that components with methyl branches in the centre of the chain have the lowest retention times, while those remote from the carboxyl group have the highest. The *iso*- and *anteiso*-isomers, i.e. with the methyl branch on the penultimate and antepenultimate carbon atoms respectively, are those most often found in nature as they are ubiquitous if minor constituents of animal lipids, and they are easily separated from each other. The *iso*-compound is eluted first, and the C_{17} fatty ester with this structure has an ECL value of approximately 16.5 on packed columns of

15 per cent EGSS-X™ and EGSS-Y™, while the related *anteiso*-compound has an ECL value of 16.7 on these columns (author, unpublished). Further data for WCOT columns are included in Table 5.3. Relatively small changes are seen for these compounds with stationary phases differing widely in polarity.

As with unsaturated fatty acids, FCL values obtained for monomethyl-branched fatty acids have been used to estimate ECL values for multibranched acids with some success [9,11]. Similarly, Jacob [423,425] has tabulated FCL data for branching of various kinds in fatty acyl chains. However, GC-mass spectrometry (see Chapter 7) is probably the only reliable means of identification. Diastereoisomers of branched-chain acids have been separated on WCOT columns [9,20], and the problems of such analyses have been reviewed by Smith [858]. Among fatty acids of this kind, most attention has been given to phytanic acid, 3,7,11,15-tetramethylhexadecanoic acid, a metabolite of phytol which can accumulate in the tissues of ruminants and of humans under certain conditions [562]; it tends to emerge very close to 17:0 on stationary phases of moderate polarity, and with 18:2(n-6) on Apiezon L™.

The difficulties involved in the analysis of complex mixtures of branched-chain fatty acids are perhaps best illustrated by some selected examples of actual analyses, and the reader may find illuminating those studies of such components in human milk [238], *Vernix caseosa* (mono-, di- and trimethylbranched) [668], ruminant tissues [232,853], avian uropygial gland secretions (reviews) [423,425], the bacteria *Streptomyces* R61 and *Actinomadura* R39 [136] and the bacterium *Desulfovibrio desulphuricans* (*iso*- and *anteiso*-methylbranched and monoenoic) [117].

5. Cyclopropane, cyclopropene and other carbocyclic fatty acids

Cyclopropane fatty acids are common constituents of bacterial fatty acids and sometimes accompany cyclopropene fatty acids in certain seed oils. The ECL values of the complete series of methyl esters of C_{19} isomeric cyclopropane fatty acids have been recorded on polar (NPGS, DEGS and PEGA) and non-polar (Apiezon L™) liquid phases; DEGS and PEGA were in packed columns and the remainder were in WCOT columns [176]. In each instance, the ECL values are approximately one unit greater than those of the corresponding C_{18} monoenoic esters from which they are derived, synthetically or biosynthetically. Therefore, as with the monoenes, a cyclopropane ring in the centre of the chain has less effect on retention time than one at either extremity of the molecule. The 9,10- and 11,12-methyleneoctadecanoates, which are occasionally found together in bacterial lipids, were separable on the non-polar WCOT column. Synthetic fatty acids containing a cyclopropane ring with a *trans*-configuration were eluted before the corresponding *cis*-isomers on packed columns of EGS and Apiezon L™ [179]. The chromatographic properties of cyclopropane and other carbocyclic fatty acids have been reviewed [547].

Methyl esters of cyclopropene fatty acids are less easily analysed by gas chromatography as they tend to decompose or rearrange on GC columns to give spurious peaks. As cautioned in the previous Chapter, it is necessary to avoid acidic conditions when preparing the methyl ester derivatives. If highly inert supports and silicone liquid phases are used, however, successful GLC of the native esters is possible [258,759]. For example, methyl sterculate eluted after 18:0 and methyl malvalate just before 18:2 (and partially overlapping with it) on a packed column of 5% SP-2100™ (2 m x 2 mm); as little as 0.3% of each could be quantified [258]. Some useful results have also been achieved on a glass WCOT column coated with Carbowax 20M™ [99,281]. The author has successfully chromatographed methyl sterculate on a well-conditioned EGSS-X™ column (packed) where it eluted just after methyl linoleate, although the same compound decomposed on a new EGSS-Y™ column. On the other hand, it might be easy to inadvertently miss a small amount of degradation on a GC column, since Conway *et al.* [194] found that some minor decomposition products co-chromatographed with other fatty acids on a Carbowax 20M™ WCOT column.

Most workers have preferred to prepare stable derivatives prior to analysis. For example, cyclopropene fatty acids can be subjected to hydrogenation, or reaction with silver nitrate [449] or methanethiol [746]. Silver nitrate in anhydrous methanol reacts with cyclopropene rings in about 2 hours at 30°C to form predominantly methoxy ether but with some enonic derivatives, which appear as twin peaks (because of reaction on either side of the ring) on analysis by GC [99,241,281]. An application of this procedure to the analysis of kapok seed oil is illustrated in Figure 5.14. Alternatively, a brief reaction with hydrazine will selectively reduce the cyclopropene compounds to the more stable cyclopropanes; by examination by GC before and after the reaction, the small amounts of natural cyclopropane components can also be identified [194].

Seed oils containing cyclopropene fatty acids have been successfully analysed by means of HPLC in the reversed-phase mode (reviewed elsewhere [168]) and this may now be the method of choice.

Other saturated and unsaturated cyclic esters are more stable to gas chromatography. Methyl 11-cyclohexylundecanoate, a minor constituent of rumen microrganisms and milk fat, is eluted with 18:2(n-6) on a packed column of PEGA and with 18:0 on one of Apiezon L™ [354]. In addition, retention data have been published for some synthetic and natural ω-alicyclic fatty acids on polar and non-polar phases [215,458].

Cyclopentenoic acids occur in seed oils of *Hydnocarpus* and related species of the genus Flacourtiaceae, and they have been subjected to GLC analysis. The main components are chaulmoogric (13-cyclopent-2-enyltridecanoic),

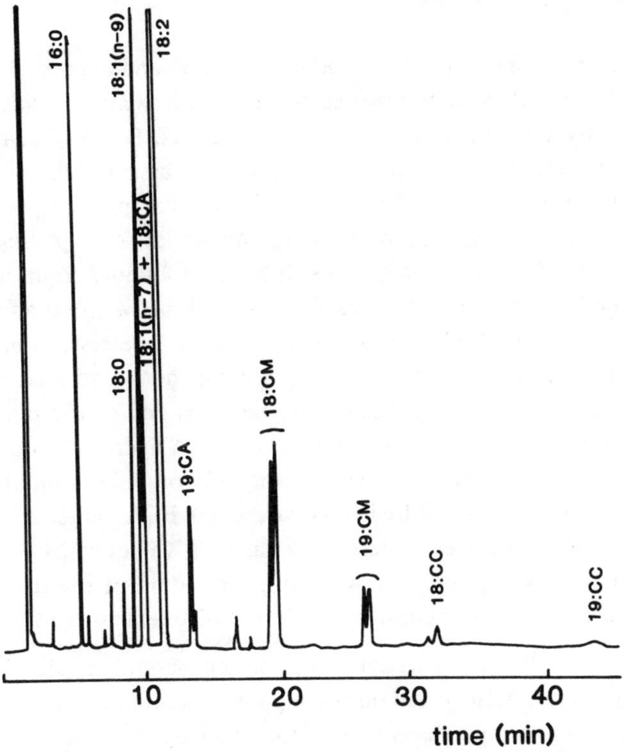

Figure 5.14 Analysis of kapok seed oil methyl esters after reaction with silver nitrate in anhydrous methanol, on a glass WCOT coated with Carbowax 20M™, maintained at 190°C with hydrogen as carrier gas [99]. Abbreviations: 18:CA, dihydromalvalic acid; 19:CA, dihydrosterculic acid; 18:CM, methoxy derivatives of malvalic acid; 19:CM, methoxy derivatives of sterculic acid; 18:CC, enone derivatives of malvalic acid; 19:CC, enone derivatives of sterculic acid. (Reproduced by kind permission of the authors and of *Analytical Chemistry*, and redrawn from the original paper).

hydnocarpic (11-cyclopent-2-enylhendecanoic) and gorlic (13-cyclopent-2-enyltridec-6-enoic) acids, but higher and lower homologues exist. On packed columns of polar phases, such as EGS, the cyclopentene ring has a substantial effect on retention time but methyl hydnocarpate and methyl oleate elute together; on non-polar phases, such as Apiezon M™, this pair are separated [75,876,1013]. By a careful use of polar and non-polar phases, it is possible to estimate all the main components. Fortunately, the principal fatty acid constituents of such seed oils are now much more readily separated on fused silica WCOT columns coated with phases such as Silar 5CP™ and Carbowax 20M™, from which methyl hydnocarpate elutes before methyl stearate as illustrated in Figure 5.15 ([587] and W.W. Christie, E.Y. Brechany and V.K.S. Shukla, *Lipids*, in the press).

6. Oxygenated fatty acids

Oxygenated fatty acids are not uncommon in plant and microbial lipids, and they are also found in animal tissues, where 2-hydroxy fatty acids especially are ubiquitous constituents of the sphingoglycolipids. Their GC characteristics have been reviewed by Vioque, who presents a substantial list of ECL data [954]. Tulloch has reported ECL values for the complete series of methyl hydroxy- and acetoxystearates on polar and non-polar liquid phases in packed columns [926]. As is usual with substituted compounds, components with central functional groups are not easily separated, but where the substituents are close to either end of the molecule, positional isomers can be resolved. In this instance, the ECL values are highest for the 4- and 5-isomers, drop as the functional group approaches the carboxyl end of the molecule and rise appreciably close to the terminal methyl end. A hydroxyl group increases the ECL value of a fatty acid greatly, especially on polar columns, and on EGS and SE-30, methyl 12-hydroxystearate has ECL values of 26.25 and 20.00 respectively. Its acetate derivative has ECL values of 24.70 and 20.55 respectively on these phases. In addition, the nature of the derivative has a profound effect on the elution profile for different isomers.

Compounds with free hydroxyl groups present several problems to the gas chromatographer. Adsorption on the support or on the walls of the column and hydrogen-bonding effects may come into play, so that unsymmetrical peaks are obtained and recoveries are incomplete. Losses can occur becuase of transesterification of the hydroxyle group with polyester liquid phases. Although such phenomena are less apparent with modern catalyst-free liquid phases and highly inert supports, and they are now much less troublesome with grass of fused silica WCOT columns coated with non-polar phases (c.f. [542], it is still advisable to chromatograph the compounds in the form of a non-polar volatile derivative such as the acetate [926], trifluoroacetate [988] or trimethylsilyl ether [995]. With the last especially, sharper peaks, better recoveries and improved resolutions of positional isomers are obtained. *n*-Butylboronate derivates are invaluable for characterising 2- and 3-hydroxy fatty acids (see Chapter 4 for details of preparation) [55,133].

The 2-hydroxy fatty acids of sphingolipids have been resolved on packed columns of Se-30™ [145] and EGSS-X™ [531] as the methyl ester/trimethylsilyl ether derivatives for identification by mass spectrometry. On the other hand, much better separations were obtained on a fused silica WCOT column (25 m x 0.25 mm) coated with OV-101, as shown in Figure 5.16 [5]. In this instance, the 2-hydroxyl acids were chromatographed simply

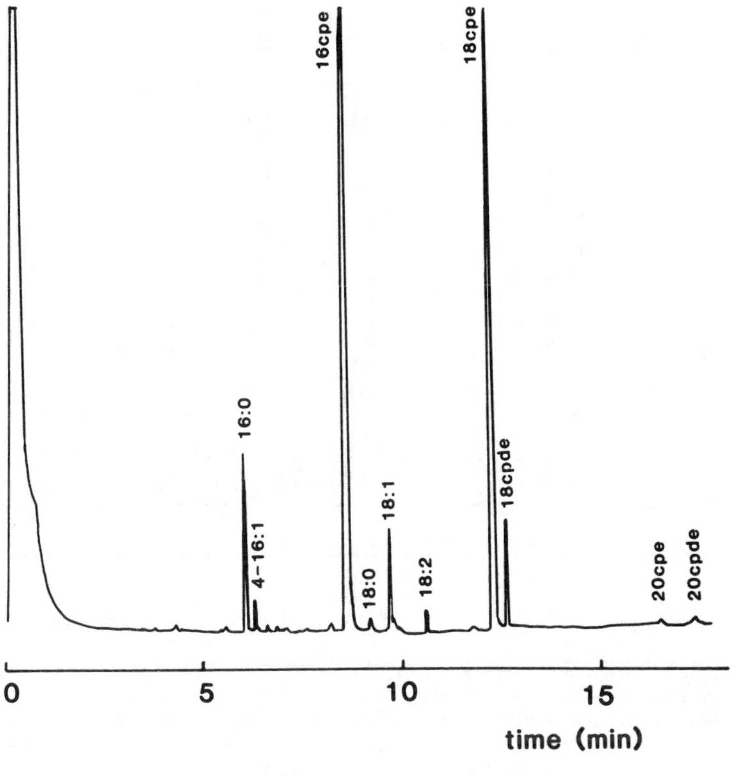

Figure 5.15 Separation of the methyl ester derivatives of the fatty acids of *Hydnocarpus anthelminticus* on a WCOT column of Carbowax^{TM} (see figure 5.8 for conditions). Abbreviations: 16cpe, 11-hendecylcyclopent-2-enoate; 18cpe, 13-tridecylcyclopent-2-enoate; 18cpde, 13-tridec-4-enylcyclopent-2-enoate; 20cpe, 15-pentadecylcyclopent-2-enoate; 20cpde, 15-pentadec-9-enylcyclopent-2-enoate.

as the methyl ester derivates together with the non-hydroxy fatty acid constituents, althought he two groups could also be resolved on their own following a TLC separation. Other natural hydroxy fatty acids (also containing isolated double bonds) occur in seed oils, and ECL data for several of these (and derivatives) have been published [4,482,907]. With fatty acids with conjugated unsaturation that also contain allylic hydroxyl groups,

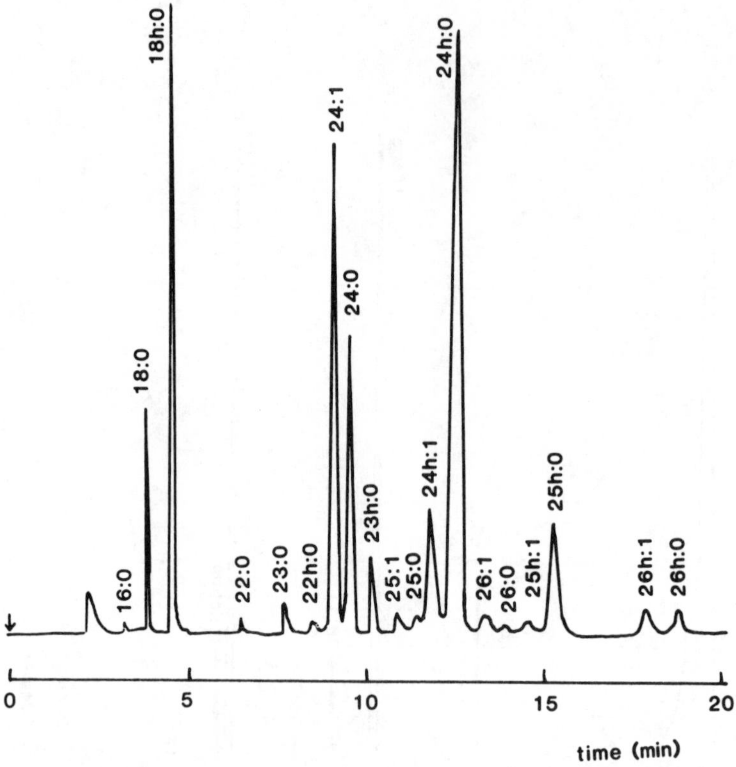

Figure 5.16 Separation of hydroxy and non-hydroxy fatty acid methyl esters from cerebrosides of bovine brain on a fused silica WCOT column (25 m x 0.25 mm) coated with OV-101™[5]. The column temperature was 280°C and the flow-rate of the nitrogen carrier gas was 0.5ml/min. (Reproduced by kind permission of the authors and of the *Journal of Chromatography*, and redrawn from the original paper).

dehydration may occur leading to spurious peaks [619]. The problem was apparently not alleviated if the acetate or methoxy derivatives were prepared, but it might be instructive to repeat this work with more modern catalyst-free liquid phases on inert supports, or better with a non-polar phase on a WCOT column of fused silica. Certainly, TMS ether derivatives of conjugated hydroxy fatty acids have apparently been subjected to GC on WCOT columns successfully [1005].

Synthetic polyhydroxy esters have been subjected to GLC in the form of the trimethylsilyl ethers [988], as trifluoroacetates [988] and as isopropylidene derivatives [985]. *Erythro*- and *threo*-forms of vicinal diols can be separated on packed columns when they are converted to either of the last two derivatives and, as these compounds can be prepared quantitatively to a high degree of stereochemical purity from *cis*- or *trans*-olefins respectively, this provides a basis for gas chromatographic separation and estimation of stereoisomers of unsaturated acids on packed columns [985]. (Epoxide derivatives can be used in a similar manner as discussed in Sections D.1 and 2 above [116,243,244]. It appears to be a general rule that derivatives with the *trans*- or *threo*-configuration are eluted before those of the *cis*- or *erythro*-configuration, especially on non-polar stationary phases. Polyhydroxy fatty acids (also with oxo and epoxyl groups) occur naturally in the form of polyesters in shellac and in plant cutins. For the latter, the composition, methods of analysis and GC retention data have been reviewed [391,929].

Optical enantiomers of the methyl esters of mono-hydroxy fatty acids can be resolved by gas chromatography if they are first converted to the (-)-menthyl, *D*-phenylpropionyl or related derivatives [348,468] (reviewed briefly elsewhere [456,858]). Particularly good results were obtained with racemic methyl 2-hydroxy palmitate and stearate, after conversion to the (-)-α-methoxy-α-trifluoromethylphenylacetate derivatives, on a WCOT column coated with OV-1™ as shown in Figure 5.17 [93]; it is probable that the *D*-form elutes first. Some limited success has also been achieved in separating diastereoisomers of polyhydroxy fatty acids [988].

Hydroperoxy fatty acids as such cannot be separated by GC as they decompose at high temperatures, and HPLC is probably the preferred method for their analysis [168]. Nonetheless, there are times when it is advantageous to convert them to the hydroxy derivatives by means of sodium borohydride reduction and then to the TMS ethers for GC analysis, for example for identification by GC-mass spectrometry; products derived from linoleic [223,341,911,946,1005], arachidonic [124,407,568,1005] and docosahexaenoic acids [948] have been examined in this way (the list is not intended to be comprehensive). Woollard and Mallet [1005], in particular, have presented a comprehensive list of ECL data for compounds of this type. In addition, GC methods were used for the determination of the absolute configuration of hydroperoxides formed by lipoxygenase reaction [124,568,947].

Retention data for all the isomeric methyl oxostearates were recorded by Tulloch [926], and the elution pattern resembles that described above for the corresponding acetoxy derivatives. GC retention data for some conjugated polyenoic fatty acids with keto groups in position 4 have been recorded [334].

GC separations of synthetic epoxy fatty acids have been referred to briefly above. Epoxy fatty acids also occur naturally, in seed oils especially, and ECL data for a number of these have been published [191,735,871].

Furanoid fatty acids are present in some seed oils, but more interestingly perhaps they are found in the reproductive tissues of fish. The chromatographic analysis of these compounds has been reviewed by Lie Ken

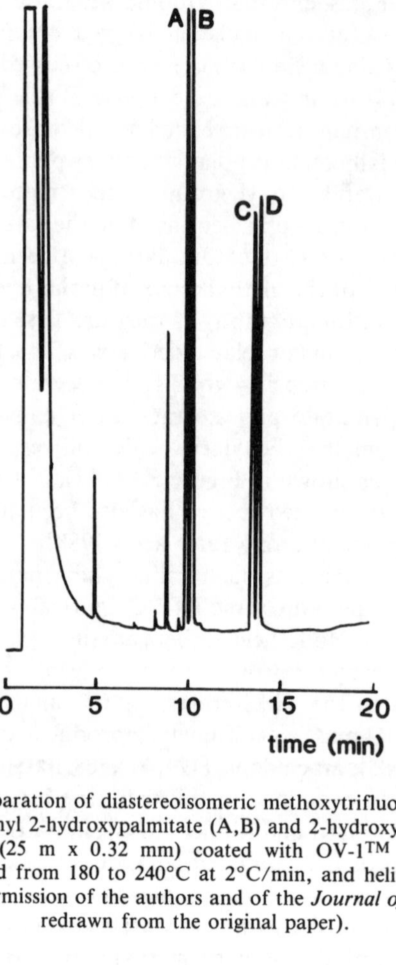

Figure 5.17 Separation of diastereoisomeric methoxytrifluoromethylphenylacetate derivates of methyl 2-hydroxypalmitate (A,B) and 2-hydroxystearate (C,D) on fused silica WCOT columns (25 m x 0.32 mm) coated with OV-1™ [93]. The column was temperature-programmed from 180 to 240°C at 2°C/min, and helium was the carrier gas. (Reproduced by king permission of the authors and of the *Journal of Chromatography*, and redrawn from the original paper).

Jie [547]. It is apparent that the methyl ester derivatives can readily be subjected to GC analysis, and retention data have been published for a number of synthetic [4,548,549] and naturally-occurring compounds [300,301,335,790,825].

7. *Other fatty acids*

Deuterated fatty acids can be separated from non-deuterated in part at least on WCOT columns coated with polar stationary phases [693,695]. *t*-Butyldimethylsilyl ester derivatives (see Chapters 4 and 7) are useful for GC-mass spectrometric estimation of these compounds [692,709,1004]. Long-chain dicarboxylic acids have been separated by GC and identified by GC-MS in the serum of patients suffering from Reye's syndrome [663] and in the lipids of royal jelly [542,543]. Brominated vegetable oils are added to soft drinks to disperse flavouring agents; after acid-catalysed methanolysis, the methyl ester derivatives of fatty acids containing two, four and six bromine atoms have been separated on packed [538] or fused silica [151] columns with non-polar silicone stationary phases.

8. Free fatty acids

There may be times when it is necessary to analyse fatty acids by GC in the free (underivatised) form, and procedures for achieving this have been reviewed in some detail by Kuksis [505]. While this is not easily accomplished on the more conventional polyester phases, it is not at all difficult if specially made packings are used, such as the FFAP™ and Carbowax 20M™-terephthalic acid phases discussed briefly in Section B.1 above, in glass or fused silica WCOT columns (see for example [211,731]). Fused silica columns containing a chemically-bonded phase of the FFAP type are now available commercially specifically for this type of analysis. It is also possible to achieve acceptable separations on fused silica columns coated with methylsiloxane polymers, provided that the column is not overloaded [131].

G. QUANTITATIVE ESTIMATION OF FATTY ACID COMPOSITION

Although this is the last Section of this Chapter, it is certainly not the least important. Ackman [10] has suggested that many analysts might consider it indelicate if asked "When did you last verify the response quantitation and linear range of your gas chromatograph?" Yet as data collection has become more automated, it has become easier to neglect this aspect because of the popular view that "computers don't make mistakes". In fact, errors or selective losses can be introduced during sample preparation, on injection into the column, while on the column, at the detector, and in measuring peak areas and relating them to sample quantity.

With modern flame ionisation detectors, there should be sufficient linear range to cope with most problems, although this can be abused if columns are grossly overloaded (a common fault with novices to the technique). It has recently been demonstrated that the effective linear response of detectors can be improved if a higher flow-rate is used than is required for optimum sensitivity [38]. As discussed from a theoretical standpoint in Chapter 3, the areas under the peaks on the GLC traces are within limits linearly proportional to the amount (by weight) of material eluting from the columns. Problems of measuring this area arise mainly when components are not completely separated, and there is no way of overcoming this difficulty entirely. When overlapping peaks have distinct maxima, the height multiplied by retention time method of area measurement is probably the most accurate manual technique in isothermal analyses, but computer analysis of peak shapes may improve the accuracy of the estimation. Where one component is visible only as a minor shoulder or broadening of a major peak, no manual or computer method is likely to give very precise results for the individual components, although electronic integration can at least give an accurate measure of the total amount of material present in a multiple peak. Wherever feasible, column conditions should be altered in an attempt to improve resolution, or another liquid phase can be tried. The magnitude of this problem can be much less with modern WCOT columns.

Errors arising through loss of components on the columns can be minimised with packed columns as discussed above, by using catalyst-free liquid phases and highly inert supports. With all column types, it is necessary to check at frequent regular intervals whether losses are occurring by running standard mixtures of accurately known composition through the columns. These standards should be similar to the samples to be analysed; for example, saturated standards should not be used to calibrate columns for the analysis of polyunsaturated fatty acids, and the calibration should be checked regularly. Methyl ester mixtures made originally to the specifications of the National Institutes of Health in the U.S.A. are available from several commercial sources for the purpose. Difficulties arise most often with polyunsaturated fatty acids, and it is possible to circumvent them in part by adding an appropriate internal standard. For example, it has been suggested that a 22:1 fatty acid be used in the determination of 22:6(n-3) in fish oils [189] and a 20:3 fatty acid for 20:4(n-6) in animal tissue lipids [294]. Of course, with fatty acids of this type a major potential source of loss is autoxidation caused by faulty sample-handling technique. One simple and convenient means of checking this aspect is to analyse a sample before and after hydrogenation to compare the relative proportions of the components of various chain-lengths (a suitable procedure is given in Chapter 4).

With WCOT columns, losses are most likely to occur through faulty injection technique (see Chapter 3 for a detailed discussion). The author is convinced of the value of "hot needle" and "cold-trapping" injection techniques, especially with split injection, and he used these in preparing the chromatograms in all those Figures in this book from his own laboratory. Factors affecting accuracy and reliability with this type of injection system with specific reference to fatty acid analysis have recently been reassessed [77]. A high speed of injection was found to be especially effective in avoiding discrimination when the sample was in the needle, while rapid vaporisation was achieved by using relatively dilute solutions of the smallest size of sample that could reasonably be analysed. In addition, an injection temperature much higher than is usually recommended, namely 375°C, improved the reproducibility appreciably and had no detectable effect on the recovery of highly unsaturated components. On-column injection techniques with WCOT columns of course present fewer problems to the analyst.

Even with packed columns, good sample injection technique can improve the recoveries of fatty acids and the resolutions attainable. The syringe needle should be inserted rapidly but steadily into the column until it reaches a predetermined depth, just below the surface of the packing material, when the plunger is pressed in firmly; the needle is left in place for about one second and is then removed rapidly.

If necessary, calibration factors may have to be calculated for each component to correct the areas of the relevant peaks in the mixtures analysed. When flame ionisation detectors are used, small correction factors can be applied, when high precision is required, to compensate for the fact that the carboxyl carbon atom in each ester is not ionised appreciably during combustion [8,10,27]. There are also small effects of this kind due to the absence of hydrogen atoms at double bonds. A list of these factors is contained in Table 5.4. The degree of correction necessary is obviously greatest for fatty acids of shorter chain-length or with a high degree of unsaturation. In a recent reassessment of this aspect of quantification in the analysis of fatty acid methyl esters on WCOT columns, it was concluded that these theoretical detector response factors were indeed valid and applied equally to the analysis of short-chain and polyunsaturated components [38,78,79,197]. If further correction factors were found necessary by analysts, it was concluded that some aspect of the chromatographic technique has not been optimised. On the other hand, a deficiency in the carbon atom response for lower esters has been found by other analysts, and it appears that empirical response factors are necessary in the analysis of milk fats say [27,72,73]. Careful calibration with pure standards is necessary in this instance.

In general, the proper approach to the generation of results of high accuracy is to optimise the equipment parameters and operational technique (sample preparation and injection) so that the true answer is obtained with a primary standard, rather than to introduce empirical correction factors to correct for faulty practice.

It is not easy to give an objective assessment of the standard of accuracy that should be possible in routine analyses of fatty acids, but in a collaborative study of IUPAC methodology for fatty acid analysis, typical coefficients of variation (%) at various concentrations were 15 (2% level), 8.5 (5% level), 7 (10% level) and 3 (50% level) [257]. As might be expected, the greatest errors were experienced in estimating butyric acid in milk fat.

When an analysis has been completed, the results can be expressed directly as weight percentages of the fatty acids for presentation in tabular form. On the other hand, it is often necessary to calculate the molar amounts of each acid as, for example, in most lipid structural studies (positional distributions and molecular species proportions). This is performed simply by multiplying the area of each peak by an arithmetic factor, obtained by

Table 5.4
Theoretical response correction factors (RCF) for flame ionisation detectors to convert to weight percent methyl ester, and molecular weight correction factors (MWF) to convert to molar percent for some common fatty acids.

Fatty acid	RCF	MWF	Fatty acid	RCF	MWF
4:0*	1.54	2.647	19:0	0.99	0.865
6:0*	1.31	2.077	20:0	0.98	0.827
8:0*	1.077	1.709	20:1	0.98	0.833
10:0	1.12	1.452	20:2	0.97	0.838
12:0	1.08	1.262	20:3	0.97	0.843
13:0	1.06	1.184	20:4	0.96	0.849
14:0	1.04	1.116	20:5	0.95	0.854
14:1	1.04	1.125	21:0	0.98	0.794
15:0	1.03	1.055	22:0	0.97	0.763
16:0	1.02	1.000	22:1	0.97	0.767
16:1	1.01	1.007	22:2	0.96	0.771
16:2	1.00	1.015	22:3	0.96	0.776
16:3	1.00	1.023	22:4	0.95	0.780
16:4	0.99	1.031	22:5	0.94	0.784
17:0	1.01	0.951	22:6	0.94	0.789
17:1	1.00	0.957	23:0	0.97	0.734
18:0	1.00	0.906	24:0	0.96	0.707
18:1	0.99	0.912	24:1	0.96	0.711
18:2	0.99	0.918	24:5	0.94	0.726
18:3	0.98	0.925	24:6	0.93	0.730
18:4	0.97	0.931			

* The theoretical response factors may not apply here [73].

dividing the weight of a selected standard ester (say 16:0) by the molecular weight of the component, followed by renormalising to 100%. For convenience, these factors are also listed in Table 5.4. It should be noted that if fatty acid compositions are calculated on a weight percent basis, it is not always necessary to positively identify each compound; this cannot be avoided if molar proportions are required.

CHAPTER 6

ISOLATION OF FATTY ACIDS AND IDENTIFICATION BY SPECTROSCOPIC AND CHEMICAL DEGRADATIVE TECHNIQUES

A. INTRODUCTION

As cautioned in the previous Chapter, GC alone cannot give unequivocal identifications of the compounds separated. Ideally for this purpose, individual pure fatty acids (usually in the form of the methyl ester derivatives, prepared as described in Chapter 4) should be isolated by a combination of complementary chromatographic methods, and examined first by non-destructive spectroscopic techniques before chemical degradative procedures are applied. For example, adsorption chromatography will separate normal fatty acids from those containing polar functional groups. Silver ion chromatography can be used to segregate fatty acids according to the number and geometrical configurations of their double bonds; a portion of each fraction should be hydrogenated so that the lengths of the carbon chains of the components can be confirmed. Finally some form of partition chromatography must be utilised to separate components of different chain-lengths in each, so that the position and configuration of the double bonds may be determined by spectroscopic (principally IR and NMR spectroscopy and mass spectrometry) or by chemical oxidative degradation procedures. Appropriate spectroscopic and chemical techniques can also be used to detect and locate other functional groups in the fatty acyl chains. With many of these techniques, GC will be necessary to monitor separations or as an aid to identification of the products of reactions. It is rarely possible for one technique on its own to give all the structural information necessary on an unknown.

Of these techniques, gas chromatography coupled with mass spectrometry has become of such importance that it merits a separate treatment, and this forms the subject of the next Chapter. As the interface between the GC column and the mass spectrometer and the need to prepare derivatives of high molecular weight may limit the resolution attainable by GC alone, silver ion chromatography and HPLC in the reversed-phase mode especially are important complementary techniques. They permit isolation of simpler fractions, more readily analysed by GC-MS and other techniques. It should, however, be noted that mass spectrometry cannot normally give evidence as to the stereochemistry or configuration of functional groups. Other spectroscopic and chromatographic techniques together with GC retention data are required in this instance.

The chromatographic procedures used for the isolation of fatty acid derivatives may be based on either TLC or HPLC, and the latter is increasingly being favoured. Certainly, the author has now virtually abandoned TLC, because of the convenience and cleanliness of HPLC and the high reproducible resolution attainable. Nevertheless, TLC still has much to offer in terms of flexibility and economy, and an enormous amount of data has been obtained by this means; it is not neglected in this Chapter. Reversed-phase TLC and preparative GLC are now little used and are not discussed here; they have been reviewed elsewhere [163].

B. ISOLATION OF INDIVIDUAL FATTY ACIDS FOR STRUCTURAL ANALYSIS

1. Concentration techniques

(i) Urea Adduct Formation

When urea is permitted to crystallise in the presence of certain long-chain aliphatic compounds, it forms hexagonal crystals with a channel, into which the aliphatic compounds may fit, provided they do not contain functional groups that increase their bulk, and thence they are removed from solution. Such crystals are known as *urea inclusion complexes*. Saturated straight-chain acids (as the methyl ester derivatives) form complexes readily. On the other hand, the double bonds of unsaturated fatty acids increase their bulk so that monoenoic fatty acids do not form complexes easily, but tend to form them more readily than dienes, which in turn form them somewhat more easily than compounds with three or more double bonds. Fatty acids with double bonds of the *trans*-configuration form complexes before the analogous compounds with *cis*-double bonds. The effect of double bond configuration and number on urea complex formation has been studied systematically [883]. Unfortunately, the separations are complicated by the fact that shorter chain-length compounds do not complex as readily as do higher fatty acids; methyl oleate, for example, is adducted with approximately the same facility as methyl laurate [421]. For this reason among others, the procedure has never been developed as an analytical technique *per se*. The following method can be applied to obtain a concentrate of, for example, polyunsaturated fatty acids from a natural mixture (as the methyl esters).

"The esters (10g) are dissolved in methanol (100 ml) to which urea (20 g) is added. The mixture is warmed until all the urea is in solution, when it is allowed to cool to room temperature with occasional swirling. After a minimum of 4 hours, the material is filtered through a Buchner funnel to remove the urea complexes, which are washed twice with 2.5 ml portions of methanol saturated with urea. The solution, which is greatly enriched in polyunsaturated esters, is then poured into 1 % aqueous hydrochloric acid (60 ml) and extracted alternately with hexane (50 ml) and diethyl ether (50 ml). The combined organic layers are washed with water (2 x 50 ml) and dried over anhydrous sodium sulphate, before the solvent is removed under reduced pressure."

The procedure can be scaled up or down considerably. As an example, a GLC trace of material obtained in this way from 0.25 g of the standard mixture described in Chapter 5.C above (Fig. 5.1) is illustrated in Figure 6.1.

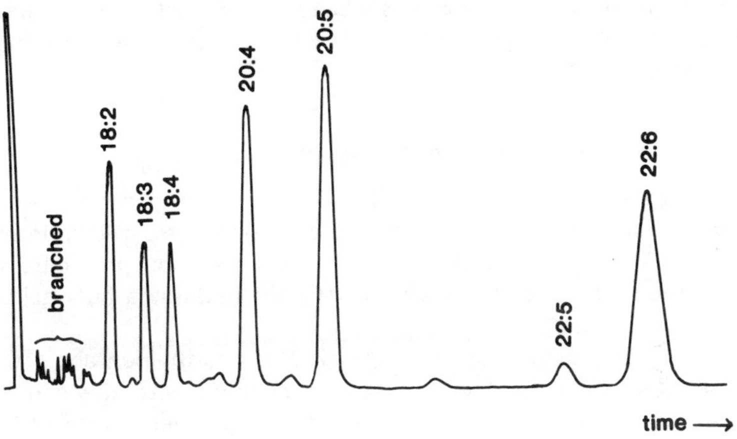

Figure 6.1 GLC recorder trace (the EGSS-X™ column of Table 5.1 and Fig 5.1) of methyl esters that did not form urea adducts from the natural mixture illustrated in Fig. 5.1

The fraction obtained which represents 20 % of the original esters is enriched in the polyunsaturated components, but also contains some branched-chain esters, not readily apparent earlier, and some constituents of shorter chain-length.

The adducted esters can be recovered, when required, by breaking up the complexes with water and extracting the esters into hexane or diethyl ether.

Another particularly valuable application consists in the isolation of concentrates of branched-chain and cyclic esters from natural mixtures, and many useful separations of the former especially have been described [232,238,668,853,854]. *Iso*-branched esters complex more readily than the corresponding *anteiso*-compounds, so some change in the ratio of these may occur during processing. Urea fractionation procedures have also been used to separate ω-hydroxy fatty acids (after acetylation to increase their bulk) from related compounds with the substituent elsewhere in the chain [178].

Attempts to use urea in thin-layer adsorbents or in columns on an analytical scale have not been entirely convincing, and the following simplified procedure is of value when small amounts only of esters are available [552].

> "The methyl esters (up to 100 mg) are dissolved in hexane (4ml), and urea (1.5 g) moistened with methanol (15 drops) is added. After standing overnight, the solid is filtered off and thoroughly washed with hexane; the washings and the hexane filtrate are combined, washed with water, dried over anhydrous sodium sulphate and evaporated, yielding a branched-chain and/or polyunsaturated fraction."

The main advantage of methods using urea are that large quantities of esters can be separated and that with care there is little chance for harm to come to polyunsaturated esters.

(ii) Partition in the form of metal ion complexes

Although silver ion complexation is usually used in conjunction with chromatography to separate unsaturated compounds (see below), Peers and Coxon [702] have described a simple solvent partition procedure which permits the isolation of a concentrate of polyunsaturated fatty acids.

> "The fatty esters (up to 1 g) in 2,2,4-trimethylpentane (10 ml) are partitioned with vigorous shaking with the same volume of ethanol-water (1:1) containing silver nitrate (2.5 g). The upper organic layer, which contains mainly saturated and monoenoic components, is removed. The lower layer is diluted with water (10 ml), and is extracted with hexane (3 x 10 ml), which is dried over anhydrous sodium sulphate and evaporated to yield the polyunsaturated esters."

Take care to prevent spillages, and do not allow the silver ion solution to come in contact with the skin.

Concentrates of polyunsaturated fatty acids (as the methyl esters) can be obtained from suitable starting materials, by preparing the mercuric acetate adducts (see Chapter 4 for practical details of the preparation of adducts and regeneration of the original double bonds) and partitioning them between methanol and pentane; the methanol layer retains the adducts of the more unsaturated esters which can be regenerated unchanged. For example, methyl linoleate of 95 % purity and methyl linolenate of 90 % purity have been produced on the 50-100 g scale by this method from the esters of safflower and linseed oils respectively. The method could no doubt also be adapted to the preparation of concentrates of other polyunsaturated fatty acids.

(iii) Low temperature crystallisation

Fractions enriched in polyunsaturated fatty acids are readily obtained by low temperature crystallisation, a technique that demands little by way of expensive equipment. Most methods employ acetone for crystallisation of samples in the form of the methyl ester derivatives, or acetone, diethyl ether or hexane for the free acids, which are generally preferred for the purpose, at temperatures down to -70°C (attainable with solid carbon dioxide as refrigerant).

Three fractions enriched in saturated, monoenoic and polyenoic fatty acids respectively can be prepared [137]. As an example, the free acids are taken down to the lowest working temperature (about -50°C generally), at a concentration of 1 g per 10 ml of acetone, and are held there for up to 5 hours with gentle swirling until equilibrium is reached. The solution is then filtered through a Buchner funnel, cooled to just below the solution temperature, and the crystals are washed with a small amount of cold solvent. The material in solution consists mainly of polyunsaturated fatty acids, although it is always contaminated by small amounts of saturated and monoenoic components. If the volume of solvent is reduced until a 10 % solution is again attained, the process can be repeated and improved separations obtained. The crystals may be redissolved in fresh solvent and further fractionated at slightly higher temperatures into components enriched in monoenoic and then in saturated fatty acids. In the method's favour, large quantities of fatty acids can be processed in a single operation, and very little harm can come to polyunsaturated acids at the low temperatures employed.

2. Silver ion chromatography

Since its introduction by Morris in 1962, TLC on silica gel impregnated with silver nitrate has been of enormous value to the lipid analyst. It is sometimes termed "argentation" chromatography. The basis of the separation is the facility with which the double bonds in the alkyl chains of fatty acids form polar complexes reversibly with silver compounds. Fatty acids can be separated according to both the number and the configuration of their double bonds and sometimes, with care, according to the position of the double bonds in the alkyl chain. HPLC has been slow to make a mark in this area, because of problems in preparing stable columns, but many of the major difficulties now appear to have been resolved. However, most of the data on the elution characteristics of silver complexes of unsaturated fatty acids has been obtained by TLC.

Usually, 3 to 5 % by weight of silver nitrate relative to the weight of silica gel is incorporated into the slurry used to make the plates, which are then activated in the normal way. They must be stored in the dark and are stable thus for a month or so. Plates with better keeping properties have been prepared by incorporating 30% ammonia into the slurry, but a well-ventilated area is needed for activation [998]. Excellent results have recently been claimed for pre-coated alumina plates, impregnated by immersion in a 10 % solution of silver nitrate, followed by drying and reactivation [127]. On exposure to light, silver-impregnated plates darken rapidly, and it is important that they be handled and developed in a darkened room or cupboard whenever possible. Fatty acids (as methyl esters) on the plate can be visualised under UV light after spraying with 2′, 7′-dichlorofluorescein in 95 % methanol (0.1 % w/v), when they appear as yellow spots on a red-purple background.

It is reportedly possible to separate fatty acid methyl esters with zero to six double bonds into distinct fractions on a single plate with a double development in hexane-diethyl ether-acetic acid (94:4:2 by volume), provided that the atmospheric relative humidity is below 50% [231]. It is more usual, however, to attempt to separate methyl esters of fatty acids with zero to two double bonds on one plate, and those with three to six double bonds on another as illustrated in Figure 6.2, although complete separation of

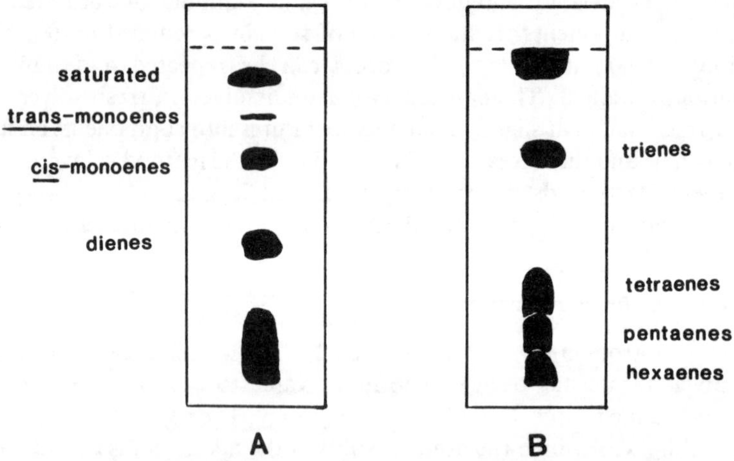

Figure 6.2 Separation of methyl ester derivatives of unsaturated fatty acids by TLC on silica gel G impregnated with 10 % (w/w) silver nitrate.
Plate A: mobile phase hexane-diethyl ether (9:1, v/v). Plate B: as A but solvents in the ratio 2:3.

components with four or more double bonds is never easy. Hexane-diethyl ether (9:1, v/v) will separate components with up to two double bonds, and the same solvents in the ratio 2:3 will separate polyunsaturated esters [163]. After visualising with the spray reagent, components with zero to two double bonds are eluted from the adsorbent with diethyl ether or chloroform; chloroform-methanol (9:1, v/v) may be necessary for complete recovery of polyunsaturated compounds. Unwanted silver ions contaminating fractions can be eliminated from the extracts by washing them with dilute ammonia (about pH 9). Up to 5 mg of esters can be separated on a 20 x 20 cm plate coated with a layer 0.5 mm thick of silica gel containing 5 % (w/w) silver nitrate.

The elution characteristics of a wide variety of unsaturated esters have been studied. For example, the complete series of methyl *cis*- and *trans*-octadecenoates [323] and methylene-interrupted *cis,cis*-octadecadienoates [157] have been subjected to silver nitrate TLC. When run in order on a single TLC plate, each series migrates in the form of a sinusoidal curve similar to that observed with isomeric polar esters on silica gel alone (see below), with a minimum at the 6-18:1 isomer (or at 6,9-18:2 in the case of the dienes) and a maximum at the 13-18:1 isomer. *Trans*-isomers migrate consistently ahead of the corresponding *cis*-isomers, with the exception of the *cis*-2 component, which not only migrates ahead of its *trans*-analogue but also ahead of methyl stearate. With natural samples, monoenoic fatty acids containing *trans*-double bonds can be estimated by separating them from the *cis*-compounds by means of silver nitrate TLC with hexane-diethyl ether (9:1 v/v) as developing solvent and eluting them, together with the band containing the saturated components, from the adsorbent. If the samples are analysed by GLC before and after the separation, the amount of the *trans*-acids in the mixture can be determined [181]. Alternatively, the *cis*- and *trans*-components can be eluted individually and quantified by GLC with an internal standard.

With care, it is possible to separate positional isomers of unsaturated fatty acids by silver nitrate TLC. The most consistent and successful separations of this kind were achieved by Morris *et al.* [622], who utilised silica gel impregnated with up to 30 % silver nitrate, developing the plates several times in the same direction if necessary, with toluene as developing solvent at temperatures as low as -5 to -25°C (complex formation is stronger at low temperatures). With this system, the methyl 6-, 9- and 11-*cis*-octadecenoates can be cleanly separated from each other. Layers containing such high proportions of silver nitrate are very friable, but are not required for more routine separations.

An acetylenic group is less polar than a *cis*-double bond on silver nitrate TLC, so methyl stearolate migrates just ahead of methyl oleate [620] and methyl crepenynate ahead of methyl linoleate [325]. The allenic ester, methyl labellenate, also migrates ahead of methyl oleate [326]. Esters with conjugated double bonds are less strongly retained than similar compounds with isolated double bonds; for example, methyl 9-*cis*,ll-*trans*-octadecadienoate migrates with methyl oleate when hexane-diethyl ether (9:1 v/v) is the developing solvent, and just ahead of it when toluene is used [161]. It is worth noting that compounds which are not resolved with hexane-diethyl ether solvent systems are frequently separable with aromatic solvents such as toluene and *vice versa*. Cyclopropane and saturated branched-chain esters co-chromatograph with normal saturated straight-chain compounds.

It appears likely that silver ion HPLC will soon begin to supplant TLC procedures. As the author has reviewed the technique in some detail elsewhere [168], some salient references only together with important developments of a more recent date are given here. Several groups have reported successful separations of unsaturated fatty acid derivatives on microparticulate silica gel impregnated with silver nitrate [88,372,378,379,685,814]. It appears that the most suitable grade of silica gel for the purpose is one with a large pore size. Unfortunately, silver nitrate leaches rapidly from such columns, contaminating samples and reducing the working life of the column.

A better approach consists in binding the silver ions to a cation-exchange resin, as the silver ions are thus retained more strongly on the column. Some of the first attempts showed great promise, but did not make full use of the benefits of HPLC technology [33-36,246]. Better results were obtained with icosanoids on columns containing sulphonic acid-bonded phases loaded with silver ions [736], and the author has had considerable success with a column of this type in the separation of the more conventional fatty acids. A prepacked column (250 x 4.6 mm i.d.) of Nucleosil™ 5SA was loaded with silver ions simply by injecting small aliquots of silver nitrate, via a Rheodyne™ injector, into a mobile phase of water [169]; when the excess silver ions began to emerge from the end of the column, the aqueous phase was flushed out with organic solvents of decreasing polarity.

To illustrate the power of this column, an application to the separation of the fatty acids of the sea snail *Rapana thomasiana* is illustrated in Figure 6.3 [175]. The column was eluted with a gradient of acetonitrile into methanol, and a mass (light-scattering) detector was used (UV detection at low wavelengths would also have been possible). The major part of the eluent was diverted by a stream splitter at the end of the column for manual collection of fractions as they were seen to emerge. Fractions with zero to six double bonds were clearly resolved, with no cross-contamination and no contamination with silver ions, in less than 25 minutes (and on about the 0.5 to 1 mg scale). After evaporating the solvents, the fractions were examined by GC on a WCOT column coated with Silar 5CP™ as shown in Figures 6.4 and 6.5. The GC trace of the unfractionated methyl esters was so complex that even the more common fatty acids were not easily distinguished (published elsewhere [175]).

After fractionation, some relationships between various components are immediately apparent. Both the chain-length and the number of double bonds in each fatty acid derivative in the chromatograms are ascertained, while relative retention times can be correlated with those of constituents of the unfractionated material. By careful measurement of ECL values, it is then

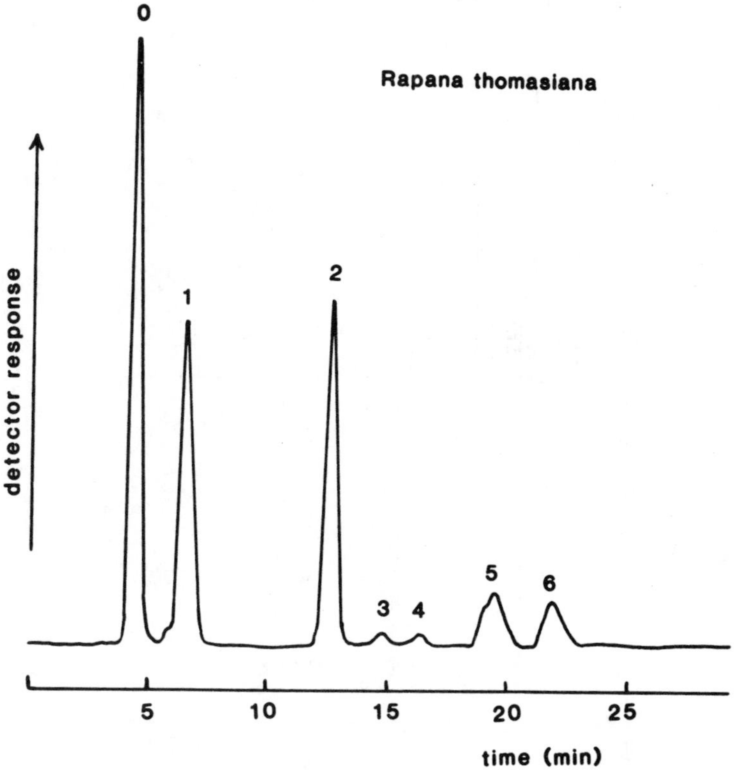

Figure 6.3 Separation of the methyl ester derivates of the fatty acids of *Rapana thomasiana* by silver ion HPLC. A column of Nucleosil 5SA™ (250 x 4.6 mm), loaded with silver ions, was eluted with a gradient of methanol to methanol-acetonitrile (9:1, v/v) over 40 min at a flow-rate of 0.75 ml/min. A mass detector was used with a stream splitter at the end of the column. Further details are given elsewhere [175].

frequently possible to assign double bond positions. In this sample, the identities of nearly fifty different fatty acids in the fractions were confirmed by GC-mass spectrometry following conversion to the picolinyl ester derivatives (see next Chapter).

The saturated fraction contained the usual range of straight-chain even- and odd-numbered fatty acids, and many branched-chain constituents including the isoprenoid 4,8,12-trimethyl-13:0. In the monoenes, there were a number of positional isomers of the C_{16} (2), C_{18} (5), C_{20} (4) and C_{22} fatty acids, not to mention the odd-chain constituents, with 7-20:1 (20:1(n-13)) surprisingly being the single most abundant component. The diene fraction presented real problems of identification, since it contained trace amounts only of the conventional methylene-interrupted compounds, with a number of fatty acids with several methylene groups between the double bonds, i.e. 5,11- and 5,13-20:2 and 7,13- and 7,15-22:2. After this the polyene fractions

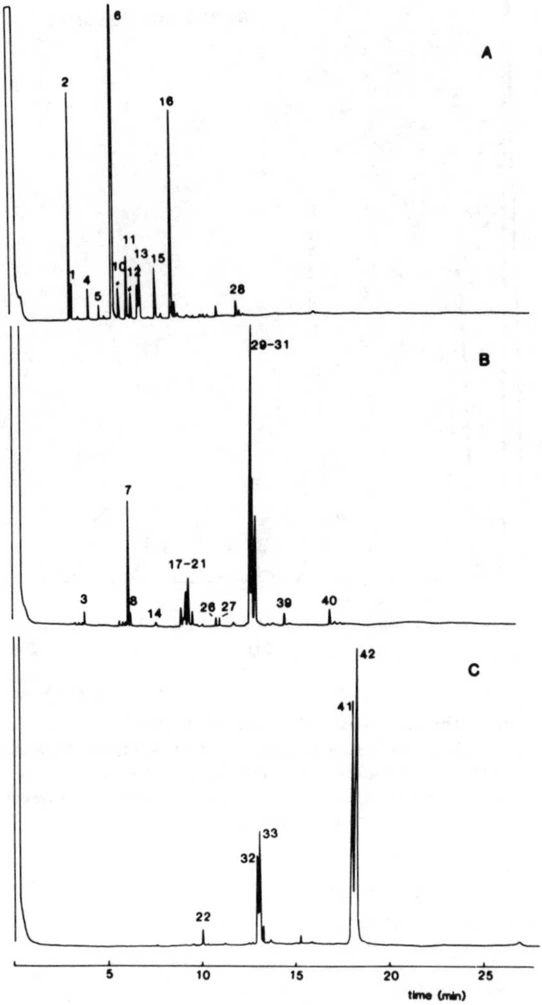

Figure 6.4 GC separations of the fractions separated by silver ion chromatography (illustrated in Figure 6.3) [175]. (A), saturated; (B), monoenes; (C), dienes. A WCOT column (25 m x 0.22 mm i.d.) coated with Silar 5CP™ was used with hydrogen as carrier gas at a flow-rate of 1 ml/min; the column was maintained at 155°C for 3 minutes, then the temperature was raised at 4°C/min to 195°C, and was maintained at this for a further 17 min. Peak designations: 1, 4,8,12-trimethyl-13:0; 2, 14:0; 3, 14:1(n-5); 4, 15:0; 5, *iso*-methyl-15:0; 6, 16:0; 7, 16:1(n-7); 8, 16:1(n-5); 10, C_{17}-cyclic (unidentified); 11, *iso*-methyl-16:0; 12, *anteiso*-methyl-16:0; 13, 17:0; 14, 17:1(n-8); 15, 14-methyl-17:0; 16, 18:0; 17, 18:1(n-13); 18, 18:1(n-11); 19, 18:1(n-9); 20, 18:1(n-7); 21, 18:1(n-5); 22, 18:2(n-6); 26, 19:1; 27, 19:1(n-8); 28, 20:0; 29, 20:1(n-13); 30, 20:1(n-9); 31, 20:1(n-7); 32, 5,11-20:2; 33, 5,13-20:2; 39, 21:1(n-14); 40, 22:1(n-15); 41, 7,13-22:2; 42, 7,15-22:2.

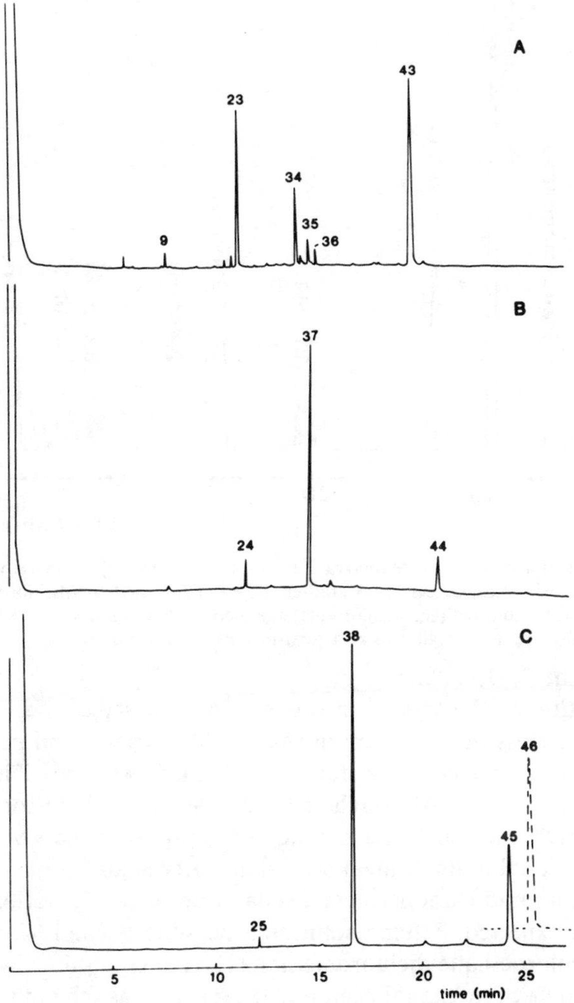

Figure 6.5 As Figure 6.4 except - (A), trienes; (B), tetraenes; (C), pentaenes with hexaene (dotted line) superimposed. Peak designations: 9, 16:3(n-4); 23, 18:3(n-3); 24, 18:4(n-3); 25, 18:5(n-3); 34, 20:3(n-9); 35, 20:3(n-6); 36, 20:3(n-3); 37, 20:4(n-6); 38, 20:5(n-3); 43, 22:3(n-6) (?); 44, 22:4(n-6); 45, 22:5(n-3); 46, 22:6(n-3).

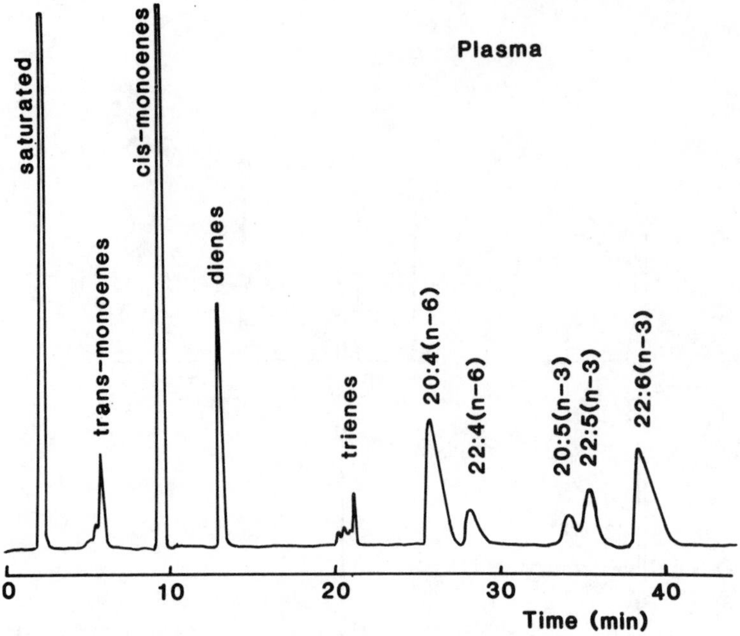

Figure 6.6 Separation of methyl ester derivatives of fatty acids by HPLC on a silver ion column (as in Figure 6.3) with mass detection. A gradient of 1,2-dichloroethane-dichloromethane (1:1) was changed to 1,2-dichloroethane-dichloromethane-methanol-acetonitrile (45:45:5:5 by volume) over 40 min at a flow-rate of 1.5 ml/min.

may in fact be 7,13,16-22:3).

Fractionation of the sample in this way compensated to a considerable degree for any loss of resolution in the GC-MS system, and permitted the identification of many more fatty acids than was possible with the unfractionated sample. Although the example chosen here for illustrative purposes is rather unusual, the technique has proved equally valuable with more conventional animal, algal and plant fatty acids.

With such a rapid elution scheme, separation of configurational isomers would not be expected. Better resolution is possible by employing a gradient from 1,2-dichloroethane-dichloromethane (1:1, v/v) to this solvent mixture containing 10 % of methanol-acetonitrile (1:1, v/v) as the mobile phase, as shown in Figure 6.6. Here the fatty acid esters of human plasma lipids are separated, and *cis*- and *trans*-isomers of monoenoic fatty acids are resolved as are some positional isomers of the polyunsaturated constituents. Excellent separations of positional and configurational isomers of derivatives of monoenoic fatty acids from hydrogenated oils have been accomplished by utilising dichoroethane-dichloromethane as the isocratic mobile phase (author, unpublished).

It is not hard to predict that separations of this type will prove invaluable to lipid analysts in future when identifying unknown fatty acids. One problem with the column appears to be that it is unsuited for certain applications, as residual sulphonic acid groups on the stationary phase bring about transesterification in some circumstances, although this can be avoided with care. A build-up in operating pressure tends to occur in prolonged use, that appears to be due to impurities accumulating on the column; these can be removed by elution with acetonitrile-methanol (1:1, v/v) into which a few milligrams of silver nitrate in acetonitrile is injected. In general, it is probably advisable to avoid ether-containing solvents as silver nitrate can catalyse a reaction between traces of hydroperoxides and double bonds to form epoxides [154]. Recently, a column consisting of silver-loaded mercaptopropyl silica gel was described that shows promise [234]; further results are awaited with interest. In addition, the author (unpublished) has obtained promising preliminary results with Bond Elut™ cartridges, packed with a bonded sulphonic acid medium, in the silver ion form.

3. Chromatography of mercury adducts

Before silver nitrate chromatography was developed, some similar separations of unsaturated compounds were achieved by TLC of mercuric acetate derivatives and some workers continue to find it of value. The procedure is rather tedious as the derivatives must be prepared prior to the analysis (see Chapter 4), then decomposed when the fractionation is over, before components can be analysed further by other procedures. Also, the resolutions that can be obtained are not as good as those accomplished with silver nitrate chromatography. Usually the methoxy-bromomercuri-derivatives are prepared, as these are less polar than the acetoxy-mercuri-compounds and can be fractionated according to the number of double bonds by TLC [614,827,830,972,973] or column chromatography [198] procedures. Sebedio *et al.* [827,830] utilise hexane-dioxan (3:2, v/v) as the solvent for development in TLC, and detect the bands by spraying with diphenylcarbazone in ethanol (0.2 %, w/v). One advantage of the method in some circumstances is that there is no separation of the *cis-* and *trans-*isomers, which can be collected as a single band and re-chromatographed by other procedures later.

4. Reversed-phase HPLC

Before gas chromatography was developed, liquid-liquid partition chromatography was the most useful technique for separating individual (or critical pairs of) fatty acids from natural mixtures. After a period in abeyance, the instrumentation developed for HPLC has been applied to utilise this principle to effect excellent separations of fatty acid derivatives on an analytical or semimicro-preparative scale. Again, as the procedure has been reviewed rather comprehensively by the author [168], there is no need to repeat this here. However, a brief summary of the principles and of some selected applications that complement GC analysis is worthy of discussion.

The technique involves partition of a solute between a stationary and a mobile phase as in GC, except that in the former both phases are liquids; the term "reversed-phase" implies that the mobile phase is more polar than the stationary one. By far the most widely-used stationary phase consists of octadecylsilyl ("C_{18}" or "ODS") groups, linked to a silanol surface by covalent bonds, although C_8 phases are increasingly being found to have some utility. Invariably, unsaturated fatty acids are eluted appreciably ahead of the saturated fatty acids of the same chain-length, each double bond reducing the retention time (or volume) by the equivalent of about two carbon atoms. Thus oleic acid derivatives tend to elute in the same region as palmitate; as these are always major components of plant and animal tissues, it is essential in assessing separation conditions for natural samples that these fatty acids should be adequately resolved. 14:0, 16:1, 18:2 and 20:4 fatty acids form a further troublesome group. Because of technical improvements in the production of the microparticulate phases, resolution has become less of a problem in recent years. On the other hand, due to the nature of the separation, the various fatty acids are easily confused, and it is necessary to be especially vigilant to ensure that components separated by reversed-phase HPLC are identified correctly.

As methyl ester derivatives are by far the most useful for chromatography in general and for GC in particular, it is perhaps most relevant to consider HPLC of fatty acids in this form here. The main difficulty lies in the choice of detection system, as lipids lack chromophores that facilitate spectrophotometric detection (see Chapter 2 also). Aveldano *et al.* [67] made a systematic study of the separation of methyl ester derivatives with real samples (as opposed to standard mixtures) of mouse brain fatty acids on a column (250 x 4.6 mm) of Zorbax™ ODS, maintained at 35°C; the mobile phase was acetonitrile-water (7:3, v/v) changed to acetonitrile alone, and with UV detection at 192 nm. The nature of the separation is illustrated in Figure 6.7. When the nature of the separation process is understood, the order of

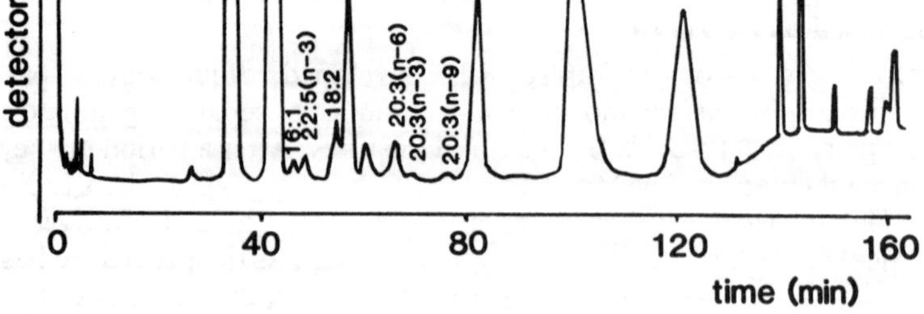

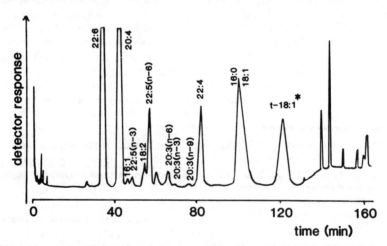

Figure 6.7 Separation of the methyl ester derivatives of fatty acids from the phospholipids of mouse brain by reversed-phase HPLC with spectrophotometric detection at 192 nm [67]. In essence, the column of Zorbax™ ODS phase, maintained at 35°C, was eluted stepwise with acetonitrile-water (7:3, v/v) then with acetonitrile alone. Methyl elaidate was added as an internal standard. (Reproduced by kind permission of the authors and of the *Journal of Lipid Research*, and redrawn from the original paper).

elution of different components is logical, but a newcomer to the technique would certainly find it puzzling. Nonetheless, a number of different components are sufficiently well resolved, for structure determination say. As isolated double bonds contribute most to absorption at low wavelengths, the response of the detector is strongly dependent on the degree of unsaturation of each constituent fatty acid.

It is noteworthy that *cis*- and *trans*-isomers are well resolved, and reversed-phase HPLC has been suggested as a means of estimating such fatty acids (as the methyl esters) in hydrogenated fish oils [894]. In this instance, an isocratic mobile phase, consisting of methanol-water (89:11, v/v), was used with refractive index detection. Some positional isomers can also be resolved, and as an example various conjugated trienoic acids have been separated by the technique [900].

Alternative procedures for reverse-phase HPLC of fatty acids have been developed, in which fatty acid esters containing strongly UV-absorbing substituents in the alcohol moiety are prepared, so that components emerging from the columns can be detected by means of UV spectrophotometry. Borch [119] was one of the first to exploit this technique and obtained some remarkable separations of fatty acids, in the form of the phenacyl esters, by means of HPLC with a C_{18}-bonded stationary phase and acetonitrile-water mixtures as the mobile phase, in conjunction with UV-detection. A wide range of components are separable, including polyunsaturated fatty acids and isomers in which the position or configuration of the double bond varies. Indeed, oleic and petroselinic acid are resolved in this system, a feat that is not readily achieved by other techniques. Innumerable separations of fatty acids in the form of derivatives such as the 2-naphthacyl, p-bromophenacyl (for example of C_{30} to C_{56} fatty acids [742]), or methoxyphenacyl esters have been recorded, generally with gradients of methanol-water or acetonitrile-water mixtures as the mobile phase [168]. Separations of geometrical and positional isomers of fatty acid derivatives of this kind have also been studied [986,994]. The author (unpublished) has observed that phenacyl esters are readily hydrolysed or transesterified by the procedures described in Chapter 4, if this is required.

One advantage of the method is that any impurities not converted to UV-absorbing derivatives are not detected, so cannot obscure the separations. On the other hand, it is likely again that the principal use of such methods will be for small-scale preparative separations of single components for structural analysis or for liquid scintillation counting. The technique may also prove invaluable for determination of the fatty acid composition of seed oils containing thermally-labile fatty acids, such as those with cyclopropene moieties [987]. GC is likely to remain the method of choice for analytical separations of most natural fatty acids in the foreseeable future.

One further reversed-phase separation worth recounting is of picolinyl ester derivatives of fatty acids, since these are particularly valuable for structural identification by GC-mass spectrometry (see Chapter 7). Figure 6.8 illustrates a separation of the picolinyl ester derivatives of the fatty acids

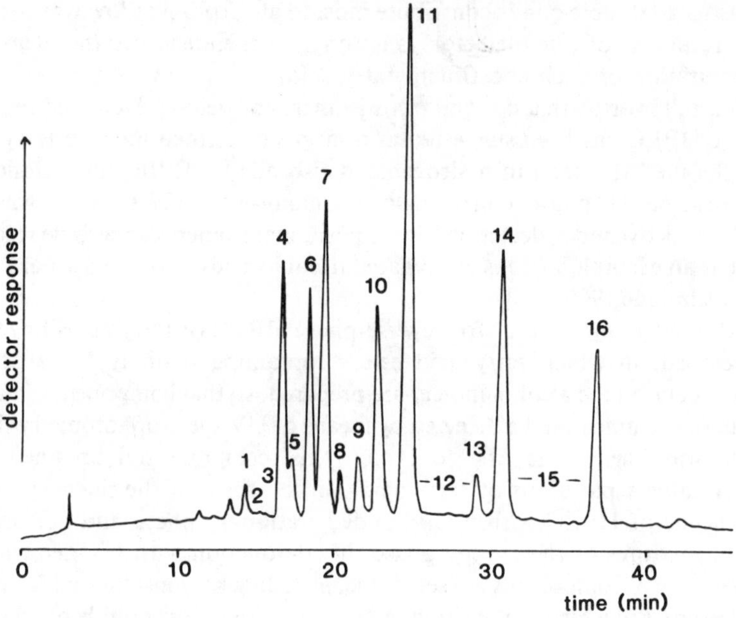

Figure 6.8 Separation of the picolinyl ester derivatives of the fatty acids of cod liver oil by HPLC in the reversed-phase mode on a column (250 x 5 mm) of Spherisorb™ C8 (see text for elution conditions) [184]. The main components in each peak are as follows: 1, 14:1; 2, 18:4; 3, 16:2; 4-5, 14:0, 20:5, 18:3; 6, 16:1; 7, 22:6, 20:4; 8, 18:2; 9, 22:5; 10, 16:0; 11, 18:1; 12, 20:2; 13, 18:0; 14, 20:1; 15, phytanic acid; 16, 22:1. (Reproduced by kind permission of the *Journal of Chromatography*, and redrawn from the original paper).

of cod liver oil on a column containing an octylsilyl bonded phase with stream-splitting for collection purposes and mass (light-scattering) detection [184]. The solvent reservoirs contained methanol (A) and water-pyridine-acetic acid (B) (98.5:1.5:0.025 by volume), and a gradient of the solvents in the ratio (A:B) from 80:20 to 92:8 was generated over 40 min at a flow-rate of 0.75 ml/min. Components which form critical pairs in this system are generally resolved well by GC. On subsequent examination of each of the fractions from the column in turn by GC-MS, 39 different constituents were positively identified.

5. Adsorption chromatography

In general, fatty acids that differ only in chain length or degree of unsaturation cannot be separated by adsorption chromatography, although short-chain and polyunsaturated fatty acids migrate more slowly on silica gel and other adsorbents than do C_{16} to C_{20} saturated or monoenoic components, so that fractions enriched in these compounds can sometimes be obtained. If polar functional groups (especially oxygenated moieties) occur in the alkyl chain, however, some useful separations are possible. It is again more usual to separate fatty acids as the methyl ester derivatives, but unesterified fatty acids can also be chromatographed if 1 % of acetic or formic acids is incorporated into the mobile phase when silicic acid is the adsorbent. Sufficient material for many structural analyses can be obtained by preparative TLC and the elution characteristics of polar fatty acid esters on thin layers of silica gel have been thoroughly examined. Inevitably, most work in this area has been done with TLC systems, but HPLC will certainly replace these in due course with the TLC experience serving as a guide.

In particular, Morris *et al.* [623] have studied the chromatographic behaviour of the complete series of methyl hydroxy-, acetoxy- and keto-stearates. When each series is arranged in order as a line of spots on a single TLC plate, which is then developed in various hexane-diethyl ether mixtures, the compounds migrate in the form of a sinusoidal curve with a minimum at the 5-substituted isomer and a maximum at the 13- or 14-substituted isomer. Isomers with functional groups in positions 2 to 7 can often be separated from each other and from the remaining compounds. This sinusoidal effect is exhibited by other isomeric series of polar aliphatic compounds. As an example of a practical separation, methyl 6-, 14- and 2-hydroxy-stearates were resolved by TLC on silica gel layers, with hexane-diethyl ether (85:15, v/v) as mobile phase [955]. Figure 6.9 illustrates the elution characteristics of some hydroxy esters with this solvent system. Positional isomers of dihydroxy esters, i.e. methyl 6,7- and 9,10-dihydroxy-stearates have been separated by TLC, with diethyl ether-hexane (4:1, v/v) as developing solvent [621]. Up to 20 mg of esters can be separated on a 20 x 20 cm plate coated with a layer of silica gel 0.5 mm thick. As with silver ion TLC, bands are detected by means of a 2′,7′-dichlorofluorescein spray, and they are recovered by elution from the adsorbent, after scraping it into a small column, with diethyl ether or a chloroform-methanol mixture (about 9:l, v/v).

Column chromatography on silicic acid has also been used to isolate polar fatty acid esters, and as examples, methyl 9- and 13-hydroxy-stearate [224] and the acetate derivatives of methyl 9,10-dihydroxy-myristate and 9,10-dihydroxy-palmitate [178] have been resolved by this technique. A small-scale column procedure, with Florisil as adsorbent, has been described for separating normal from 2-hydroxy fatty acid esters, i.e. the component fatty acids of sphingoglycolipids [122].

Threo- and *erythro*-isomers of vicinal dihydroxyesters can be separated on thin layers of silica gel impregnated with boric acid (10 % w/w) as a complexing agent, with hexane-diethyl ether (60:40, v/v) as the developing solvent (the *threo*-isomer migrates more rapidly); the compounds cannot be separated on silica gel alone [618]. Sodium arsenite-impregnated layers give

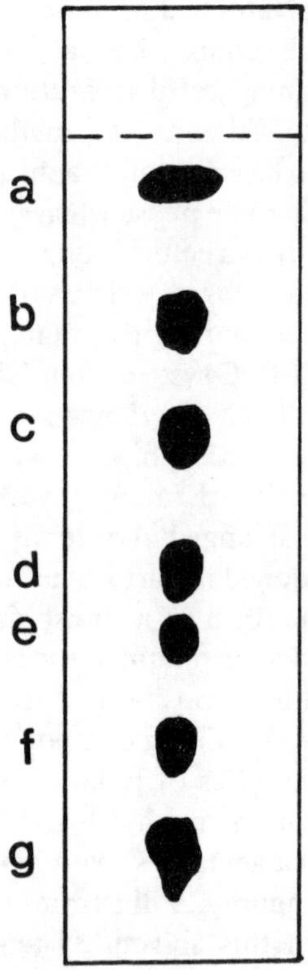

Figure 6.9 TLC of methyl esters of oxygenated fatty acids on silica gel G, with a mobile phase of hexane-diethyl ether (85:15, v/v). Abbreviations: a, 16:0; b, methyl 9,10-epoxystearate; c, methyl 12-ketostearate; d, methyl 2-hydroxystearate; e, methyl 14-hydroxystearate; f, methyl 6-hydroxystearate; g, methyl 9,10-dihydroxystearate.

even more remarkable separations of isomeric polyhydroxy fatty acids.

A great deal of work has been done on the separation of hydroperoxides of unsaturated fatty acids by HPLC and columns of silica gel (reviewed by the author [168]). As the separation is carried out anaerobically at room temperature, little harm can come to the samples. Conjugated double bonds usually present in such fatty acids permit sensitive and specific detection by UV spectrophotometry at 235 nm. The technique has been used with other oxygenated fatty acids, and as an example, epoxy fatty acids were isolated from the total fatty acids of a seed oil by HPLC with a column of Partisil™ silica gel with hexane-diethyl ether (9:1, v/v) as the mobile phase [871].

Where the fatty acids contain double bonds in addition to more polar groups, silver ion chromatography may be of further assistance in achieving separations to complement those by GC.

C. SPECTROSCOPY OF FATTY ACIDS

A detailed discussion of the principles of spectroscopy is beyond the scope of this text, but it is hoped that sufficient information is given to point the reader in the right direction. Most spectroscopic methods are based on empirical collations of vast amounts of data obtained with model compounds of known structure, and in the interpretation of the spectra of unknowns, some knowledge of these data is required. It will only be possible to reproduce a few selected examples here. Applications of spectroscopic procedures in lipid analysis have been reviewed in some detail elsewhere [153,483]. The mass spectra of fatty acid derivatives are described in Chapter 7.

1. Infrared absorption and Raman spectroscopy

Infrared (IR) spectra are obtained when energy of light in the infrared region at a given frequency is absorbed by a molecule, thereby increasing the amplitude of the vibrations of specific bonds between atoms in the molecule. The most useful and conveniently measured region of the IR spectrum (limited by sodium chloride optics) is over a range of wavelengths of 2.5 to 15 μm (equivalent to wave numbers of 4000 to 667 cm^{-1}).

Fatty acids may be subjected to IR spectroscopy in the free (unesterified) state, bound to glycerol or as the methyl ester derivatives, although an esterified form is to be preferred as a band due to the free carboxyl group between 10 and 11 μm may obscure a number of other important features in the spectra. Most information on the chemical nature of fatty acid derivatives can be obtained when they are in solution and Figure 6.10 illustrates the IR spectrum of soyabean oil in carbon tetrachloride solution. The sharp band at 5.75 μm is due to the esterified carbonyl function, which is also responsible for a band at 8.6 μm. With free fatty acids, the first of these bands is displaced to 5.9 μm and there are also broad bands at 3.5 μm and 10.7 μm. *cis*-Double bonds give rise to small bands at 3.3 μm and 6.1 μm, that are useful as diagnostic aids and are considered sufficiently distinct for use in quantitative estimations in some circumstances [43,59]. Most of the remaining bands are absorption frequencies of the hydrocarbon chain.

Many other functional groups give rise to characteristic absorption bands which can be used to identify or to estimate the amount of a given acid. The most important of these is a sharp peak manifested by *trans*-double bonds at 10.3 μm (967 cm^{-1}), and IR spectroscopy has long been one of the principal methods for estimating compounds with this structural feature in the fatty acid chain. In the recommended procedure [414], a calibration curve for the absorbance at 10.3 μm is drawn up from standard solutions of methyl elaidate in carbon disulphide with equivalent methyl stearate solutions run simultaneously to improve the background correction; the IR spectrum of the unknown is scanned in the region 9.5 to 11 μm and the *trans*-double bond

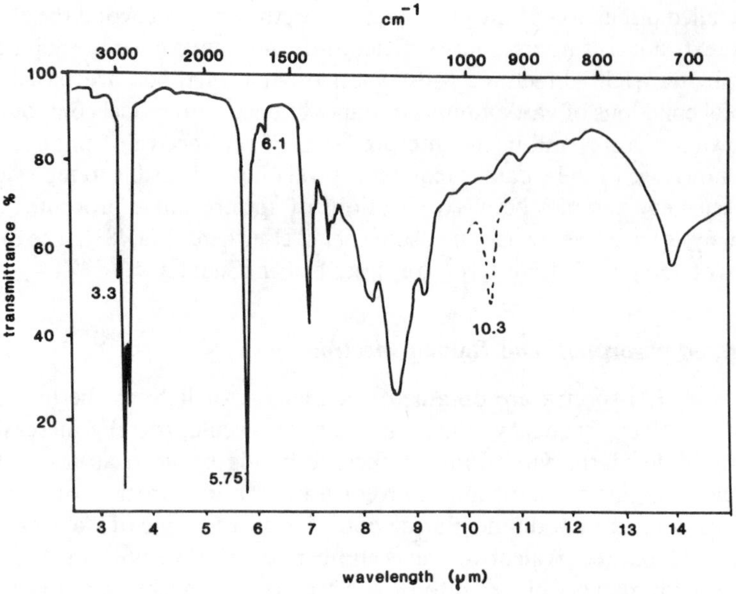

Figure 6.10 Infrared spectrum of soyabean oil in carbon tetrachloride solution. The insert (dotted line) at 10.3 μm illustrates an absorption band due to a *trans*-double bond.

content is derived by reference to the standard curve. If a *trans*-double bond is part of a conjugated system, the band maximum may be shifted to about 10.1 μm.

Free hydroxyl groups give rise to bands at 2.76 and 10.9 μm, epoxy groups produce a double band with maxima at 11.8 and 12.1 μm, allenes give a small band at 5.1 μm, a cyclopropene ring produces a small band at 9.9 μm and a cyclopropane ring produces two small bands at 3.25 and 9.8 μm. Some of the absorption frequencies caused by the more unusual functional groups have been discussed by Wolff and Miwa [982], and more detailed general information can be found in a number of reviews [153,266,490]. Characteristic frequencies are not altered markedly by the position of a group in the aliphatic chain, unless it is at either extremity of the molecule or immediately adjacent to another functional group. IR spectroscopy is therefore a valuable method for detecting the presence and, on occasion, for estimating the amount of certain functional groups in fatty acids, especially when these are in esterified form in a natural lipid mixture, but other spectroscopic or chemical procedures must be used to fix the exact position of the group in the alkyl chain.

Fourier transform IR (FTIR) spectroscopy appears to have been little used by lipid analysts to date, but this situation will no doubt change soon.

Two applications of GC-FTIR to some standard fatty acids and to cyclic monomers produced in heated sunflower oils have been published, and may serve to illustrate the power of the technique [589,931].

Structural features in unsaturated fatty acid methyl esters (in carbon tetrachloride solution) can give rise to distinctive bands in Raman spectroscopy [208-210]. For example, characteristic Raman bands are found for *cis*-double bonds (1656 cm^{-1}), *trans*-double bonds (1670 cm^{-1}) and triple bonds (2232 and 2291 cm^{-1}); a terminal triple bond gives a single band at 2120 cm^{-1} (this group does not give a distinctive band in IR spectroscopy). Again, the position of a double bond does not affect the spectrum significantly unless it is at either extremity of the molecule.

2. *Ultraviolet spectroscopy*

The ultraviolet (UV) spectrum of a compound is generally measured over the range 220 to 400 nm. It is nowadays used principally to detect or to confirm the presence of fatty acids containing conjugated double bond systems in natural oils, or to observe chemical or enzymatic isomerisation of fatty acid double bonds in which conjugated systems are formed. With such acids, series of broad bands of increasing intensity the greater the number of double bonds in the conjugated system, are found in the UV region at successively higher wavelengths. For example, with conjugated dienes, λ_{max} is 232 nm (ϵ = 33,000), with conjugated trienes (e.g. α-eleostearic acid), λ_{max} is at 270 nm (ϵ = 49,000) and with conjugated tetraenes (e.g. α-parinaric acid), λ_{max} is at 302 nm (ϵ = 77,000). Different geometrical isomers have slightly different spectra; the greater the number of *trans*-double bonds, the higher the extinction coefficient but the shorter the wavelengths of the band maxima. Isolated *cis*-double bonds exhibit a specific absorbance at 206 nm, but with a relatively low extinction coefficient so this feature is of little use as a diagnostic aid. Conjugated triple bonds also affect the spectra. A peak of absorbance at 234 nm is characteristic for the *cis,trans*-conjugated double bond system produced by the action of lipoxygenase, a property that is utilised in the estimation of polyunsaturated fatty acids with methylene-interrupted *cis,cis*-double bonds in lipids [840] and more widely to detect and quantify lipid peroxidation in general [476,726]. A review of applications of UV spectroscopy in lipid analysis has been published [153].

3. *Nuclear magnetic resonance spectroscopy*

The nuclei of certain isotopes are continuously spinning with an angular momentum, which can give rise to an associated magnetic field. If a very powerful external magnetic field is applied to the nucleus and made to oscillate in the radio frequency range, the nucleus will resonate between different quantised energy levels at specific frequencies, absorbing some of the applied energy. Such very small changes in energy can be detected, amplified and displayed on a chart. The trace obtained, of the variation in the intensity of the resonance signal with increasing applied magnetic field, is the nuclear magnetic resonance (NMR) spectrum. In organic compounds, the isotope of hydrogen, ^{1}H, displays this phenomenon whereas the main isotopes of carbon, oxygen and nitrogen do not, so the resonance frequencies of hydrogen atoms in molecules are those most often measured and the technique in this instance is referred to as proton NMR spectroscopy. One of the less abundant (naturally) isotopes of carbon, ^{13}C, also exhibits the phenomenon but until comparatively recently, the inherent 6000 fold loss of sensitivity relative to proton NMR limited the value of the technique. Developments in instrumentation and data processing have made ^{13}C NMR much more accessible and since all carbon atoms in organic compounds give distinctive signals, whether or not they are linked to protons, a great deal of structural information can be obtained from the spectra. As the power of modern instruments has increased, the size of sample needed has decreased to about 1 mg for an analysis time of 4 hours at 300 MHz.

Compounds must be in solution for analysis and for proton NMR analysis, the solvent should preferably not contain the isotope ^{1}H. Carbon tetrachloride is suitable for non-polar lipids, but deuterochloroform may also be used and deuterated methanol has been added to this to effect solution of phospholipids. Chemical shifts are not measured in absolute units but are recorded as parts per million of the resonance magnetic field. Tetramethylsilane is added to the solvent as an internal standard, and in the conventional system it is given the arbitrary value zero on the so-called δ (delta) scale. (On the usual charts, values increase from right to left with the increasing strength of the magnetic field). In an alternative system (the τ scale), tetramethylsilane is given the value 10. To convert:

$$\delta = 10 - \tau$$

Fortunately, the same solvents can usually be used for both ^{1}H and ^{13}C NMR, so that both spectra can be obtained from a single preparation. With ^{13}C NMR, chemical shifts are reported as upfield or downfield from tetramethylsilane. Whereas the spectral width of a proton NMR spectrum is about 10 ppm, that of a ^{13}C spectrum is approximately 200 ppm, and this gives a corresponding improvement in the resolution attainable.

In recent years, NMR spectroscopy (especially proton NMR but also, ^{13}C NMR to some extent) has been increasingly applied to the identification of lipid structures, and in particular to the detection, and often the location, of double bond systems in fatty acid chains (again in the form of methyl ester derivatives). The topic has been reviewed most recently by Pollard [725] and by Chapman and Goni [153]. The NMR spectrum of methyl linoleate obtained at 60 MHz was illustrated and described previously by the author

[163], and it has been published elsewhere [945]. When the same spectrum is obtained on a 300 MHz instrument, each of the main features is of course much better resolved [725]. Here, the proton NMR spectrum of methyl linolenate (9,12,15-octadecatrienoate) obtained at 300 Mhz, is illustrated in Figure 6.11. There are eight main features: a multiplet at 5.33 δ for the

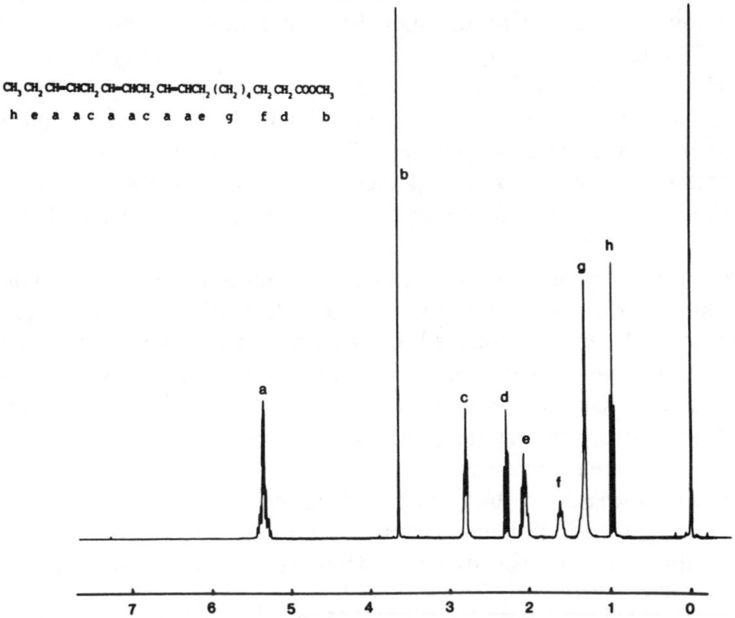

Figure 6.11 Proton NMR spectrum of methyl linolenate in deuterochloroform solution at 300 MHz (tetramethylsilane as internal standard). Features in the spectrum and corresponding protons in the fatty acid are labelled "a to h". The spectrum was kindly obtained on a Bruker instrument by Professor F.D. Gunstone.

olefinic protons, a sharp singlet at 3.64 δ for the protons on the methoxyl group, a triplet at 2.78δ for the methylene groups between the double bonds, a triplet at 2.28 δ for the protons on the carbon atom adjacent to the carboxyl group, a multiplet at 2.06δ for the methylene groups on either side of the double bond system, a triplet at 1.6δ for the protons on carbon atom 3, a broad peak at 1.3δ for the protons in the chain (carbons 4 to 7) and a triplet at 0.98δ for the terminal methyl protons (this signal is at 0.88δ in the spectrum of methyl linoleate). Integration of the signals assists in confirming the assignments to particular protons.

A wide range of unsaturated fatty acids have been subjected to proton NMR spectroscopy on 60 to 100 MHz instruments. They include the complete series of methyl *cis*-octadecenoates (2- to 17-18:1) [322], all the methylene-interrupted *cis,cis*-octadecadienoates [180], several *cis,cis*-octadecadienoates with more than one methylene group between the double bonds [328] and very many natural polyunsaturated fatty acids [320,399]. From studies with such compounds, it is now known how variations in the positions of double bonds affect the NMR spectrum of a long-chain fatty acid. For example, the 2- to 5- and 14- to 17-18:1 isomers can be distinguished by this technique, largely because of small changes in the signal associated with the olefinic protons, but *cis*- cannot be distinguished from the corresponding *trans*-isomers. The terminal methyl group of polyunsaturated fatty acids of the (n-3) series produces a well-separated triplet at a slightly lower field than that for the (n-6) family, and this feature has been utilised in the estimation of such compounds in natural mixtures [299]. NMR spectra of some natural conjugated trienoic acids (α- and β-eleostearic acid) have been determined; the olefinic protons give rise to a complex multiplet at approximately 6.0δ [399].

More informative spectra are obtained with more powerful instruments (220 to 300 MHz), as is seen in Fig. 6.11, and data are available for a large number of different unsaturated fatty acids, including *cis*- and *trans*-monoenes, polyenes and acetylenic compounds [268,269]. Only the 10- and 11-isomers of the methyl octadecenoates cannot be distinguished, for example, while all the corresponding acetylenic compounds have unique features in their spectra. Some evidence as to the configuration of the double bond can be adduced, especially in the presence of lanthanide shift reagents, which also permit the location of centrally-placed double bonds [270]. The use of chemical shift reagents for structure determination with lipids has been reviewed [707].

Free hydroxyl groups in fatty acid chains give rise to two separate signals; that due to the -OH proton is indistinct, and its intensity and position may vary because of hydrogen bonding effects, but that due to the -CHO- proton at 3.6δ is quite characteristic. All the isomeric hydroxy stearates have been examined by NMR spectroscopy, and all can be distinguished from each other by this technique when quinoline is used as the solvent [927]. In addition, much valuable information on the structures of hydroxy acids can be obtained if lanthanide shift reagents are added [977,978]. When the hydroxy esters are acetylated, the acetoxy protons give rise to a sharp signal at 2.1δ. Keto groups influence the *alpha*-methylene protons, which produce a signal similar to that for protons adjacent to a carboxyl group. Many other functional groups give rise to distinctive signals: epoxide ring protons at 2.8δ [401], cyclopropene ring protons at 0.8δ [401] (suggested as a means of quantification [699]), cyclopropane ring protons at 0.6 and -0.3δ [561], and olefinic protons in a cyclopentene ring at 5.7δ [400].

Methyl branches on aliphatic chains do not give signals that are helpful in locating their positions, unless the branch is immediately adjacent to either end of the molecule; *iso*-compounds can, therefore, be recognised, but *anteiso*-compounds cannot be distinguished from fatty acids having the methyl group in a central position [401].

There has recently been great interest in natural-abundance ^{13}C NMR spectroscopy of fatty acids, as features in the spectra can be assigned to virtually every carbon atom in fatty acids differing widely in structure. Model compounds are used to obtain chemical shift assignments, and these are essentially additive for particular functional groups in an aliphatic chain. Most effort has gone into an understanding of the spectra of conventional polyunsaturated fatty acids [141,142,332,448], but much work has also been done with saturated fatty acids [84,448,930], and those containing conjugated double bonds [933,937,939], acetylenic bonds [141,142,331], hydroxyl groups [748,931] and oxo moieties [930]. The ^{13}C spectrum of methyl linolenate is illustrated in Figure 6.12, and the detailed assignments of the signals to

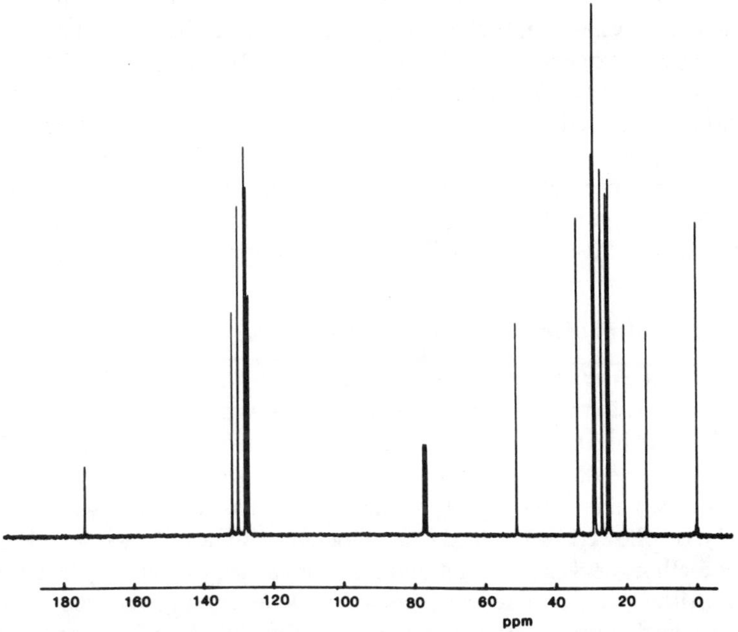

Figure 6.12 ^{13}C NMR spectrum of methyl linolenate in deuterochloroform solution at 75.5 MHz (tetramethylsilane as internal standard). Chemical shifts: C-1, 174.16; C-16, 131.92; C-9, 130.24; C-12/13, 128.29; C-10, 127.8; C-15, 127.18; -O-CH3, 51.36; C-2, 34.11; C-7, 29.63; C-4 to 6, 29.21 to 29.18; C-8, 27.25; C-11, 25.68; C-14, 25.58; C-3, 24.99; C-17, 20.60; C-18, 14.29. The spectrum was kindly obtained on a Bruker instrument by Professor F.D. Gunstone.

particular carbon atoms are listed in the legend. It can be seen that nearly every carbon atom has a distinct signal, in essence only those from carbons 4 to 6 overlapping.

Cis- and *trans*-isomers of unsaturated fatty acids are readily distinguished, an effect that is enhanced by lanthanide shift reagents [564], and ^{13}C NMR spectroscopy has been suggested as a means of quantification of *trans*-unsaturation in lipid mixtures [708,776]. Unfortunately, ^{13}C NMR spectroscopy is relatively insensitive for many purposes, requiring approximately 1 to 10 mg of fatty acid for a usable spectrum, so it is only likely to be used to complement mass spectrometric analysis with unknowns, for example.

D. IDENTIFICATION OF FATTY ACIDS BY CHEMICAL DEGRADATIVE PROCEDURES

1.. Chain-length determination

One of the first steps in the determination of the structure of an unsaturated fatty acid is to establish its chain-length, and this can be ascertained simply by catalytic hydrogenation to form the saturated compound, which is then identified positively by gas chromatography. A procedure suitable for the purpose is described in Chapter 4. Ideally, the reaction should be carried out on the single fatty acid of interest, but valid results can be obtained in some circumstances with natural mixtures or with fractions isolated by silver ion chromatography, for example. A procedure suitable for determining the chain-length of oxygenated fatty acids is described in Section D.3 below.

2. Location of double bonds in fatty acid chains

The positions of double bonds in alkyl chains are generally determined by oxidative fission across the double bond, followed by gas chromatographic identification of the products. Lipid analysts have largely accepted two procedures as suitable for the purpose: oxidation with permanganate-periodate reagent (frequently termed "von Rudloff oxidation"), and ozonolysis followed by oxidative or reductive cleavage of the ozonide. The former method yields mono- and dibasic acids as the products, while the latter can give either these or aldehydes and aldehydo-esters. Both procedures have been reviewed in some detail elsewhere [26]. Ozonolysis techniques have the advantages that over-oxidation and spurious by-product formation are negligible, recovery of short-chain fragments is less of a problem, and it can be used when other functional groups (e.g. hydroxy- or epoxy-) are present. It is of course necessary to have equipment to generate ozone (although this is probably a worthwhile investment if many samples must be analysed). The permanganate-periodate procedure, on the other hand, uses readily-available inexpensive chemicals and over-oxidation is minimal. In the author's opinion, the drawbacks of this method are sometimes painted in too dark a light. It is especially valuable when only an occasional sample must be analysed.

(i) Permanganate-periodate oxidation

In this procedure, the methyl ester of the unsaturated fatty acid in *tert*-butanol solution is oxidised by a solution containing a small amount of potassium permanganate together with a larger amount of sodium metaperiodate, which continuously regenerates the permanganate as it is reduced, while the whole is buffered by a solution of potassium carbonate. When the reaction is complete, the solution is acidified, excess oxidant is destroyed by addition of sodium bisulphite or preferably by passing sulphur dioxide into the solution, and the products are extracted thoroughly with diethyl ether before being methylated for GLC analysis. Very little over-oxidation occurs with this reagent if the reaction is carried out properly, but it is not easy to achieve quantitative isolation of short-chain mono- and dibasic acids or half-esters of these for GLC analysis. The problem has been partially resolved by injecting the free short-chain fatty acids directly on to a GLC column of Porapak Q [343], although Carbowax™ 20M-terephthalic acid might be a better choice of liquid phase, or by pyrolysing the tetramethylammonium salts of the acids, thus converting them to methyl esters, in the heated injection port of the gas chromatograph [227]. Where the compound to be oxidised consists of more than one positional isomer, only the longer-chain fragments are obtained in reproducible yields, and wherever possible, these alone should be considered when determining the amounts of individual positional isomers [373]. The recommended procedure is as follows (only the highest purity reagents should be used) [958,959].

> "A stock oxidant solution of sodium metaperiodate (2.09 g) and potassium permanganate (0.04 g) in water (100 ml) is prepared. This solution (1 ml) together with potassium carbonate solution (1 ml; 2.5 g/l) is added to the monoenoic ester (1 mg) in *tert*-butanol (1 ml) in a test-tube, and the mixture is shaken thoroughly at room temperature for 1 hour. At the end of this time, the solution is acidified with one drop of concentrated sulphuric acid, and excess oxidant is destroyed by passing sulphur dioxide into the solution, which is then extracted thoroughly with diethyl ether (3 x 4 ml). The organic layer is dried over sodium sulphate, before the solvent is removed carefully on a rotary evaporator or in a stream of nitrogen at room temperature. The products are methylated for GLC analysis, preferably by reaction with diazomethane, freshly-prepared by the procedure of Schlenk and Gellerman [800] (see Chapter 4)."

The procedure can be scaled up for polyunsaturated fatty acids, but the proportion of water to *tert*-butanol should be kept as close to 2:1 (v/v) as possible. Malonic acid, formed by oxidation of methylene-interrupted double bonds, is oxidised further and is not detected. Ambiguity may result when polyunsaturated fatty acids with more than one methylene group between the double bonds are oxidised, as it is then not possible to state which dibasic fragment contained the original carboxyl group. However, this difficulty can be resolved by reducing the carboxyl group to an alcohol prior to the analysis [329].

(ii) Ozonolysis and reductive or oxidative cleavage

Ozone attacks olefins rapidly and quantitatively to form ozonides with no over-oxidation and very few side reactions, if the reaction is carried out at low temperature in an inert solvent such as pentane. The ozonide can be cleaved reductively to aldehydes and aldehydo-esters by a number of reagents of which the most convenient are dimethyl sulphide in methanol [751] and tetracyanoethylene [487], although triphenyl phosphine has been more widely used [880]. Reduction of the ozonides with sodium borohydride to yield products which are alcohols or alcohol esters, is reported to be less subject to interference in GC analysis with packed columns [451]. The following procedure is generally satisfactory [737].

> "A solution of ozone in pentane is prepared by bubbling oxygen containing ozone through purified pentane at -70°C until a blue colour is obtained. The unsaturated ester (1 mg) is dissolved in pentane (1 ml) and cooled to 0°C, when the ozone solution is added dropwise until the blue colour persists. After 1 min, the reagents are removed in a stream of nitrogen. Methanol (0.3 ml) then dimethyl sulphide (0.5 ml), both precooled to -70°C, are added and the mixture is maintained at this temperature for 20 min, before it is allowed to warm up to room temperature. The excess reagents can be removed in a gentle stream of nitrogen, provided that the shortest-chain aldehydes are absent, and the products are dissolved in pentane or hexane for injection into the gas chromatographic column for analysis."

With larger amounts of ester, it may be necessary to add more methanol to the reaction mixture to assist the reaction. There may be difficulties in recovering short-chain aldehydes or aldehyde esters quantitatively for analysis. Dimethyl sulphoxide is the other product of the reaction, but does not interfere with the GLC analysis as it tends to elute ahead of dodecanal.

Oxidative ozonolysis became a more attractive technique when Ackman [12,826] demonstrated that ozonolysis could be carried out in a solution of 7

% boron trifluoride-methanol reagent in such a way that that acid ozonolysis products were formed, with a minimum of secondary reaction, and they were then methylated in a one-pot reaction. The reaction is carried out as follows:

> "A stream of 2-4 % ozone in oxygen (120 ml/min) is bubbled into a solution of the fatty acid or its ester (1 mg) in 7 % boron trifluoride-methanol (2 ml) for 1 min. The tube is sealed tightly and heated at 100°C for 1 hour, then the tube is cooled, water (2 ml) is added, and the methyl esters of the acidic fragments are extracted with two portions of methylcyclohexane."

The products of the two types of reaction can be identified on almost any GC column or phase; difunctional compounds are more rapidly eluted on silicone liquid phases, but are less likely to be confused with monofunctional compounds on polyester phases. With both types of liquid phase, temperature-programmed analysis is generally necessary if all the fragments must be determined. Dicarboxylic acid standards are readily available for comparison, but aldehydo-esters are not and it may be necessary to make up a suitable standard by ozonolysis of monoenoic esters of known structure that are commercially available, e.g. methyl petroselinate, methyl oleate and methyl vaccenate. A GLC trace of the ozonolysis products of such a mixture on an EGS column is illustrated in Figure 6.13.

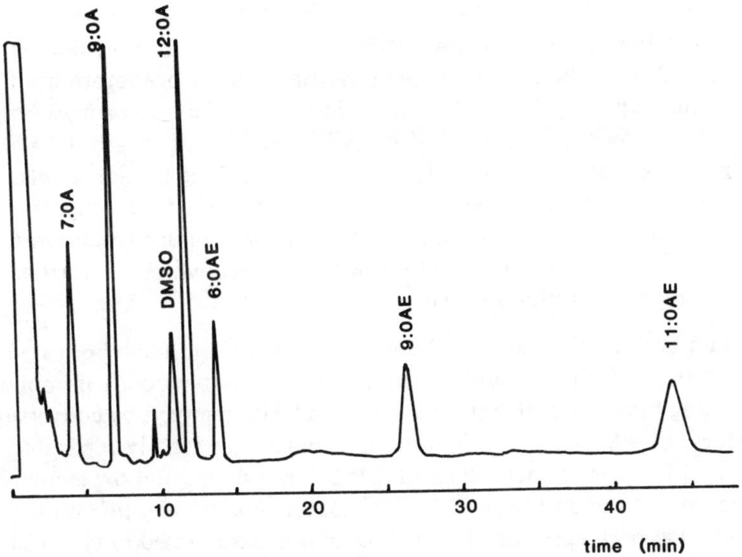

Figure 6.13 GC recorder trace of the products of reductive ozonolysis from methyl esters of 18:1(n-12), 18:1(n-9) and 18:1(n-7) fatty acids on a packed column of EGS [751]. (Reproduced by kind permission of the authors and of the *Journal of Lipid Research*, and redrawn from the original paper).

With oxidative ozonolysis, standards are freely available for identification purposes. The procedure has been applied successfully to monoenoic fatty acids [12] and, with some modification, to polyunsaturated fatty acids with other than methylene-interrupted double bonds [26,756].

(iii) Special procedures for configurational isomers or for polyunsaturated fatty acids

The procedures described above give optimum results with mono- and dienoic fatty acids. When the structures of polyunsaturated fatty acids must be determined, especially when they contain double bonds of both the *cis*- and *trans*-configurations or are part of conjugated double bond systems, more positive identification can be obtained by partially reducing the compounds prior to oxidation [738]. Hydrazine, a reagent that does not cause any double bond migration or stereomutation, is used for the purpose, under conditions such that a high proportion of monoenoic compounds are formed; isomers are then found with double bonds in each of the positions in which they were present in the original polyunsaturated fatty acid. *cis*- and *trans*-Monoenes are separated by silver ion chromatography (see above), and their structures are determined separately by one of the above methods so that the original compound is identified fully. Hydrazine reduction, which is better performed on the free acid than on the ester, is carried out as follows (c.f. Chapter 4):

> "The free fatty acid is heated in air at 35°C with 100 volumes of 10 % hydrazine hydrate in methanol for a predetermined time (typically 1.5 to 2 hours), found by trial and error with a standard polyunsaturated acid, such that there is an approximately 50 % yield of monoenes. Excess methanolic hydrogen chloride (6 % w/w) is then added to stop the reaction, and the mixture is refluxed for 2 hours to convert the acids to the methyl esters, which are recovered for further study as described earlier."

A number of variations on the procedure exist and have been reviewed by Privett [737]. For example, the positions of double bonds in monoenes containing both *cis*- and *trans*-isomers, can be determined by conversion to epoxides, of which the configurational isomers are readily separable, and cleavage of these with periodic acid [245]. Similarly, partial oxymercuration of the double bonds followed by mass spectrometric identification of the products has been applied to the analysis of polyunsaturated fatty acids [719].

3. Location of other functional groups in fatty acids

Spectroscopic aids to the recognition and location of functional groups in fatty acid chains are discussed above and in the next Chapter (mass spectrometry). Chemical methods for characterization are also of value, sometimes as an aid to mass spectrometric identification.

Isolated triple bonds in fatty acids are not easy to recognise, as they do not exhibit particularly distinctive features when examined by any of the spectroscopic techniques other than Raman spectroscopy, but a specific TLC spray consisting of 4-(4^1-nitrobenzyl)-pyridine (5 %) in acetone, that gives a violet colour with acetylenes, has been described [819]. The position of a triple bond can be located by mass spectrometry (Chapter 7), and triple bonds are readily cleaved by the permanganate-periodate reagent. Ozone reacts with triple bonds, although more slowly than with double bonds, and the products of reductive cleavage of the ozonide are mono- and dibasic acids rather than aldehydes and aldehydo-esters, so that double and triple bonds in a single fatty acid can be differentiated [734]. In addition, double bonds are hydroxylated by peracids while triple bonds remain unchanged [74].

Allenic groups have distinctive IR and NMR spectra, and all the natural fatty acids containing this functional group exhibit a marked optical activity. The position of the group in the fatty acid chain can be determined by partial reduction and oxidation of the monoene fragments as described above [74].

A number of spectroscopic procedures are of value for the detection and location of oxygenated functional groups, such as keto, hydroxyl or epoxyl, in fatty acid chains. In addition, several chemical techniques are available, and for example, the presence of hydroxyl groups can be confirmed by GLC analysis before and after the preparation of volatile derivatives, such as the acetates or trimethylsilyl ethers, as described in Chapters 4 and 5. The first step in identifying an acid of this type is to determine its chain-length. To accomplish this, the acid is first hydrogenated to eliminate any multiple bonds before free hydroxyl or epoxyl groups are converted to iodides by the action of iodine and red phosphorus. The iodine atom in the aliphatic chain is then removed by hydrogenolysis with zinc and hydrochloric acid in methanol, when the resulting saturated straight-chain compound is identified by GC [605]. Keto groups should be reduced to hydroxyl groups by the action of sodium borohydride, prior to analysis by the above procedure [855]. A chemical procedure is available for determining the position of the hydroxyl group [177], but mass spectrometry would be favoured by most analysts.

Epoxyl groups are cleaved directly with periodic acid in halogenated solvents [100] or in diethyl ether [530], and the position of the ring is established by GC analysis of the products. Furanoid fatty acids, separated on TLC plates, can be detected by spraying with a 2 % solution of tetracyanoethylene in acetone, which reacts to give blue spots on a yellow background [300].

The chemistry of cyclopropane and cyclopropene ring-containing fatty acids has been reviewed elsewhere [159]. The presence of such functional groups can be detected by a number of spectroscopic procedures and also by various chemical techniques. For example, cyclopropene rings give a pink coloration with carbon disulphide (the Halphen test) and a brown coloration with silver nitrate. The ring is disrupted by ozonolysis or permanganate-periodate oxidation, and a *beta*-diketo compound is formed which can be identified by mass spectrometry [398]. Similarly, the reaction with silver nitrate can be used with GC to quantify individual cyclopropene fatty acids in seed oils (see Chapter 5) and in conjunction with mass spectrometry for identification purposes (see Chapter 7).

Cyclopropane fatty acids behave as normal saturated fatty acids on silver ion chromatography, but react with bromine and so can be distinguished by this means [128]. They form methoxy-derivatives with boron trifluoride-methanol reagent [610], and they are converted to methyl-branched fatty acids by vigorous catalytic hydrogenation [571]. Both types of derivative can be characterized by mass spectrometry.

Methyl branches in fatty acids are not easily located by chemical means as such compounds are comparatively inert to most chemical reagents. However, their positions can be determined chemically by oxidising the fatty acids vigorously with acidic potassium permanganate; homologous series of normal and branched-chain acids are formed together with a neutral keto compound that identifies the point of branching [639,669]. Both the acidic and keto products can be isolated and identified by GC. These and related procedures have been reviewed by Polgar [723].

CHAPTER 7

GAS CHROMATOGRAPHY - MASS SPECTROMETRY AND FATTY ACIDS

A. INTRODUCTION

In recent years, gas chromatography in combination with mass spectrometry (MS) has become one of the most powerful tools in the hands of lipid analysts. Simple GC-MS systems, such as the Finnigan MAT "ion trap" detector and the Hewlett Packard "mass selective detector", have become less costly, more reliable and simpler to use; the author has made use of the latter instrument in his own research and in preparing many of the figures for this chapter. It is therefore likely that even more laboratories will acquire this facility. The basic principles of an electron-impact (EI) mass spectrometer are described in Chapter 3. For fatty acid identification especially, it could be argued that this basic system will have the capacity to provide answers to most problems. The main difficulties arise with compounds with labile functional groups, which may not always be subjected safely to GC and where it may not be easy to recognise the molecular ion, because excessive fragmentation has occurred. Reducing the ionization potential from the standard 70 eV can sometimes help.

Other "soft" ionisation procedures are then often the preferred alternative. For example, in chemical ionisation (CI) procedures, the sample is mixed with a large excess of a reagent gas such as methane or butane, which is ionised. The reagent ions interact with the sample, transferring the charge and some energy, and bringing about fragmentation; the resulting ions are separated and analysed. Field ionisation and field desorption are similar methods in which the molecules are subjected to a powerful electrical field, which brings about ionisation with a relatively small transfer of energy, so that the amount of fragmentation is reduced. In the field desorption method, the sample is ionised without prior volatilisation and this is the mildest method of all. Some of these ionisation procedures are more compatible with GC separation than are others. Fast atom bombardment (FAB) MS is a solid-phase ionisation technique, utilising production of charged particles from the surface of a liquid matrix, and it has proved of great value for structure determination of complex ionic glycolipids and phospholipids; it is not appropriate to discuss it further in the context of this chapter. This is also true of tandem mass spectrometry of lipids, which has been reviewed elsewhere [30].

Methyl ester derivatives of fatty acids are not always the most useful for identification purposes using MS, and pyrrolidides and picolinyl esters especially have advantages in many circumstances (see Chapter 4 for methods of preparation). For example, stable ions may be produced with the latter that assist in locating double bonds in unsaturated fatty acids. It is also advantageous in some instances to prepare specific adducts of double bonds and derivatives of other functional groups to assist both with the GC separation and with identification by MS. Often this results in a considerable increase in mass with a corresponding change in GC retention characteristics. On fused silica WCOT columns coated with polar and non-polar phases, picolinyl esters tend to elute at a temperature 5°C higher than the corresponding pyrrolidides, which in turn elute 40 to 50°C higher than the methyl esters [173]. Non-polar stationary phases, such as cross-linked methyl- or methylphenylsilicones, stand up to the required conditions better, but some loss of resolution is inevitable. Analogous effects are seen with many other types of derivative, especially those used in the location of double bonds. In this chapter, most emphasis is on practical GC-MS procedures, and mechanistic aspects of the fragmentation process are not treated in depth. Perhaps even more than with other spectroscopic techniques, interpretation of mass spectra is dependent on experience gained with model compounds, and it is not possible to draw up simple rules to guide the reader. This chapter should not be read in isolation; it should be remembered that other chromatographic procedures can give simplified fractions that are more easily analysed subsequently by GC-MS than are the intact samples (see Chapter 6). GC-MS, as opposed to MS with sample introduction by other means, gives the analyst information of two kinds on a given compound - its mass spectrum and GC retention time. The latter can often be of crucial importance as an aid to identification with closely related isomers. Aspects of mass spectrometry of fatty acid derivatives have been reviewed by several authors [45,446,569,612,704,809].

B. SATURATED STRAIGHT-CHAIN FATTY ACIDS

The methyl ester derivatives of long-chain saturated fatty acids are easily identified by electron-impact MS, and their spectra are characterised by a prominent molecular ion (M^+), and other significant ions equivalent to m/z = M-31 (loss of methanol) and M-43 (loss of C_2, C_3 and C_4 as a result of a complex rearrangement), together with a series of ions of general formula - $[CH_3COO(CH_2)_n]^+$, often with intensity maxima at m/z = 87, 143 and

199 [781]. The base ion at m/z = 74 is often termed the "McLafferty rearrangement ion", and is formed after cleavage of the parent molecule *beta* to the the carboxyl group [574]. (The mechanism of its formation is one of the most widely studied processes in mass spectroscopy). CI mass spectra, in contrast, have a prominent quasi-molecular ion (MH^+), and minor ions only at $[MH\text{-}32]^+$ and $[MH\text{-}32\text{-}18]^+$ [631].

Mass spectra of pyrrolidine derivatives of saturated fatty acids have prominent molecular ions, and a base peak at m/z = 113, equivalent to the McClafferty ion [48]. Similarly, picolinyl ester derivatives have spectra with distinctive molecular ions, in addition to abundant ions containing the pyridine ring at m/z = 93, 108, 151 (the McLafferty ion) and 164 [359]. The spectrum of picolinyl palmitate is illustrated in Figure 7.1. In the high mass

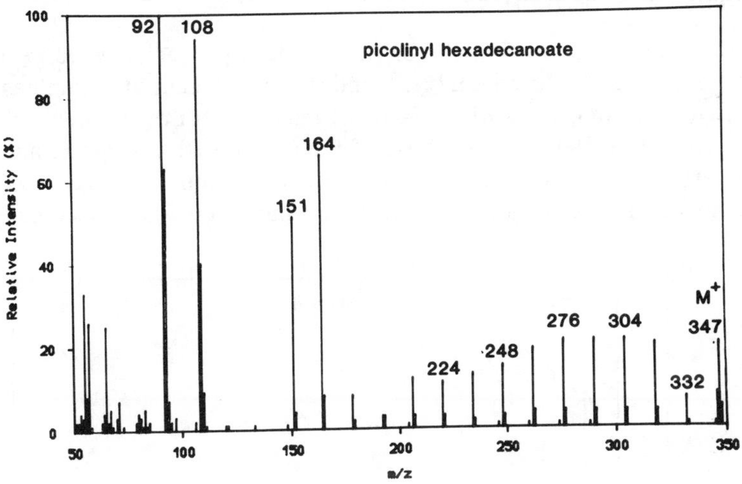

Figure 7.1 The mass spectrum of picolinyl hexadecanoate.

range, there is a regular series of ions 14 atomic mass units (amu) apart, i.e. at m/z = 332, 318, 304, 290, 276 and so forth, representing cleavage between successive methylene groups.

Mass spectrometry has long been the favoured method for the analysis of deuterated fatty acids (reviewed elsewhere [570,934]). Methyl ester derivatives and electron impact MS are not the best combination for locating deuterium atoms on specific carbon atoms in saturated fatty acids, because of the complex nature of the rearrangements that happen. Pyrrolidine derivatives of deuterated fatty acids are better in some circumstances [48,50], but isotope effects can occur within the mass spectrometer, and elimination reactions take place that confuse the results [484]. Conversely, specifically deuterated fatty acids are excellent substrates for the study of ion fragmentations, and the trimethylsilyl (TMS) esters of deuterated decanoic acids have been used for this purpose, for example [935].

In measurements of isotopic purity, field desorption MS (of the free acids) has considerable potential, since the molecular ion is the base peak [539]. It is also possible to use *t*-butyldimethylsilyl esters of fatty acids (see Chapter 4 for details of preparation) with EI-MS, since these give a particularly prominent $[M - C_4H_9]^+$ ion from which the precise molecular weight can be determined [692,709,1004].

C. UNSATURATED FATTY ACIDS

1. Methyl ester derivatives

Procedures for locating double bonds in fatty acids were reviewed comprehensively by Minnikin [612], and the more recent literature has been assessed by Schmitz and Klein [809]. EI mass spectra of unsaturated fatty acids are very different from those of their saturated analogues, and they also vary a little according to degree of unsaturation. The mass spectrum of methyl oleate is illustrated in Figure 7.2. Again, there is a distinct molecular

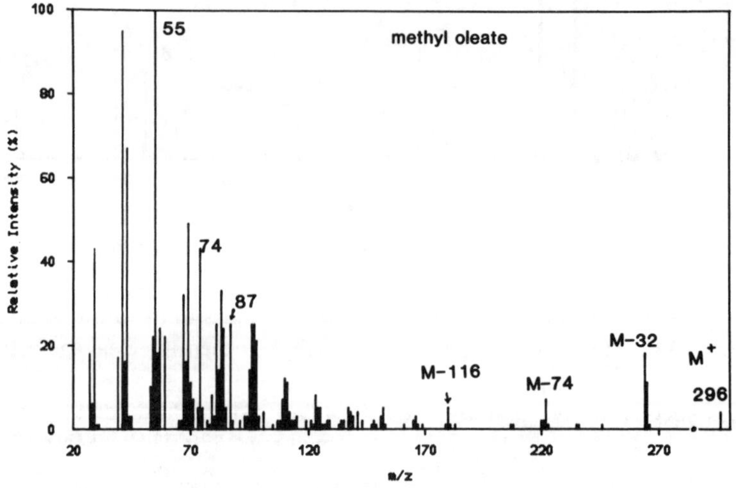

Figure 7.2 The mass spectrum of methyl octadec-9-enoate.

ion with prominent ions for loss of methanol ($[M-32]^+$, the base peak with some instruments [338]) and of a methoxyl radical $[M-31]^+$. $[M-74]^+$ and $[M-116]^+$ ions stand out in the high mass range. In the spectra of dienes and trienes, the molecular ion is more pronounced, while those representing losses of 32, 74 and 116 amu are less so.

Unfortunately, there are no ions that serve to indicate the location or stereochemistry of the double bonds in positional isomers. This is believed to be because double bond migration occurs when the molecular ion is formed, resulting in a range of common intermediate products, which in turn give common fragment ions. The only exception is the 2-isomer, which has a characteristic ion at m/z = 113, the base ion for the *cis*-isomer [780]. Similarly, in the mass spectra of the series of isomeric methylene-interrupted octadecadienoates, only that of the 2,5-isomer differs from the remainder in having the ion at m/z = 113 together with one at m/z = 139 [180]. With the spectra of polyenoic fatty acids, the intensities of some ions vary, but they are not readily interpretable in terms of the positions of the double bonds. Nonetheless, it has been claimed that there are specific ions in the spectra of the main families of polyenes that may be of diagnostic value [253]. Methyl esters of a number of different fatty acids of the (n-3) family are reported to give a characteristic fragment at m/z = 108 (27 to 66% of the base peak), while those of the (n-6) series give a prominent ion at m/z = 150 (9 to 30% of the base peak). These ions are believed to represent fragments from the terminal region of the molecule. Fatty acids of the (n-9) family would therefore be expected to have an abundant ion at m/z = 192, and this was indeed present in the spectrum of 20:3(n-9) (11% of the base peak). Features of this kind were used for the provisional identification of long-chain (C_{26} to C_{38}) fatty acids of the (n-6) family with four or five double bonds in tissues of patients with the Zellweger syndrome [733,838], and of unusual fatty acids of the (n-3) series in ram spermatozoa [732]. In these examples, it is necessary to assume that all the double bonds are methylene-interrupted.

While it is obviously disappointing not to be able to obtain complete information on a fatty acid from its EI mass spectrum, the capacity to determine accurate molecular weights together with GC retention time data can be of considerable value to the analyst. Two GC-MS studies selected from many to have been published testify to this [39,655].

Acetylenic fatty acids, as the methyl ester derivatives, have complex and distinctive EI mass spectra, which vary with the position of the triple bond and can be used to distinguish isomers [480].

With CI procedures, the molecular ion is more readily seen with polyunsaturated components [631], and in some circumstances more information can be obtained on the location of double bonds. Usually, however, it is necessary to have the free acid or a salt form. For example, collision-activated decomposition spectra of negative ions obtained by chemical ionisation gave distinct fragments corresponding to cleavage allylic to the double bond of elaidic acid [918]. Analogous methods have been described that make use of iron, lithium and other alkali metal salts of monoenoic acids as the substrates [32,447,700]. Similarly, by making use of a suitable reagent gas, it may be possible to locate isolated double or triple bonds [139]. Such techniques are not of course always compatible with gas chromatography.

2. Pyrrolidine derivatives

Various types of amide derivatives of fatty acids give distinctive EI mass spectra from which many functional groups, including double bonds, can be located. It appears that the reason for this is that the charge on the molecular ion is sited on the nitrogen-containing functional group predominantly, rather than being localised at the double bond. Acyl pyrrolidines were the first to be employed extensively for the purpose, and their use has been reviewed by Andersson [45]. The method of preparation is described in Chapter 4. On electron impact, pyrrolidides fragment to produce a series of ions as a result of successive cleavage of carbon-carbon bonds induced by radicals. In comparison to methyl esters, there is less scrambling of hydrogen atoms.

The EI mass spectrum of *N*-octadec-9-enoylpyrrolidine is illustrated in Figure 7.3 [45,51]. In addition to the base peak at m/z = 113 (observed with

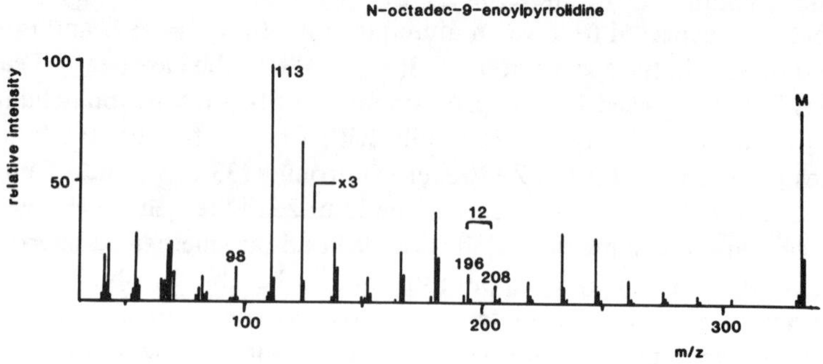

Figure 7.3 The mass spectrum of *N*-octadec-9-enoylpyrrolidine.

saturated derivatives (Section B)) and the molecular ion, there is a series of ions 14 amu apart in general, except in the vicinity of the double bond, where there is an interval of 12 amu., i.e. occurring between m/z = 196 and m/z = 208, and corresponding to fragmentation between carbons 8 and 9 in the aliphatic chain. For the spectra of the pyrrolidides of the isomeric 5- to 15-octadecenoic acids, the following rule was formulated:

> "If an interval of 12 amu, instead of the regular 14, is observed between the most intense peaks of clusters of fragments containing *n* and *n+1* carbon atoms in the acid moiety, a double bond occurred between carbons *n* and *n+1* in the molecule".

The pyrrolidides of the remaining octadecenoic acid isomers also have distinctive spectra, but they do not fit the rule. The technique has been utilised with vicinal dideuteriated monoethylenic compounds [49].

Methylene-interrupted polyunsaturated fatty acids, as the pyrrolidides, have characteristic spectra that can usually be interpreted in terms of the positions of the double bonds with some modification to the above rule [45,47]. Similarly, a conjugated diene, 9,11-octadecadienoate, has been identified by this means in the lipids of human tissues [417]. It must be admitted, however, that it is not easy to identify unknowns from first principles using such spectra. Nonetheless, a good molecular ion is almost always obtained; with access to the spectra of model compounds and to natural mixtures the fatty acids of which have been identified by other means, it is possible to make good use of pyrrolidides in the analysis of unknown samples. For example, they have been used to identify fatty acids in the bovine lens [941], marine organisms [70,207,452,534,843], bacteria [909] and algae [770].

Isomeric mono-acetylenic fatty acids in the form of the pyrrolidides have distinctive mass spectra which can be used for identification purposes [942].

3. Picolinyl ester derivatives

Harvey [359] first proposed the use of picolinyl (3-hydroxymethyl-pyridine) ester derivatives for the identification of fatty acids containing double bonds and other functional groups. Like the pyrrolidides, they fragment under electron impact by radical-induced cleavage at each carbon-carbon bond. The method of preparation is described in Chapter 4. The author recently took part in a comparative study of the utility of pyrrolidides and picolinyl esters for the recognition of unsaturated fatty acids in natural mixtures of animal and marine origin [173]. Without doubt, picolinyl esters are to be preferred and give spectra from which both the number and the positions of double bonds can be deduced; pyrrolidides have marginally better chromatographic properties. Harvey [359] obtained better results by operating his mass spectrometer at an ionisation potential of 25 eV, but the author obtained satisfactory spectra at 70 eV; instrumental differences may explain the discrepancy.

The EI mass spectrum of the picolinyl ester derivative of oleic acid is illustrated in Figure 7.4, and details of the spectra of the complete series of isomeric octadecenoate derivatives (2- to 17-) have been published [172]. That shown differs from those published by Harvey [359,361] in that the base ion is at m/z = 92 rather than at 108, possibly a consequence of the different ionisation potentials utilised in the two studies. There is a particularly prominent molecular ion (m/z = 373), and the ions containing the pyridine ring at m/z = 92, 108, 151 and 164 are all very evident as with saturated compounds (Section B). Points of cleavage where characteristic fragmentations occur that permit the location of the double bond are

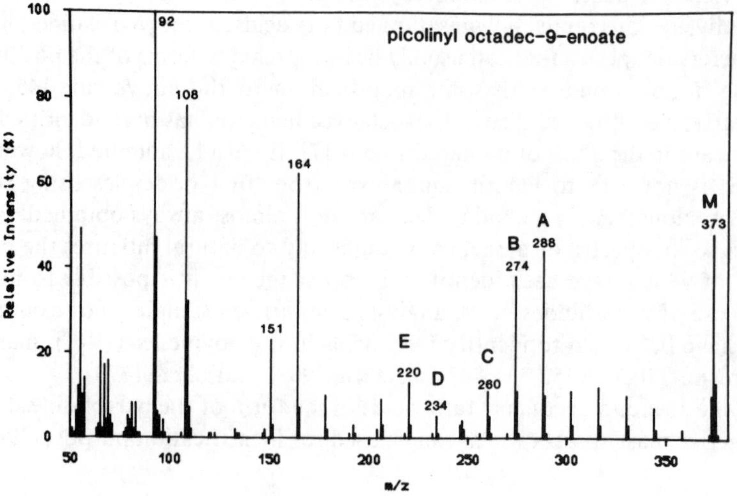

Figure 7.4 The mass spectrum of picolinyl octadec-9-enoate.

llustrated in Figure 7.5a, and data for each of the isomers is listed in Table 7.1. In nearly all the spectra, a distinctive feature is a doublet of prominent

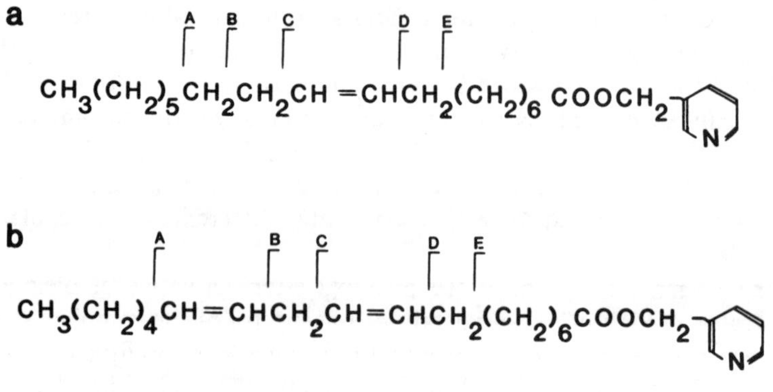

Figure 7.5 The bonds where characteristic mass spectrometric fragmentation occurs for (a) picolinyl octadec-9-enoate and (b) picolinyl 9,12-octadecadienoate.

ions 14 amu apart, representing cleavage at points A and B on the distal side of the double bond. In the spectrum of the 9-isomer, they are seen at *m/z*

TABLE 7.1

Relative Abundances[a] of the Molecular Ion and of Ions Characteristic of the Picolinyl Moiety, and m/z Values[b] and Relative Abundances of Ions Characteristic of Double Bond Positions in Mass Spectra of Picolinyl Octadecenoates [172]

Double bond position	m/z = 92	m/z = 108	m/z = 151	m/z = 164	M	M-1	Doublet		Gap of 26 amu		Gap of 40 amu
							A[c]	B	C	D	E (to C)
2	79	100	27	4	26	8	45(190)	40(177)			
3	100(93)	43	19	10	31	4	15(204)	8(190)			
4	84(93)	53	100	3	30	3	9(218)	2(204)			
5	100(93)	43	8	21	31	2	21(232)	4(218)	3(204)	1(178)	21(164)
6	100(93)	54	22	19	29	3	32(246)	5(232)	7(218)	1(192)	4(178)
7	100	70	16	52	52	8	43(260)	17(246)	4(232)	8(206)	5(192)
8	100	73	20	67	48	11	37(274)	28(260)	5(246)	10(220)	7(206)
9	100	75	22	61	37	7	35(288)	30(274)	10(260)	4(234)	10(220)
10	100	75	30	41	46	10	37(302)	39(288)	9(274)	2(248)	8(234)
11	100	72	36	46	43	9	32(316)	43(302)	10(288)	4(262)	7(248)
12	100	72	28	47	53	10	41(330)	49(316)	13(302)	4(276)	12(262)
13	100	67	28	50	50	10	33(344)	48(330)	13(316)	2(290)	11(276)
14	100	76	29	42	54	12	10(358)	52(344)	12(330)	3(304)	15(290)
15	100	88	38	59	54	12	12(372)	14(358)	18(344)	3(318)	17(304)
16	100	90	35	58	50	17		17(372)	5(358)	5(332)	28(318)
17	100	74	29	44	35	20			20(372)	1(346)	37(332)

[a]Expressed as a percentage of the base ion.
[b]In parentheses.
[c]Points of cleavage of ions A to E can be seen by reference to Figure 1a.

= 274 and 288; such ions in exactly the same place are found in the spectra of the picolinyl derivatives of 9-14:1, 9-16:1, 9-17:1, 9-20:1 and 9-22:1 fatty acids [173,175,184]. Their formation was rationalised in terms of an initial abstraction of allylic hydrogens on either side of the double bond with production of conjugated diene systems, which are relatively stable [361]. When natural samples containing fatty acids in which the double bonds are two carbon atoms apart are examined by GC-MS in the form of the picolinyl ester derivatives (see Figure 6.4 for example), this doublet can usually be picked out for diagnostic purposes even when the isomers are imperfectly resolved. When the double bonds are near either end of the molecule, these ions can be less obvious (Table 7.1) but there are then other distinguishing features.

Distinctive fragmentations at the double bond are also apparent for each isomer. In the spectrum of picolinyl oleate again, there is a gap of 26 amu between m/z = 260 and 234 for fragmentations adjacent to the double bond. This gap is seen in the spectra of most of the isomers, although a gap of 40 amu between points C and E (i.e. between m/z = 260 and 220 in this instance) is often easier to locate. Difficulties arise mainly with the 3- and 4-isomers, but there are other diagnostic features with these. On either side of the double bond, regular series of ions are found 14 amu apart for cleavage between successive methylene groups. Thus for oleate once more, these are seen at m/z = 358, 344, 330, 316, 302, 288 and 274 on the distal side of the double bond, and at m/z = 220, 206, 192, 178 and 164 on the carboxyl side. Analogous features are seen in the spectra of the other isomers.

Harvey [361] also demonstrated distinctive features in the spectrum of the picolinyl ester of linoleic acid, but that illustrated in Figure 7.6 is from the

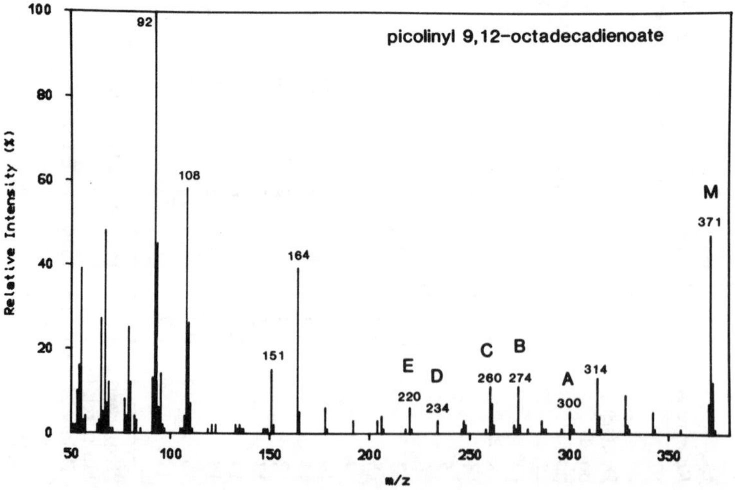

Figure 7.6 The mass spectrum of picolinyl 9,12-octadecadienoate.

author's laboratory (again it differs significantly only in that the base peak is at m/z = 92 rather than at 108). Data for the complete series of isomeric methylene-interrupted octadecadienoates (2,5- to 14,17-) have been published [172], and some of this is listed in Table 7.2 as it has proved invaluable in the author's laboratory for the interpretation of the spectra of unknown polyenoic fatty acid derivatives. Returning to Figure 7.6, ions corresponding to cleavage on either side of the terminal double bond are seen at m/z = 300 and 274 (26 amu apart), there is then a gap of 14 amu to 260, then one of 26 amu for the internal double bond to 234. When examining the spectra of unknowns, however, it is often easier to locate gaps of 40 amu between points A and C and C and E (Figure 7.5b), i.e. in this instance between m/z = 300 and 260, and between 260 and 220. Regular series of ions 14 amu apart on either side of the double bond confirm that there are no further functional groups in these regions of the molecule. The spectra of the remaining isomers have corresponding ions or other distinctive features.

With dienoic acids with more than one methylene group between the double bonds, the spectra of the picolinyl ester derivatives are less easy to interpret in terms of the positions of the double bonds [171]. The terminal bond can be identified with relative ease, and there are ions diagnostic of the internal bond provided that model spectra are available for comparison purposes. With such assistance, the author was able to identify the dienes shown in Figure 6.4 in the previous Chapter [175].

Interpretation of the spectra of the picolinyl ester derivatives of tri- and tetraenoic fatty acids containing methylene-interrupted double bonds presents little problem. Gaps of 26 amu for each of the double bonds are usually seen as with the dienes, although again it is often easier to detects gaps of 40 amu between the terminal end of each double bond and the methylene group on the carboxyl side. Full spectra of 18:3(n-3) [361], 20:3(n-9), 20:3(n-6) and 20:3(n-3) [173], 16:4(n-1) [175], 20:4(n-6) and 22:6(n-3) [361] have been published, together with brief details of many more. With the last and with isomers with 5 double bonds, the spectra are much more complex and have to be examined with care, but the spectrum of the picolinyl ester derivative of 22:5(n-6) is readily distinguished from that of 22:5(n-3) and both are as expected [173]. As an example, the mass spectrum of picolinyl 5,8,11,14,17-eicosapentaenoate (20:5(n-3)) is illustrated in Figure 7.7. There is an adequate molecular ion, a gap of 15 amu for the terminal methyl group to m/z = 378 one of 14 to m/z = 364; the gaps of 26 amu for the double bonds between m/z = 364 and 338, 324 and 298, 284 and 258, 244 and 218, and 204 and 178 are not always easy to pick out, but the associated gaps of 40 amu for each double bond and the attached methylene group on the carboxyl side, i.e. for m/z = 364 to 324 to 284 to 244 to 204 to 164, are also diagnostic.

The picolinyl ester of a monoacetylenic fatty acid gave a spectrum which resembled that from its monoethylenic analogue except that there was a gap

TABLE 7.2

Relative Abundances[a] of the Molecular Ion and of Ions Characteristic of the Picolinyl Moiety, and m/z Values[b] and Relative Abundances of Ions Characteristic of Double Bond Positions in Mass Spectra of Picolinyl Octadecadienoates [172]

Double bond position	m/z = 92	m/z = 108	m/z = 151	m/z = 164	M	M-1	Gap of 26 amu (terminal bond)		Gap of 26 amu (proximal bond)		Gap of 40 amu
							A[c]	B	C	D	E (to C)
2,5	100	75	34	3	72		1(202)				
3,6	100(93)	40	46	3	54	4	2(216)	48(190)	4(176)	46(151)	
4,7	100	74	53	12	59	5	2(230)	11(204)	1(190)	12(164)	53(151)
5,8	100(93)	49	13	17	58	6	2(244)	8(218)	2(204)	2(178)	17(164)
6,9	100	50	17	21	37	4	2(258)	8(232)	3(218)	3(192)	2(178)
7,10	100	55	15	43	66	8	8(272)	13(246)	9(232)	5(206)	4(192)
8,11	100	51	14	47	62	9	7(286)	12(260)	8(246)	5(220)	7(206)
9,12	100	58	15	39	47	7	5(300)	11(274)	11(260)	3(234)	6(220)
10,13	100	58	24	42	55	10	10(314)	14(288)	27(274)	4(248)	9(234)
11,14	100	59	20	41	57	13	9(328)	14(302)	32(288)	4(262)	10(248)
12,15	100	68	24	48	55	15	13(342)	13(316)	42(302)	5(276)	14(262)
13,16	100	67	22	48	31	10	2(356)	4(330)	32(316)	3(290)	13(276)
14,17	100	58	20	41	39	18	18(370)	1(344)	48(330)	2(304)	23(290)

[a]Expressed as a percentage of the base ion.
[b]In parentheses.
[c]Points of cleavage of ions A to E can be seen by reference to Figure 1b.

of 24 amu for fragmentations on either side of the triple bond [172]. A series of diacetylenic fatty acids gave spectra which were rather complex, but the terminal triple bond generally exhibited a fragmentation with a gap of 24 amu [174].

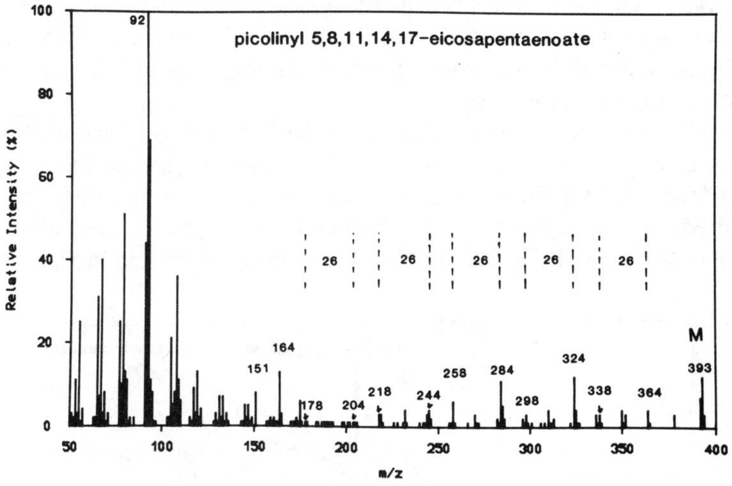

Figure 7.7 The mass spectrum of picolinyl 5,8,11,14,17-eicosapentaenoate.

It is the author's opinion then that picolinyl esters are the best general purpose derivative for locating double bonds by GC-MS. They have been used in analyses of fatty acids of animal tissues [173,184,363,365,979], a fish oil [173,184], marine invertebrates [175] and bacteria [960].

4. Addition compounds

While the EI mass spectra of pyrrolidine and picolinyl ester derivatives of unsaturated fatty acids can often yield sufficient information for structure determination, there are circumstances when the result is not clear or where confirmation by a different method is desirable. The preferred method then is to form some addition compound with the double bonds that gives a distinctive fragmentation pattern. Very many different methods have been described for the purpose, and as they have been reviewed in detail elsewhere [612,809] only the more important are considered here. Methods of preparation are described in Chapter 4.

Potentially the simplest method consisted in deuteration of polyunsaturated fatty acids with deuterodiimide, followed by conversion to the pyrrolidine derivative for examination by GC-MS [472]. The positions of the original double bonds are deduced from the ions containing deuterium atoms. In a critical examination of the method, Klein and Schmitz [484] concluded that good results could be obtained with fatty acids containing up to six double bonds, although great care was necessary in interpreting the spectra because of the occurrence of unexpected elimination reactions. There were also problems with some instruments. Nonetheless, it is one of the few methods of this type that is easily applied to polyunsaturated fatty acids. It might be of interest to ascertain how picolinyl esters would fare in this approach. One disadvantage of the procedure is that it can only be applied to pure fatty acids or to rather simple mixtures.

If simplicity is a virtue, then a further excellent method for locating double bonds, especially in monoenoic acids, consists in preparing the dimethyldisulphide adducts, since a single reagent and a one step reaction is required for the preparation (see Chapter 4) [261]. The EI mass spectrum of the dimethyldisulphide adduct of methyl oleate is illustrated in Figure 7.8.

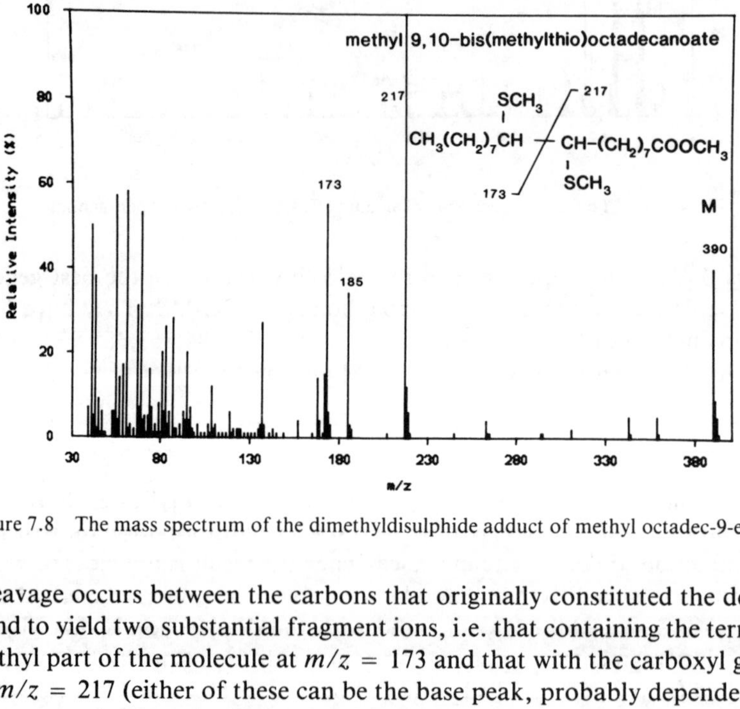

Figure 7.8 The mass spectrum of the dimethyldisulphide adduct of methyl octadec-9-enoate.

Cleavage occurs between the carbons that originally constituted the double bond to yield two substantial fragment ions, i.e. that containing the terminal methyl part of the molecule at $m/z = 173$ and that with the carboxyl group at $m/z = 217$ (either of these can be the base peak, probably dependent on instrumental factors). There is also a prominent ion at $m/z = 185$, corresponding to the latter fragment with the loss of methanol. In addition, these derivatives give good molecular ions. As the adduct adds substantially to the molecular weight of the original ester, it tends to elute at a temperature about 40°C higher than the former from a GC column containing a non-polar silicone phase. Adduct formation is entirely stereospecific, presumably by trans addition, so that *erythro*- and *threo*-derivatives are formed from *cis*- and *trans*-isomers respectively [143]. Although the different geometrical isomers have virtually identical spectra, they are eluted separately from GC columns containing either polar or non-polar phases, that derived from the *cis*-isomer eluting first [143,540]. The procedure has been used with a variety of monoenoic fatty acids [143,233,540,667,701,841,952] and with some polyenes from natural sources [841,952].

A number of workers have approached the problem of double bond location by oxidising it to a vicinal diol with alkaline permanganate or osmium tetroxide and then derivatising further to a non-polar form (see Chapter 4). The stereochemistry is also retained in this way, so that geometrical isomers can be resolved. Isopropylidene derivatives of the diols were used first for the purpose [572], but more use has been made of the TMS ethers [57,146,239,240]. The former is more useful for establishing the configuration of the original double bond, while the latter gives more intense fragments for locating it. Figure 7.9 illustrates the EI mass spectrum of the TMS ether

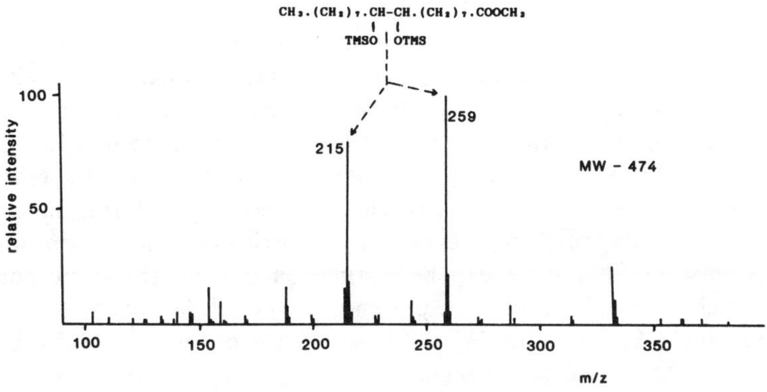

Figure 7.9 The mass spectrum of the trimethylsilyl ether derivative of the vicinal diol prepared from methyl octadec-9-enoate.

of the vicinal diol prepared from methyl oleate. Ions in the high mass range are of very low intensity and the molecular ion is not seen, but two ions at m/z = 215 and 259, which represent cleavage between the carbon atoms of the original double bond, are particularly prominent. Indeed these ions are sufficiently intense to be of use in estimating mixtures of positional isomers of unsaturated fatty acids. When softer chemical ionisation procedures are used, a small molecular ion is then seen and the ions for fragmentation between the functional moieties are still sufficiently intense for structure determination both with monoenes and dienes [58,632,717]. Indeed the geometry of the original double bond can be determined from differences in the intensities of the fragment ions. Although there are difficulties in handling polyhydroxy compounds, prepared from polyenoic fatty acids, the TMS ether derivatives give good spectra from which the original structures can be deduced [225,442,443,637,808,810].

Because so many TMS ether groups add substantially to the molecular weight of a fatty acid and can in prolonged use lead to deterioration of the ion source, other workers have preferred to prepare polymethoxy derivatives [534,893]. These give excellent mass spectra when chemical ionisation is employed.

One further procedure worth mentioning briefly here consists in preparing methoxy- or methoxyhalogeno-derivatives, via the mercuric acetate adducts, for identification by MS [611-613,834]. Oxymercuration has also been employed to convert triple bonds to keto groups, which can be located in this form or after reduction to hydroxyl derivatives with sodium borohydride [140,480].

D. BRANCHED-CHAIN FATTY ACIDS

Mass spectrometry has been used more than any other technique for the location of alkyl branches in fatty acids. The methyl ester derivatives have been used extensively, although they are certainly not the best in all circumstances, as a branch-point does not provide a centre for charge localisation. Mass spectral data for a wide range of simple methyl-branched isomers, as the methyl esters, have been published [6,56]. Unfortunately, the hardest to identify are the most commonly occurring *iso*- and *anteiso*-isomers. An *iso*-methyl fatty ester can be distinguished from the corresponding straight-chain compound by the presence of a small ion at $[M-65]^+$ and a doublet at $[M-55]^+$ and $[M-56]^+$(all less than 1% of the base peak), and by its rather different GC retention time. In the spectrum of an *anteiso*-isomer, the ion for $[M-29]^+$ is greater than that for $[M-31]^+$. Small ions at $[M-61]^+$, $[M-60]^+$ and $[M-79]^+$ are also of diagnostic value. Related esters with a single double bond in the aliphatic chain have characteristic spectra [118]. Chemical ionisation [953] and other soft ionisation techniques [445] may simplify identification.

When branching occurs near the centre of a chain, an ion of the type $[CH_3OOC(CH_2)_n]^+$, designated the *a* ion and representing cleavage adjacent to the carbon carrying the methyl group, together with *a+1* and *a+2*, and ions representing a similar fragment but with cleavage on the remote side of the methyl carbon, designated the *b* ion, are of diagnostic value. As the principal ions are also present in the spectra of the corresponding straight-chain compounds, interpretation is based mainly on changes in relative intensity and comparison with spectra of model esters. However, the triplet of *a*, *a+1* and *a+2* is often distinctive, and in the spectrum of methyl

10-methylstearate for example, they occur at m/z = 171, 172 and 173. In addition, ions derived from *a* and *b* by loss of the elements of water or methanol can assist with interpretations. When the methyl branch is on carbons 2, 3 and 4, shifts in the ions at m/z = 74 (to 88 in the 2-isomer) and 87 (to 101 in the 2- and 3-isomers), and equivalent to $[M-43]^+$ (to $[M-57]^+$ in all three isomers) and $[M-29]^+$ (to $[M-43]^+$ in the 2- and 3-isomers) serve for identification. Similar mass spectral features are used to identify dimethyl- and ethyl-branched fatty acids. Applications of GC-MS to isoprenoid fatty acids have been reviewed [562]. Those studies cited here are representative of many applications to natural samples [232,238,423,425,668,852,853].

Pyrrolidine [52] and picolinyl ester [359] derivatives of branched-chain fatty acids give especially distinctive spectra, and are certainly to be preferred with samples that are relatively simple in composition, i.e. such that the bulky nitrogen-containing moiety does not impair GC resolution excessively. Again, it is the author's experience that the latter derivative has some advantages. With both types of derivative, the spectra superficially resemble those of the corresponding straight-chain compounds, but there are characteristic fragmentations on either side of the carbon atom linked to the methyl group. The result is that a diagnostic gap of 28 amu appears in the spectrum. Thus in the spectrum of the picolinyl ester of *iso*-methylheptadecanoate (M^+ = 375), there is a gap from m/z = 332 to 360; in the spectrum of the corresponding *anteiso*-isomer, the gap is from m/z = 318 to 346. This feature is also seen in spectra from isoprenoid fatty acids, and that of the picolinyl ester derivative of phytanic acid (3,7,11,15-tetramethylhexadecanoate) is illustrated in Figure 7.10 (W.W.Christie and E.Y.Brechany, unpublished).

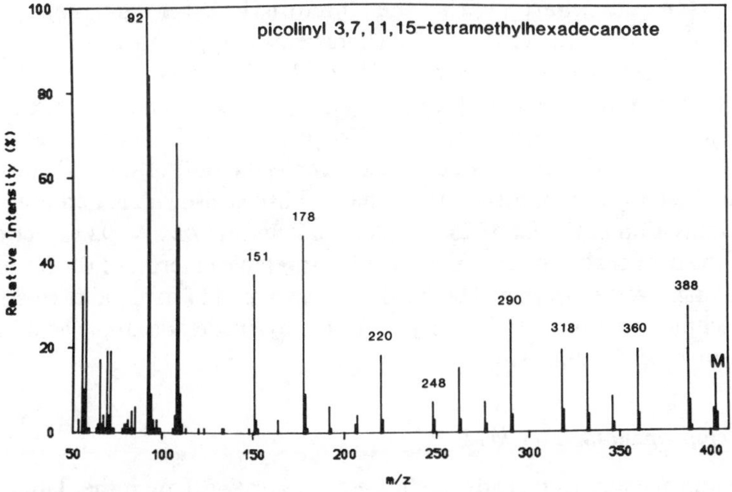

Figure 7.10 The mass spectrum of picolinyl 3,7,11,15-tetramethyl-hexadecanoate.

There is a good molecular ion at m/z = 403, then gaps of 28 amu are seen between m/z = 151 and 178, 220 and 248, 290 and 318, and 360 and 388 for each of the methyl groups. When the methyl group is in position 2, the typical ions at m/z = 151 and 164 are shifted to 165 and 178 (the base peak) respectively [361]. Both pyrrolidides [95,147,206,296,770] and picolinyl esters [96,173,175,184,363,365,979] have been used for the identification of branched-chain esters in natural mixtures.

E. CARBOCYCLIC FATTY ACIDS

1. *Cyclopropanoid fatty acids*

Mass spectrometric procedures for the location of cyclopropane rings were reviewed by Minnikin [612]. The methyl ester derivatives of cyclopropanoid fatty acids are not readily distinguished from mono-unsaturated fatty acids with a similar total number of carbon atoms by MS, apparently because on ionisation the cyclopropane ring opens up to form such a compound [179,997]. If necessary, methyl esters can be used if the ring is first opened by vigorous catalytic hydrogenation (with formation of two methyl-branched fatty acids) [571] or by reaction with boron trifluoride-methanol (with formation of two methoxymethyl fatty acids) [610].

Pyrrolidine derivatives of cyclopropanoid fatty acids give more useful spectra, although the diagnostic ions are not as clear as with the analogous monoenes [45,292,293]. The technique was used to identify 17-methyl-*cis*-9,10-methyleneoctadecanoic acid in a protozoan, for example [397].

Much better results are obtained with picolinyl ester derivatives. Harvey [360] showed that, with the picolinyl ester derivative of *cis*-9,10-methyleneoctadecanoic acid, characteristic fragments from either side of the ring are found, i.e. at m/z = 234 and 274. These can be seen in Figure 7.11 (spectrum obtained at 70 eV), together with an even more distinctive ion at m/z = 247, which represents a fragment containing carbon 9 in the ring, together with the remainder of the molecule on the same side as the picolinyl ester group (it differs from that published elsewhere, and obtained at an ionisation potential of 25 eV [360], only in that m/z = 92 rather than 108 is the base peak). Spectra which are interpretable in terms of the positions of the rings are also obtained with *bis*-cyclopropanoid fatty acid derivatives. The technique was used to identify 11,12-methyleneoctadecanoic acid in the lipids of a bacterium [960].

2. *Cyclopropenoid fatty acids*

Cyclopropenoid fatty acids (methyl ester derivatives) are rather labile and not easily subjected to GC (see Chapter 5). Nonetheless, mass spectra of the methyl ester derivatives of some naturally occurring fatty acids of this kind have been published [698]. These are rather different from the spectra of

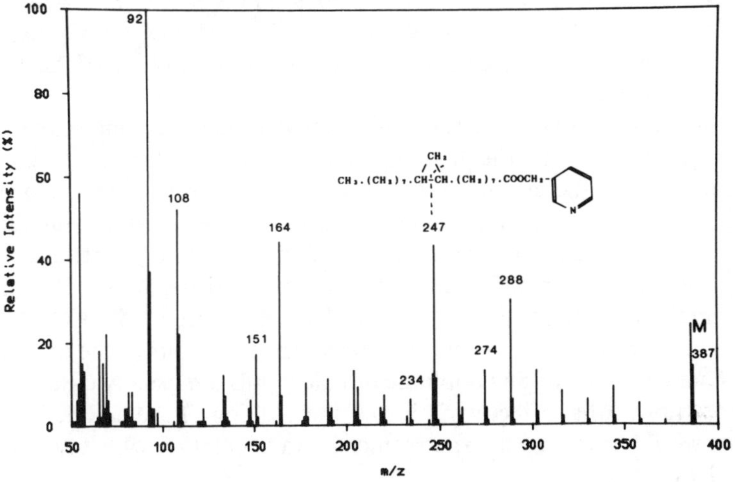

Figure 7.11 The mass spectrum of picolinyl 9,10-methylene-octadecanoate.

most other unsaturated esters, but it is doubtful whether they would assist greatly in the identification of an unknown. One of the most popular methods for the analysis of cyclopropenoid fatty acids involves GLC of the methoxy- and keto-derivatives formed by reaction with silver nitrate. Such derivatives produce rather complex spectra which can with care be interpreted in terms of the position of the original ring [37,241]. One of the first procedures in which MS was utilised involved ozonolyis or permanganate-periodate oxidation of the cyclopropene ring to form a diketo fatty acid, which has a distinctive mass spectrum [398]. On the other hand, a simpler method is the addition of methanethiol to the cyclopropene ring to yield non-polar derivatives with good GC [746] and MS [398] properties.

3. Cyclopentenyl and related fatty acids

The methyl ester derivatives of cyclopentenyl fatty acids give electron-impact mass spectra with important structural information [183]. The base peak in methyl hydnocarpate (11-cyclopent-2-enylundecanoate) is at *m/z* = 67, presumably for the cyclopentene ring itself, while there are significant fragments at *m/z* = 82 and 185 for cleavage *beta* to the cyclopentene ring. The McLafferty rearrangement ion is not prominent, although this is the base ion in the spectrum of the corresponding cyclopentyl (i.e. saturated) fatty ester. With the latter, there is a prominent ion at *m/z* = 199 for cleavage *alpha* to the cyclopentyl ring.

When in addition to a cyclopentenyl double bond there is a further double bond in the aliphatic chain, as with methyl gorlate (13-cyclopent-2-enyltridec-6-enoate), the base ion is still at $m/z = 67$, and there is another abundant ion at $m/z = 80$, formally equivalent to a dihydrofulvene ion; there are no ions diagnostic for the position of the double bond in the chain.

Pyrrolidine derivatives give good spectra with conspicuous molecular ions, and distinctive ions in the high mass range for the loss of fragments corresponding to the cyclopentenyl ring and successive methylene groups [842]. In the spectrum of the gorlic acid derivative, there are small but significant ions diagnostic for the aliphatic double bond. Even better spectra are obtained with the picolinyl ester derivatives (Christie, W.W., Brechany, E.Y. and Shukla, V.K.S, *Lipids*, in the press). The base ion in this instance is again at $m/z = 67$ for the cyclopentenyl ring, and ions in the high mass range characteristic for any aliphatic double bonds are usually clearly seen. It is not known what effect variation of the position of the double bond in the ring would have on mass spectrometric fragmentation with any of these derivatives.

Mass spectra of fatty acids with terminal cyclohexyl [215,812], cyclobutyl and cycloheptyl rings [215] have been described. Cyclic fatty acids are produced in oxidised oils, and these also have been subjected to structural analysis by GC-MS [368,705].

F. OXYGENATED FATTY ACIDS

1. Hydroxy fatty acids

The methyl ester derivatives of hydroxy fatty acids give good EI mass spectra in which there are ions diagnostic of the position of the hydroxyl group [783]. The most characteristic ions are associated with cleavages *beta* to the oxygen atom (or *alpha* to the carbon carrying this atom); they are accompanied by ions 32 amu lower formed by the elimination of methanol. Unfortunately. there is rarely a significant molecular ion, although ions equivalent to $[M\text{-}OH]^+$ and $[M\text{-}OCH_3]^+$ are usually prominent. Chemical ionisation, with ammonia as the reagent gas, gives a good $[M+1]^+$ ion, however [717].

Because of the high polarity of such esters, it is more usual to convert the hydroxyl group to a trimethylsilyl ether or other derivative for separation by GC, and these also affect the fragmentation patterns observed. This has been discussed in part in relation to addition compounds for the location of double bonds in Section C.4 above. Perhaps the most useful study of this type is one of the acetoxy and trimethylsilyl ether derivatives of all the positional isomers of the methyl hydroxypalmitates [670]. As with the hydroxy esters, fragmentation occurs *alpha* to the carbon containing the functional group. With the acetoxy derivatives, the main diagnostic ion generally represents a cleavage on the side remote from the carboxyl group, while with the trimethylsilyl ethers, the main cleavage is on the side adjacent to the carboxyl group. Neither derivative gives a good molecular ion, but there is usually an acceptable ion equivalent to $[M\text{-}CH_3]^+$ for the TMS ether. The spectrum of the trimethylsilyl ether derivative of methyl 12-hydroxystearate is illustrated in Figure 7.12 as an example. No molecular ion (at m/z = 386)

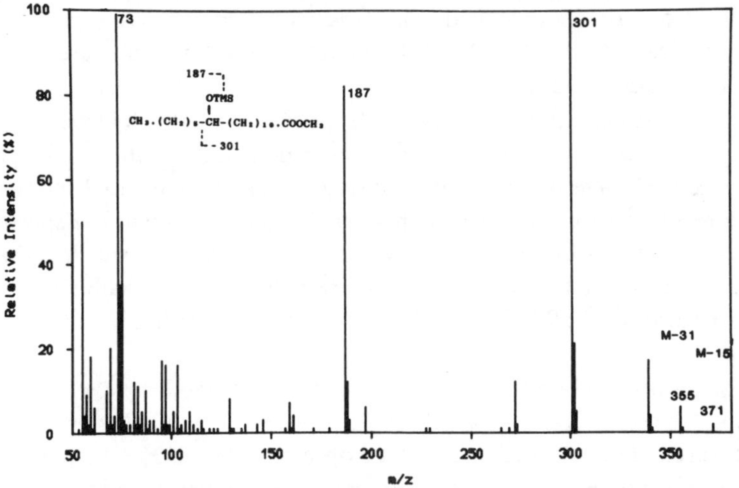

Figure 7.12 The mass spectrum of the TMS ether derivative of methyl 12-hydroxystearate.

is evident, although the molecular weight can be ascertained from a characteristic ion at m/z = 371 (M-15, loss of a methyl group) and = 355 (M-31, loss of methanol). The base peak at m/z = 301 is the fragment formed by cleavage of the molecule between carbons 12 (carrying the TMS ether group) and 13 and containing the carboxyl moiety. The abundant ion at m/z = 187 is formed by cleavage between carbons 11 and 12, and contains the terminal region of the molecule including the TMS ether group. The McLafferty rearrangement ion at m/z = 73 is almost as large as the base peak. (The CI mass spectrum of this compound has been published elsewhere [717]). CI-MS with methane or isobutane as the reagent gas is reported to give better spectra of the TMS ethers of polyhydroxy fatty acid methyl esters than does EI-MS [878].

Among a large number of reports, TMS ethers have been used with EI GC-MS to identify the 2-hydroxy fatty acids from sphingolipids [145,531], for the hydroxy fatty acids in royal jelly [543], for mono-, di- and trihydroxy fatty acids from plant cutins [239,932] and for mycolic acids (3-hydroxy) [87,272,457,919]. Cyclic di-*tert*-butylsilene derivatives have been used for GC-MS identification of 2- and 3-hydroxy fatty acids [132]. In addition, natural 2-acetoxy [68] and 2-methoxy [69] fatty acids have been identified in marine organisms using GC-MS.

The positions of hydroperoxy groups formed during the oxidative deterioration or degradation of lipids are most commonly determined by reduction of the hydroperoxide to a hydroxyl group and hydrogenation of any double bonds, followed by methylation and conversion to the TMS ether derivative for GC-MS. This approach is also of value with the eicosenoids derived from arachidonic acid. Those papers cited here are representative of many [223,263,604,833,879,911,984,1005]. Similar results are obtained with phenyl esters [371].

Pyrrolidine derivatives of the complete series of methyl hydroxystearates and of their TMS ether derivatives have been prepared and their mass spectra described [936]. The principal mode of fragmentation is enhanced and is again *alpha* to the carbon carrying the hydroxyl group, although the spectra of the 2-, 3- and 4-isomers are very different from the rest. Additional complications result when keto groups are present [932]. Picolinyl esters of hydroxy acids as the TMS derivatives also have very distinctive mass spectra [361].

When a double bond is present in a molecule in addition to a hydroxyl group, it is essential to prepare the TMS ether of the methyl ester derivative if interpretable mass spectra are to be obtained [481,833]. The position of the double bond relative to that of the TMS ether greatly affects the fragmentation pattern. With esters in which the two groups are allylic, no fragmentation occurs between them but cleavage does occur on either side of the system of functional groups; a similar effect is seen with conjugated diene or enyne systems allylic to a hydroxyl group. When there is one methylene group between the double bond and the carbon linked to the TMS ether, the ions caused by fragmentation *alpha* to the TMS group on the side of the double bond are most abundant. Where the functional moieties are separated by two methylene groups, the two *alpha*-cleavage ions are of approximately equal intensity. Procedures of this kind have been used for naturally-occurring fatty acids from seed oils [481] and for hydroxy fatty acids derived from peroxidation products [406]. In these circumstances, the preparation of pyrrolidine derivatives is not helpful, because oxygen-containing ions are so enhanced that there is little fragmentation adjacent to the double bonds [235].

Hydroperoxides from unsaturated fatty acid derivatives have been examined by direct-probe chemical ionisation MS [716,717].

2. Keto fatty acids

The methyl ester derivatives of keto fatty acids give characteristic EI mass spectra with fragmentations *alpha* and to some extent *beta* to the keto group [473]. Mass spectra of some keto-hydroxy esters [932] and of derivatives of unsaturated keto fatty acid from natural sources [481,717,833,968] have been published. In addition, mass spectra for the pyrrolidides of the complete series of oxo-stearates [936] and of some hydroxy-oxo fatty acids [932] have been described. With the former, the most abundant ion tends to be that representing cleavage *beta* to the keto group and containing the pyrrolidine moiety.

3. Cyclic oxygen-containing fatty acids

Epoxy fatty acids (as the methyl esters) give EI mass spectra which are not easily interpreted in terms of the position of the oxygen atom (especially when double bonds are also present), although it is possible with some effort if suitable model compounds are available [116,473,481]. It is generally recommended that the epoxide be isomerised to a mixture of keto compounds [473], or that the ring should be opened with boron trifluoride-methanol reagent to a mixture of methoxy-hydroxy derivatives [481,833] or with lithium aluminium hydride to hydroxy derivatives followed by trimethylsilylation [940], for identification by GC-MS. MS was used to study the incorporation of ^{18}O into the oxirane ring of 9,10-epoxyoctadecanoic acid in wheat tissues [488]. If derivatisation is undesirable, more useful spectra are obtained with chemical ionisation [474,717].

Methyl esters of furanoid fatty acids give good EI mass spectra with characteristic ions being produced by fragmentations *alpha* to the furan ring [790].

Cyclic hydroperoxides have been characterised by chemical ionisation MS with a direct exposure probe [262].

G. SOME MISCELLANEOUS FATTY ACIDS

Methyl esters of dibasic acids give EI mass spectra which are more complex than those of the monobasic equivalents [782]. The molecular ion is not easily found, but the molecular weight can be obtained from an ion representing $[M-31]^+$. There is usually a prominent and diagnostic ion at $[M-73]^+$, and a series of ions at $m/z = 84 + 14n$ appears to be typical. Straight-chain, branched-chain and unsaturated dibasic acids have been identified by this means in royal jelly [543]. Other workers have preferred to use TMS ester derivatives for identification by GC-MS, although rearrangement ions complicate the picture [229,663]. Once more, picolinyl ester derivatives would appear to be preferable for the purpose, since they give good molecular ions and diagnostic fragments in the hydrocarbon chain [361].

Two novel brominated fatty acids, (5*Z*,9*Z*)-6-bromo-25-methyl-5,9-hexacosadienoic acid and the isomeric 24-methyl compound, were identified in the form of the pyrrolidine derivatives by GC-MS, although it was necessary to replace the bromine atom in each with deuterium to establish its position [975].

PART 3

THE ANALYSIS OF LIPIDS OTHER THAN FATTY ACIDS

CHAPTER 8

GAS CHROMATOGRAPHIC ANALYSIS OF MOLECULAR SPECIES OF LIPIDS

A. INTRODUCTION

In nature, lipid classes do not exist as single pure compounds, but rather as complex mixtures of related components in which the composition of the aliphatic residues varies from molecule to molecule. In some lipids, such as cholesterol esters, only the single fatty acid moiety will vary; in others, for example triacylglycerols, each position of each molecule may be esterified by a different fatty acid. Sphingolipids contain a number of different long-chain bases which may be linked selectively via an amide bond to specific fatty acids. A complete structural analysis of a lipid therefore requires that it be separated into molecular species that have single specific alkyl moieties (fatty acids, alcohols, ether-linked aliphatic chains, long-chain bases, and so on) in all the relevant portions of the molecule. With those lipids that contain only one or two alkyl groups, this is now often technically feasible. With lipids which have more than two alkyl groups, means have yet to be developed for physically separating all the possible species that may exist. However, if stereospecific enzymatic hydrolyses are performed on fractions separated by the available methods, it may be possible to at least calculate the amounts of a very high proportion if not all of the molecular species that are present. The analyst must at the moment be content to isolate simpler molecular fractions rather than single molecular species in such instances. Alkyl-, alkenyl- and acyl-forms of a given lipid are strictly-speaking not molecular species of it and can themselves be fractionated into molecular species, and they should therefore be isolated separately before an analysis of this kind is begun.

Ideally, it would be preferable if lipids could be separated into individual molecular species without being modified in any way so that, for example, the biosynthesis or metabolism of each part of the molecule could be studied with isotopically-labelled components. The technical problems of the analysis can often be greatly reduced, however, if the polar parts of complex phospho- and glycolipids are rendered non-polar by the formation of suitable derivatives, or if they are removed entirely by chemical or enzymatic means.

Whatever approach is adopted, it is frequently necessary to apply combinations of different chromatographic procedures to achieve effective separations.

The chromatographic methods used for the analysis of molecular species of lipids differ little in principle from those used for simpler aliphatic molecules such as the fatty acids. When they are applied to the isolation of molecular species of more complicated lipids, the separations achieved depend on the combined physical properties of all the aliphatic residues. If triacylglycerols are considered to illustrate the magnitude of the analytical problem, a triacylglycerol with only five different fatty acid constituents may consist of 75 different molecular species. High-temperature GC has until recently been used largely to separate molecular species simply according to the combined chain-lengths of the fatty acid moieties. However, improvements in technology have led to separations according to degree of unsaturation also. In essence, high temperature GC is simply an analytical technique, but one which is capable of a high degree of precision. It can be married well with mass spectrometry.

Of the alternatives to GC, adsorption chromatography will permit the separation of molecules containing three normal fatty acids from those containing two normal fatty acids and one fatty acid with a polar functional group in the chain, from those containing one normal fatty acid and two polar fatty acids, and so forth. Silver ion chromatography will separate those molecules containing three saturated fatty acids from those with two saturated fatty acids and one monoenoic acid, and these are in turn separable from other distinct fractions containing molecules with fatty acids of a progressively higher degree of unsaturation, thus complementing separations by high-temperature GC particularly well. HPLC in the reversed phase mode is used to separate triacylglycerols by their partition number, a double bond reducing the effective chain-length of a fatty acid by the equivalent of about two carbon atoms. The principles of these alternative methods as they apply to molecular species separations are described further in Chapter 9. Such procedures can be applied on either an analytical or a semimicro preparative scale, but quantification is not always easy or convenient.

It should be remembered that it is almost always advisable to calculate molar rather than weight proportions (or percentages) of any molecular species isolated. Precautions should be taken to minimise the effects of autoxidation (see Chapter 2). Procedures for the analysis of molecular species of lipids have been dealt with in several monographs [163,168,506,510] in addition to many shorter reviews on specific topics (see below).

In the discussion that follows, ether analogues of specific lipids are discussed in the same sections as the acyl forms. Separations of individual pure lipid classes are treated first as distinct topics, before analyses of complex natural mixtures are described in the final section. Methods for the preliminary isolation of lipid classes are described in Chapter 2 and elsewhere

[163,168].

B. HIGH-TEMPERATURE GAS CHROMATOGRAPHY OF TRIACYLGLYCEROLS

1. Separations on Packed Columns

As triacylglycerols are the main component of virtually all the fats and oils of commercial importance, a great deal of effort has been applied to their analysis. Because of the low volatility of intact lipids, GC is beset with a number of difficulties and, indeed for some time, it was thought that molecules such as triacylglycerols, with molecular weights up to 900, would pyrolyse at the temperatures required to elute them from GLC columns. Work from the laboratories of Kuksis and Litchfield principally showed that this need not occur. Useful separations of such compounds are now achieved routinely, although the conditions necessary to elute them from the columns approach the limits of thermal stability both of the stationary phases and of the compounds themselves. The technique has been the subject of several review articles [342,500,503,504,507,509,553,590,642]. Here the separation of the pure lipid class is described; separations of total lipid extracts containing triacylglycerols are discussed in Section I below.

For packed-column use, any modern gas chromatograph should be suitable, but it is essential that it have a flame ionisation detector for maximum sensitivity and facilities for accurate temperature-programming up to at least 350°C. In addition, it should be of a construction such that on-column injection is possible, although a preheater is necessary to warm up the carrier gas before it reaches the column packing. The dead volume between the end of the packing and the detector flame should be as small as possible, and the flame jet should preferably be wider than normal so that comparatively high flow-rates of the carrier gas can be used. Automatic flow controllers for the carrier gas are a useful accessory as the flow-rate in short columns can change markedly during temperature-programming. As bleeding of even those phases which are most thermally-stable occurs at high temperatures, better results are obtained with dual-compensating columns than with single column instruments, unless an integrator with comprehensive base-line correction facilities is available. The analyst should not be discouraged from using equipment that does not meet all these criteria, especially for separating compounds of intermediate molecular weight such as diacylglycerols or their derivatives, as patience and skill can compensate for many instrumental deficiences.

Helium or nitrogen may be used as the carrier gas with packed columns and excellent recoveries of triacylglycerols have been obtained with both, although the former is to be preferred when feasible, as better resolutions are attainable at high flow-rates. Nitrogen is used most often for reasons of cost, but it is essential that it be of very high purity as traces of oxygen or water will destroy liquid phases at elevated temperatures.

Better results are obtained with glass columns than with those of other materials, but there can be technical difficulties in obtaining effective seals at each end of the column; O-rings of Viton™ or other materials generally become brittle and crack if used at temperatures above 250°C for any length of time, but graphite seals will stand up to such conditions for long periods. It is also possible to avoid this difficulty by using columns with direct glass to metal seals, and these are available for specific instruments. All-metal columns are certainly not recommended. Narrow-bore columns give the best resolutions, and those of 2 mm (i.d.) diameter are preferable to those of 4 mm; the length of the column selected will vary with the nature of the sample, but it is usually necessary to compromise resolution by using short columns (50-100 cm) in order that compounds are eluted in a reasonable time.

The most useful liquid phases are silicone elastomers of high thermal stability such as SE-30™, JXR™, OV-1™ and Dexsil 3000™, with which separations are achieved solely on the basis of molecular weight. Silanised solid supports (80-100 or 100-120 mesh) are essential and they are generally coated with low levels (1-3 %) of stationary phase. Care is necessary in preparing the columns, which must be packed firmly to ensure adequate resolution, but not too tightly, otherwise higher temperatures than are advisable may be necessary to elute samples in an acceptable time, and losses of components of higher molecular weight may occur. Finally, the column is sealed with a plug of silanised glass wool and is conditioned for four hours at a temperature at least 25°C higher than that at which it is to be used. It may be necessary to attempt the preparation of suitable columns several times before success is achieved, and the analyst should not be discouraged by initial failures. A low bleed-rate is especially important if the technique is to be used in conjunction with mass spectrometry.

The precise operating conditions and the resolutions attainable will vary with the nature of the samples to be analysed. As cautioned above, short columns are essential and the column temperature must be programmed in the range 180 to 350°C at 2 to 5°C/min, depending on the nature of the sample. Slow temperature-programming rates give improved resolution generally. The optimum flow-rate of the carrier gas will vary with the dimensions of the column and the amount of stationary phase on the packing material, but will normally be of the order of 100 ml/min. By means of a syringe, a solution of the sample should be injected directly on to the column packing at point within the oven (rather than in the flash heater), and at a temperature about 40°C below that at which the first component emerges from the column. In this way, all the sample is vaporised, but it remains as a narrow band at the top of the column until temperature-programming is under way. Samples containing about 20 μ grams of the most abundant component provide the optimum load. The entire analysis should be completed in 25 to 45 min. With samples of intermediate molecular weight such as diacylglycerol acetates, wax esters or cholesterol esters, longer columns can be used to improve the resolutions attainable and the upper temperature limit in the analysis will be lower than with triacylglycerols.

The silicone stationary phases, which must be employed for GC analysis of high molecular weight compounds, do not in general permit the separation of saturated from unsaturated components of the same chain length. Separations are then based solely on the approximate molecular weights of the compounds and for example, tripalmitin and myristopalmitoolein elute together. Components that differ by two carbon atoms in the combined chain-lengths of the alkyl moieties must be separable before the columns are considered satisfactory. This can usually be achieved with column efficiencies of 500 to 1000 theoretical plates per foot of packing material, and indeed, with well-packed columns, components differing in molecular weight by one carbon atom can often be resolved completely. Two examples of analyses of natural triacylglycerol samples from the author's laboratory are shown in Figure 8.1.

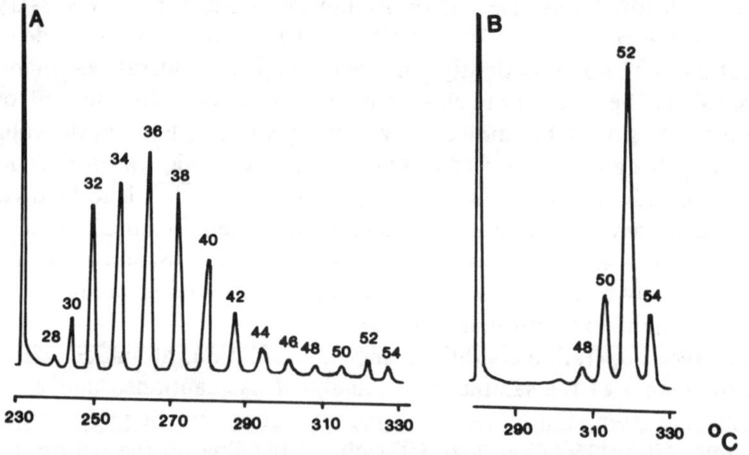

Figure 8.1 The separation of intact triacylglycerols of (A) coconut oil and (B) pig adipose tissue on a glass column (50 cm x 4 mm i.d) packed with 1% SE-30™ on Chromosorb W™ (acid-washed and silanised; 100-120 mesh). Nitrogen at 50 ml/min was the carrier gas and for separation (A), the oven was temperature-programmed from 230 to 330°C at 2°C/min, while for separation (B) it was programmed from 280 to 330°C at 2°C/min.

A shorthand nomenclature is in common use to designate simple glycerides separated in this way; the total number of carbon atoms in the aliphatic chains of the compounds (but not in the glycerol moiety) are calculated and this figure is used to denote the compound. As an example, tristearin, triolein and trilinolein are referred to as C_{54} triacylglycerols or as having a *carbon number* of 54.

It is perhaps invidious to select a single example from the wide literature on the subject, but in one of the better published separations a glass column only 30 cm in length by 2 mm i.d. was packed with 3% SE-30™ on Chromosorb G™ (100-120 mesh; acid-washed and silanised); the temperature was programmed from 275 to 350°C at 4°/min, while nitrogen at a flow-rate of 25 ml/min was the carrier gas [236].

During isothermal operation, as discussed in Chapter 3, there is a logarithmic relationship between the retention times and carbon numbers of components of a homologous series. When linear temperature-programming is used, there is a rectilinear correlation between the logarithm of the retention time and the reciprocal of the absolute temperature for short series of homologues [965]. As a result, the elution temperature rather than the elution time is sometimes quoted to described the retention characteristics of a given compound. With longer homologous series, the relationships begins to break down, and improved resolutions are obtained with non-linear (concave) temperature-programming profiles [500]. As few commercial gas chromatographs are equipped with this facility, the refinement has not been widely adopted.

When flame ionisation detectors are used, the detector response is, within limits, proportional to the weight of material eluting from the columns (see Chapters 3 and 5), and the amount of each component can be calculated from the areas of the peaks on the GLC recorder trace. There is no simple relationship between area, retention time and peak height for temperature-programmed analyses, so it is necessary to measure the area of each peak by means of an electronic digital integrator. Because of the high temperatures necessary for GC of intact lipids, there is always a danger that losses will occur on the columns as a result of pyrolysis, of reaction with the column materials and of condensation. It is, therefore, essential to check that acceptable reproducible recoveries are obtained and to calibrate the columns to compensate for any losses. The absolute recoveries from columns are not easily checked, as this requires a preparative collection facility or means of counting radioactive samples as they elute; it is, however, possible to check that recoveries are linearly related to the amount of material injected by inserting known quantities of standards (say 1-20 μg) into the columns and measuring the detector response. Alternatively, it can be assumed that in all but the very worst of columns, recoveries of tricaprin (C_{30}) or trilaurin (C_{36}) will be essentially complete, so that standard mixtures of these triacylglycerols and the higher molecular weight compounds can be analysed and the losses relative to the standard determined. The losses that can be accepted will vary with the degree of difficulty of the analysis but, as a rough guide, recoveries of the highest molecular weight component of a mixture of triacylglycerols (usually C_{54}), for example, should be at least 90 % relative to tricaprin. In practice, the most efficient way of optimising the chromatographic conditions is to vary each of them in turn, especially the flow-rate and the rate of temperature-programming, selecting those which give the maximum responses [594]. When pyrolysis occurs on the column, peaks for pure standards are often preceeded by broad humps of decomposed material. Such effects can be minimised by adding to the sample a triacylglycerol of higher molecular weight than is normally present and this presumably decomposes preferentially; triarachidin (C_{60}), for example, can be used in many circumstances [595]. By this means, recoveries and quantification of minor components are greatly improved.

As a wide range of unsaturated compounds is unlikely to be available for standardisation purposes, there are a number of advantages to be gained by hydrogenating all samples prior to analysis by high-temperature GC in packed columns. If this is done, there are no selective losses of unsaturated relative to saturated components by degradation on the column, peaks on the recorder trace are sharper, resolutions are improved and quantification of components is made easier. In the analysis of molecular species of lipids, it is necessary to know the molar proportions of all components separated, and the weight responses of the detector must be corrected by multiplying by appropriate arithmetic factors obtained from the molecular weights of the compounds, as described earlier for methyl esters of fatty acids (Chapter 5). A further advantage of hydrogenation prior to analysis then is that it greatly simplifies the range of factors required and removes any dubiety about their numerical values. This is of course particularly important with samples containing polyunsaturated fatty acids. A suitable hydrogenation procedure is described in Chapter 4.

Samples may be injected on to columns in carbon disulphide, diethyl ether or hexane solution; chloroform has also been used on occasion but tends to strip the stationary phases from the packing and damages the flame ionisation detector. While hydrogenated lipids tend to be less soluble than the unsaturated compounds in most solvents, they will usually dissolve on warming.

High-temperature GC with packed columns is then used to obtain separations of triacylglycerols simply on a molecular weight basis. Although species with carbon numbers up to 68 have been successfully resolved [357], there is little margin for error in the preparation of the column and, in most laboratories, it is considered a sufficient achievement to separate components with carbon numbers up to 56 or 58. Individual peaks are recognised by their carbon numbers relative to those of authentic standards. As single acid triglycerides (C_{42}, C_{48} and C_{54}, for example) only are available commercially, intermediate points are found by mathematical interpolation. Separation into homologues differing in carbon number by two units is achieved with relative ease, but more care is necessary to resolve components differing by one unit, as when odd-chain fatty acids are present in the sample [557]. Returning to Figure 8.1, the kind of separation obtainable with two natural triacylglycerol samples is illustrated, i.e. A - coconut oil, which contains a high proportion of fatty acids of medium chain-length, and B - pig adipose tissue (lard), which contains mainly C_{16} and C_{18} fatty acids. For these analyses, a glass column (50 cm x 4 mm) was packed with 1 % SE-30™ on Chromosorb W™ (acid-washed and silanised), and nitrogen was the carrier gas at a flow-rate of 50 ml/min. Efficient separations of the lower molecular weight components of the former sample (carbon numbers 28 to 48) are readily attained, but it is less easy to separate effectively the higher molecular weight components in the latter (C_{48} to C_{54}). The author was not always able to reproduce the separation illustrated here with new columns. It is usual to see some convergence of peaks as the temperature is increased. Again, the compounds were hydrogenated prior to the analysis, to improve the resolution and recoveries. Poor resolutions are inevitable when packed columns are used with natural triacylglycerols that contain odd- and branched-chain fatty acids as well as those with wide ranges of chain-lengths and numbers of double bonds, as in ruminant milk fats or fish oils, for example.

As cautioned earlier, quantification must be checked carefully with standard mixtures to ensure that losses of high molecular weight components relative to those of lower molecular weight are as low as possible. Losses of higher molecular weight triacylglycerols such as trierucin (C_{66}) are inevitable, however, and must at present be accepted, but all such losses can be compensated for by determining calibration factors with standard mixtures, provided that the factors are checked regularly. Data should again be converted to molar proportions. The results of an international collaborative study to establish a standard method for triacylglycerol analysis have recently been published [721]. High temperature GC has been employed for the analysis of triacylglycerols from a wide variety of natural sources, e.g. marine oils [17,18,555,557], seed oils [97,98,409,558] and animal lipids [408,511,634,849,989], and those cited are selected as representatives only of innumerable papers. In addition, the technique has been recommended for the determination of cocoa butter equivalents in the confectionery industry [687,1010]. Note that silver ion chromatography complements separations of this kind particularly well, since it provides the resolution by degree of unsaturation not achievable otherwise (see Chapter 9). Used sequentially, the two techniques permit separation of many more fractions than would be possible by either on its own, and this approach was favoured in many of the papers cited above.

Some separation of triacylglycerols according to degree of unsaturation in addition to chain-length has been achieved on packed GC columns containing polar stationary phases [54,503,898]. Whether the limited resolution obtained is sufficient to be of value to lipid analysts is doubtful, however, especially as such columns will probably have a short life-time. With phases of higher polarity, WCOT columns are greatly to be preferred (see below).

Although some preparative applications have been described, GC of triacylglycerols is now used exclusively as an analytical technique. HPLC procedures are certainly much better for small-scale preparative purposes (see Chapter 9).

High-temperature GC with packed columns has also been employed to separate ether analogues of triacylglycerols, i.e. the alkyldiacylglycerols, into molecular species. This was first accomplished for such compounds isolated from tumor tissue on a short column packed with a 3% JXR™ stationary phase [1000,1002]. Alkyldiacylglycerols tend to be minor compounds in tissues of terrestrial animals, but they can be major components of the harderian gland [105] and of some marine species [18,556], and GC proved to be of value in the studies cited here. Because they have one fewer oxygen atom than the corresponding triacylglycerols, alkyldiacylglycerols tend to elute the equivalent of one methylene unit earlier.

2. Separations on WCOT Columns

The potential benefits of glass WCOT columns for the analysis of intact lipids must have been obvious from an early stage, and indeed the first such separations with Dexsil 300™ as the stationary phase were published in 1972 [675]. Most of the more common stationary phases tended to bleed rather easily from WCOT columns at elevated temperatures, however, and further applications did not appear until chemically-bonded and cross-linked phases became available together with fused silica capillaries from about 1979 onwards. The use of WCOT columns for triglyceride separations has also been reviewed elsewhere [284,591].

Many different gas chromatographs have been used for the purpose, and most modern instruments appear suitable, but it is evident that the nature of the injection system can be of crucial importance. The minimum requirement is for some form of on-column injection. Grob [305], for example, demonstrated that techniques based on sample vaporisation in the injector are not suitable for intact lipids as discrimination in favour of the less volatile constituents occurs. With splitless injection, most losses were found to be a consequence of insufficient elution from the syringe needle; split injection gave even worse results, although the reasons for this were not clear, and only cold on-column injection gave acceptable recoveries. Any involatile material ("dirt") on the column from previous analyses affected the injection because of adsorption effects [309,310]. In addition, the flow-rate of the carrier gas can have a marked effect on sample loss and discrimination, and thermal decomposition takes place to change the sample composition [306,592]. Cold on-column injection eliminates many of these problems, although other factors then come into play [307,308,312].

The principal disadvantage of on-column injection is the contamination of the stationary phase that inevitably occurs, leading to peak broadening [309]. This effect can be minimised by using effective clean-up procedures for the triacylglycerols during sample preparation, such as by preparative TLC or HPLC [168], or more conveniently by using a short column of Florisil™ or silica gel (0.3 to 0.5 g), from which the lipid is eluted with hexane-diethyl ether (4:1, v/v; 10 ml). In addition, it is possible to insert a length of deactivated fused-silica tubing (1 to 3 m), sometimes termed a "retention gap", in front of the column to collect any impurities [309]. When this precolumn is beginning to show signs of contamination, it is replaced. Unfortunately, the low starting temperature required with an injection technique of this kind means that the analysis time is lengthened and there is additional opportunity for thermal decomposition to happen. Various solutions have been suggested for the problem, including an independently-thermostatted inlet section of the column [311] and a moveable on-column injector [283,286]. With the latter, the injector and inlet part of the column are moved up out of the oven, where they cool to room temperature. The sample is injected, most of the solvent is allowed to evaporate in the stream of carrier gas, then the injector is moved down into the oven so that the column inlet region heats up very rapidly to the initial oven temperature and the sample vaporises. As an alternative, a cold on-column injection system equipped with secondary cooling has been used; the solvent evaporates in a portion of the column cooled to 70°C, then the cooler is switched off so that the sample is heated to the oven temperature in about 30 seconds [912]. This method makes use of commercially-available equipment and can be adapted for automatic injection.

Another approach has been to use a programmed-temperature vaporiser, which in essence is a split/splitless injector that is maintained at a temperature close to the boiling point of the injection solvent, so that the sample is transferred to the column in a liquid; when the solvent has evaporated, the injector is heated at a rate of 14°C/sec to the minimum column temperature required [383,384]. The manufacturers (Perkin Elmer) claim that this injector is the closest to a universal system yet to be developed.

While WCOT columns constructed from both glass and fused silica have been used for the separation of intact triacylglycerols, there is now no doubt that the latter are to be preferred. The length of column used will be a compromise between the optimum in terms of resolution with a need to limit the exposure time of the solute to high temperatures to the minimum; commonly the length is 5 to 25 m with an internal diameter of 0.2 to 0.32 mm. Columns with a strengthened outer coating are now manufactured to better withstand temperatures above 300°C.

Initially, non-polar stationary phases only (of the methyl silicone type) were used in high-temperature GC, and cross-linking and chemical-bonding improved the properties of the columns appreciably. More polar bonded phases, consisting of phenylmethyl silicones, later came into use and are available commercially. At present, these have a temperature limit of about 360°C, and while this will no doubt be improved, the ultimate limit may depend on the pyrolysis temperature of triacylglycerols. The optimum thickness of the liquid film for high-temperature GC is about 0.1 to 0.12 μm.

As discussed earlier for packed columns, the rate and shape of the temperature-programming profile can have a marked effect on column efficiency. A non-linear (concave) rate of temperature-programming is preferable whenever this is feasible. The lower the rate of temperature-programming, the lower is the elution temperature of a given compound, but the longer is its elution time. In practice, the optimum temperature limits and the rate of programming must be determined empirically for a given sample and column.

Hydrogen has important advantages as a carrier gas with WCOT columns, in that efficiency is less dependent on linear gas velocity, as discussed in Chapter 3. It permits the elution of components at lower temperatures or elution times than with other gases, so that there is less opportunity for thermal degradation to occur, especially of more sensitive components containing polyunsaturated fatty acids. It also promotes a longer working life for the column. In addition to the possible danger of explosion with hydrogen as a carrier gas (as discussed earlier), there is one report of hydrogenation of unsaturated lipids at elevated temperatures on a polar stationary phase [862]. However, this effect has not been found by others apparently, and it has been suggested that it may have been a consequence of some impurity in the sample, the carrier gas or the stationary phase [592]. As with packed columns, all air and moisture must be rigidly excluded from the carrier gas to extend column life and efficiency.

High-temperature GC of triacylglycerols with modern WCOT columns was accomplished by several research groups in different parts of the world virtually simultaneously. It is therefore not always possible to treat the subject from a historical standpoint or to establish scientific priority. When non-polar phases are used in WCOT (as with packed) columns, triacylglycerols are separated according to molecular weight essentially and there is ordinarily no useful resolution by degree of unsaturation, although some partial separations may be seen. As an example, a separation of an interesterified palm oil on a 6 m glass WCOT column coated with a methylsilicone phase is illustrated in Figure 8.2; temperature-programming was from 250 to 350°C [616]. Components varying in carbon number from 44 to 56 are clearly resolved, and there is some evidence for the presence of intermediate species containing odd-chain fatty acids. An improvement in resolution over Figure 8.1 is obvious. Similar results were obtained with other vegetable oils. In addition, comparable analyses have been reported by others with seed oils [204,305,313,951], algal triacylglycerols [768], beeswax [550], and butter

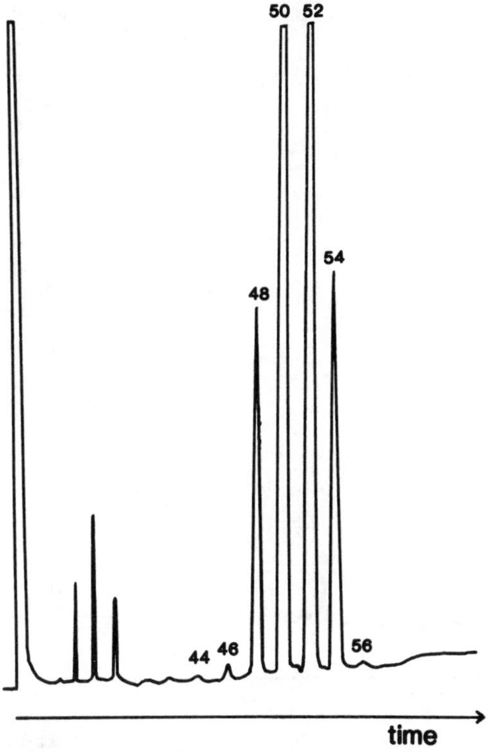

Figure 8.2 Separation of interesterified palm oil on a glass WCOT column (6 m x 0.4 mm) coated with a methylsilicone phase [616]. Helium at 6 ml/min was the carrier gas, and the oven was temperature-programmed from 250 to 350°C at 4°C/min. The numbers above each peak refer to the carbon number of the component. (Reproduced by kind permission of the authors and of *Revue Francaise des Corps Gras*, and redrawn from the original paper).

[42,313,455,641,806,961], plasma [542,592,647] and other animal lipids [313,654]. The technique has been used in comparisons of triacylglycerols and chloropropanediol diesters in milk fat from goats [514,652], and of triacylglycerols and alkyldiacylglycerols in human milk [514].

In the hands of a skilled analyst and with a good column, some partial resolution is possible according to degree of unsaturation or because of variation of the chain-lengths of fatty acids within a molecular species of a given carbon number [291,768,923,924]. A separation of coffee oil on a 15 m glass column coated with OV-101™ is illustrated in Figure 8.3; temperature-programming was from 310 to 330°C [291]. The component of carbon number 54 here, for example, is separated into three fractions according to the number of unsaturated fatty acids in the molecule, but not

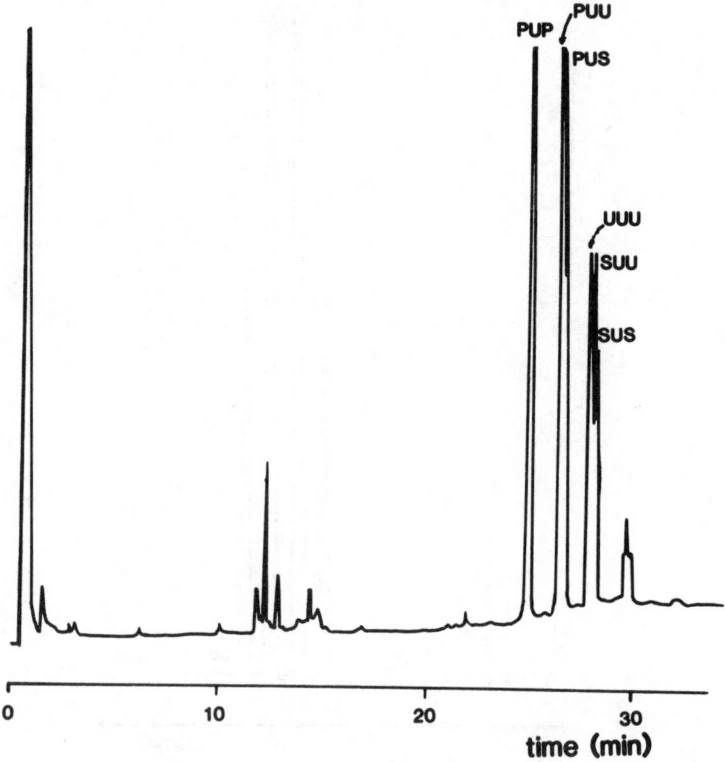

Figure 8.3 Separation of coffee oil on a glass WCOT column (15 m x 0.3 mm) coated with OV-1™ [291]. Hydrogen was the carrier gas, and the oven was temperature-programmed from 200 to 310°C at 4°C/min. Abbreviations: P, 16:0; S, 18:0; U, a C_{18} unsaturated fatty acid. (Reproduced by kind permission of the authors and of the *Journal of Chromatography*, and redrawn from the original paper).

by the number of double bonds within each acid. With care, four fractions can sometimes be seen, eluting in the order UUU, UUS, USS and SSS, where S is a saturated and U an unsaturated C_{18} fatty acyl residue, i.e. unsaturated species elute before saturated. Unfortunately, it is doubtful whether the resolution is quite good enough to be of real analytical value. In an attempt to obtain more meaningful data on the relative proportions of saturated and unsaturated molecular species, triacylglycerols were subjected to high temperature GC after ozonolyis of double bonds followed by reductive cleavage [285].

More recently, some remarkably effective separations of triacylglycerols have been achieved on WCOT columns coated with more polar (or "polarisable") silicone phases containing a high proportion of phenyl groups, mainly in the laboratories of Geeraert and Sandra, who developed the moveable on-column injection system discussed above. It is then possible to separate triacylglycerol species according to the number of double bonds in each fatty acyl residue within a given carbon number. Excellent resolutions of seed oil triacylglycerols especially have been obtained on a WCOT column (25 m x 0.25 mm i.d) of fused silica coated with a methylphenyl silicone polymer (RSL-300™), containing 50% phenyl groups [287-290]. As an example, a separation of palm oil is illustrated in Figure 8.4; temperature-

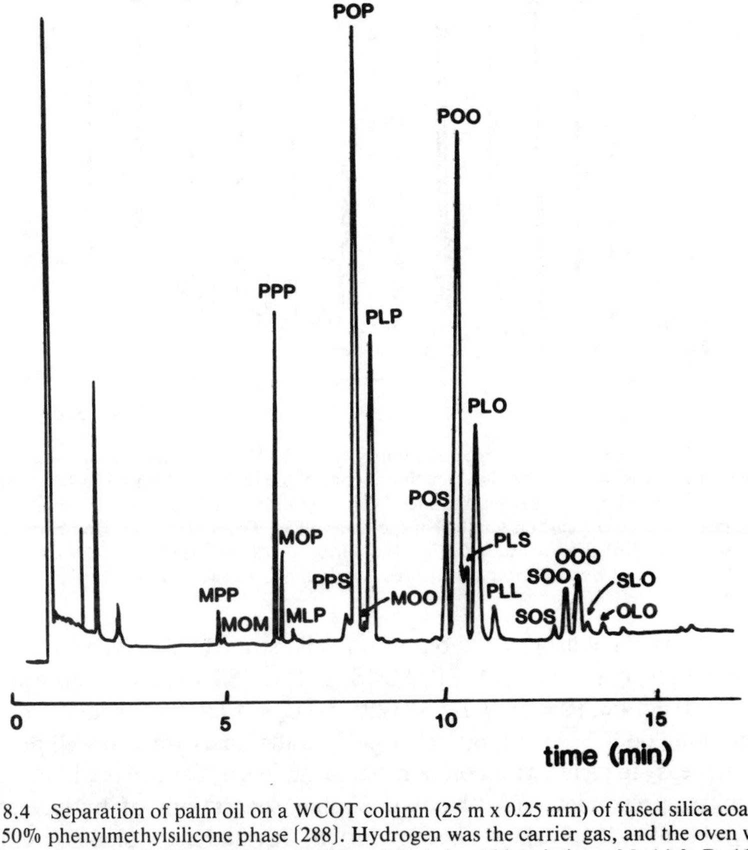

Figure 8.4 Separation of palm oil on a WCOT column (25 m x 0.25 mm) of fused silica coated with a 50% phenylmethylsilicone phase [288]. Hydrogen was the carrier gas, and the oven was temperature-programmed from 340 to 355°C at 1°C/min. Abbreviations: M, 14:0; P, 16:0; S, 18:0; O, 18:1; L, 18:2. (Reproduced by kind permission of the authors and of the *Journal of High Resolution Chromatography and Chromatography Communications*, and redrawn from the original paper).

programming was from 340 to 355°C over only 16 minutes [288]. It can be seen that the C_{52} species is separable into seven fractions, while the C_{54} species splits into six. In a more unsaturated seed oil, fractions emerge in the order - SSS, SSO, SSL, OOO, SLO, OOL, SLL, OLL, LLL and LLLn, where S = 18:0, O = 18:1, L = 18:2 and Ln = 18:3. In this instance, unsaturated fractions elute after saturated ones. The quality of the separation and the speed of the analysis will ensure that this technique is widely used, especially in the oils and fats industry, assuming that the precision of quantification is comparable to that with apolar phases.

Within a given carbon number group, some resolution is achieved for combinations of fatty acids of different chain lengths. In Figure 8.5, a

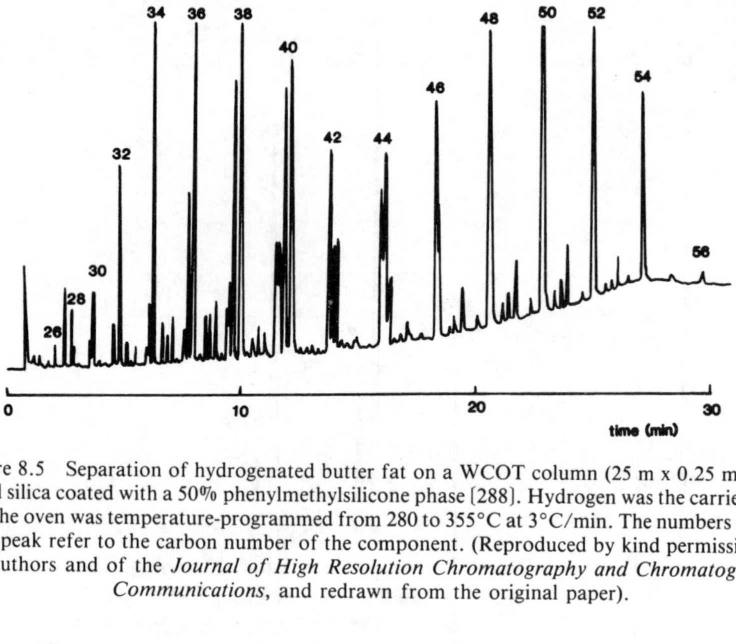

Figure 8.5 Separation of hydrogenated butter fat on a WCOT column (25 m x 0.25 mm) of fused silica coated with a 50% phenylmethylsilicone phase [288]. Hydrogen was the carrier gas, and the oven was temperature-programmed from 280 to 355°C at 3°C/min. The numbers above each peak refer to the carbon number of the component. (Reproduced by kind permission of the authors and of the *Journal of High Resolution Chromatography and Chromatography Communications*, and redrawn from the original paper).

separation of a hydrogenated butter fat is illustrated [288]. The C_{46} fraction, for example, may contain MPP, MMS, LaPS, CSS and many more species, where M = 14:0, P = 16:0, La = 12:0 and C = 10:0. Intermediate fractions containing odd-chain and branched-chain fatty acids are also well resolved. It is not easy to identify the components within particular peaks without access to mass spectrometry (see below). Similar separations of butter fat and vegetable oils [384,912] have been reported with stationary phases containing up to 65% phenyl moieties in the polymer, and analyses of other fats and oils are described in a review article [284]. In addition, triacylglycerols have been hydrolysed to diacylglycerol derivatives for analysis by high-temperature GC on polar phases in order to obtain more information on molecular structure (see Section C.2 below).

The key to a wider acceptance of high temperature GC of triacylglycerols on WCOT columns is the precision that can be attained in quantification. It is virtually essential that electronic integration be applied for peak area measurements, ideally with some form of automatic base-line correction. The response of the detector should in theory be quantitative, in that it is linearly related to the amount of material eluting from the end of the column. However, if some of the sample is selectively lost during injection (discussed above) or if losses occur through degradation on the column, the overall efficiency of the process can fall off. Good injection technique and clean samples can eliminate some of the losses. There is little that can be done to prevent thermal degradation entirely, but it can be minimised by careful optimisation of the operating conditions, and reproducible if not quantitative recoveries can be attained.

In a detailed study of the factors affecting the quantification of intact triacylglycerols with WCOT columns coated with a non-polar phase, Mares and Husek [592] demonstrated that the recovery of the higher saturated homologues was dependent on such factors as the injection technique, column quality, the flow-rate of the carrier gas, the weight of the solute and its molecular weight. Column quality is not easy to define, and the analyst is to a considerable extent in the hands of the suppliers. During use, the stationary phase begins to thin out and bare patches can appear, and there can be contamination by residues of previous samples. Such factors will inevitably lead to a loss of resolution and worsening recoveries. It is well documented that the flow-rate of the carrier gas changes during temperature-programming (see Chapter 3), so the detector response must change also. Optimisation to minimise the weight correction factor that is required for higher molecular weight species, such as triarachidin, must be carried out empirically for each new column. As long as good peak shape is maintained (no overloading), the larger the sample the better the response tends to be for triarachidin, but deleterious effects are much less apparent with C_{54} triacylglycerols. If sufficient care is taken in optimising the system and in the measurement of weight correction factors, excellent reproducibility can be achieved with non-polar phases [592]. Others appear to have a less sanguine view, while still recommending the technique [455].

Less information is available on quantification with WCOT columns coated with the more polar phases. In addition to the factors mentioned above, Mares [591] found that the response diminished with the length of the column and perhaps more importantly with the degree of unsaturation of the solute. However, the losses were reproducible so that good quantification was reportedly possible, except for highly unsaturated seed oils, with careful calibration. Others consider that the relatively greater bleed from the polarisable phases causes some quenching of the response to triacylglycerols of higher molecular weight, although reproducible results are again obtained after calibration [912]. In contrast, Geeraert [284] reported that recoveries were complete and that the detector response was directly proportional to the carbon content of each molecule, except for the most highly unsaturated species found in such samples as fish oils. With the common range of vegetable oils, such as soybean or palm oil, and for confectionery fats, such as butter, cocoa butter (and substitutes) and coconut oil, there was a uniform response factor of unity. It is possible that the quality of the last results are due in some measure to the special injection system and columns used by Geeraert, but further objective studies are obviously required.

3. Gas Chromatography-Mass Spectrometry

Such is the complexity of the GC traces of intact triacylglycerols that some additional means of identifying components is necessary, and foremost among these is mass spectrometry (MS). The topic has been briefly reviewed [386,483,642]. In most of the early work, samples were introduced into the mass spectrometer with direct probe insertion, but improvements in technology have made it possible to introduce triacylglycerols via a GC column. On the other hand, it is probably true to say that more work is now being done with HPLC interfaced to MS (reviewed elsewhere [168,515]).

Barber *et al.* [83] were the first to describe the EI mass spectrum of a triacylglycerol, and this was soon followed by other systematic studies [2,537]. The mass spectrum of 1,2-dipalmitoylolein is illustrated in Figure 8.6 [386].

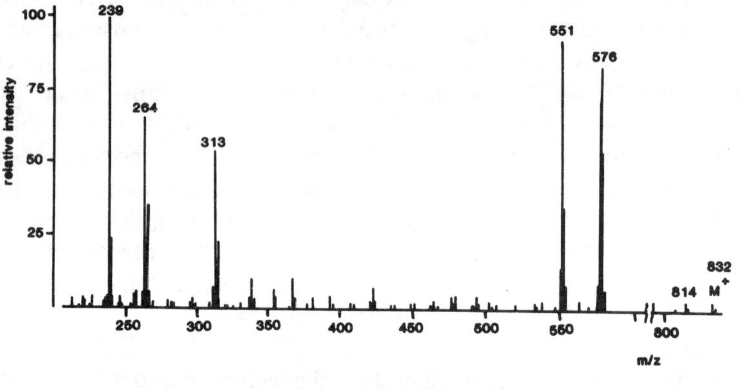

Figure 8.6 The mass spectrum of 1,2-dipalmitoylolein.

Usually, there is a very small molecular ion only, in this instance at m/z = 832, followed by a unique peak for an ester at m/z = 814 (M-18 or loss of water). There are, however, intense ions that are characteristic of the various fatty acyl residues and these fall into two classes, i.e. those containing two acyl residues and those with only one. In the first class, there is an ion equivalent to the loss of an acyloxy group, i.e. $[M\text{-}RCOO]^+$, together with a related ion but minus a further hydrogen atom. In the example shown here, the loss of a palmitoyl acyloxy group (equivalent to 255 amu) gives ions at

m/z = 577 and 576 (having one palmitoyl and one oleoyl residue), while the loss of the oleoyl moiety (equivalent to 281 amu) gives ions at 551 and 550 (having two palmitoyl residues). The relative intensities within each pair is dependent on whether the ion fragment contains an unsaturated residue (when the smaller ion is more intense). The other important class of diagnostic ions containing the individual fatty acid moieties are of the form RCO^+, though if the fatty acid group is unsaturated an additional hydrogen atom is lost. Thus in Figure 8.6, the oleoyl moiety produces an ion at m/z = 264, while that from palmitate is at m/z = 239. Related ions with an additional 74 amu corresponding to the glycerol backbone, are found at m/z = 339 and 313 respectively; in this instance, the presence of a double bond has no effect. When the triacylglycerol contains three different fatty acids, there are three ions in each class. Metastable-ion mass spectrometry can be of great value for the recognition of specific ion fragmentations [86].

EI-MS with direct probe insertion was used, for example, to determine the structure of unusual tetraacylglycerols containing an allenic acid [877] and triacylglycerols containing sorbic acid [251]. Molecular species of mixed triacylglycerols have also been analysed in this way [385]; the data were fed into a computer programmed to recognise key fragments, apply response factors and calculate the relative proportions of each fraction. One advantage of such a procedure is that large numbers of samples can be handled routinely, provided that suitable equipment is available.

When GC-MS was attempted with packed columns interfaced to mass spectrometers via molecular separators, difficulties were obtained with the recovery of triacylglycerols, but useful data were obtained from several natural samples in this way [634]. The task of analysing triacylglycerols of marine origin containing isovaleric acid was perhaps a relatively easy one [112]. Capillary columns of fused silica can now be introduced directly into the ion source of some instruments and this eliminates many of the technical problems (c.f. [196,806]). In addition, it is recognised that better spectra with substantial molecular (or quasi-molecular) ions are obtainable if soft ionisation methods, such as chemical ionisation [274,630,635] or field desorption [247,274,539,607,818] MS, are used. Only the former of these can be used in conjunction with gas chromatography, and the technique has been employed succesfully with packed [635] and with WCOT columns; with the latter, triacylglycerols from algae [768] and butter fat [961] were analysed. Ammonia was employed as the reagent gas initially [630,635], but better results appear to be attainable with methane [768,961]. By this means, it is possible to recognise and quantify individual species in a single chromatographic peak and not resolved by GC alone.

1-O-Alkyl-2,3-diacylglycerols have similar mass spectra to triacylglycerols [237]. The mass peak is small, and that for $[M-18]^+$ is smaller than in triacylglycerols. In the high mass range, a peak corresponding to $[M-(O-alkyl)]^+$ is always prominent, and there are also ions corresponding to the loss of the acyloxy groups. Further fragmentations lead to ions equivalent to $[M\text{-}(HO\text{-}acyl + acyl)]^+$.

4. *Supercritical Fluid Chromatography*

Supercritical fluid chromatography is a rapidly developing technique, which is in effect a hydrid between GC and HPLC, using much of the instrumentation of the former while the mobile phase is a liquified gas, commonly carbon dioxide. Much effort is being expended in developing the instrumentation and applications, and some interesting separations of molecular species of triacylglycerols have been described [740,969]. While there appear to be serious doubts about reproducibility at the moment, it seems probable that improvements will be made. The general technique is the subject of a recent monograph [864].

C. DIACYL GLYCEROPHOSPHOLIPIDS AND DIACYLGLYCEROLS

1. *Preparation of Diacylglycerols from Phospholipids*

Molecular species of complex lipids such as glycerophosphatides are most easily separated after the phosphorus moiety has been removed, and this is certainly true for gas chromatography. Normally, de-phosphorylation is accomplished by enzymatic hydrolysis with phospholipase C, although other methods are available. While it is technically possible to subject diacylglycerols *per se* to GC analysis, this is rarely attempted, because of problems of tailing and of acyl migration. The 1,2-diacyl-*sn*-glycerols obtained by enzymatic hydrolysis are therefore purified and acetylated (or otherwise derivatised) immediately to prevent acyl migration during subsequent separation procedures. In addition, alkyl and alkenyl-forms may have to be isolated before more detailed studies are commenced. One advantage of this approach is that diacylglycerol acetates and related derivatives from all phospholipid classes may be separated into molecular species by exactly the same procedures. GC methods for the analysis of molecular species of phospholipids in the form of the diacylglycerol derivatives have been reviewed [276,516,584,591,642,744,956]. The technique is also used in profiling lipid classes in tissues (see Section I below). In addition, diacylglycerol derivatives are of value in structural analyses of triacylglycerols, although diacylglycerols *per se* tend to be minor constituents of tissue lipids.

No satisfactory chemical method for removing the phosphorus moiety of glycerophosphatides has been developed. Acetylation with equal mixtures of acetic anhydride and acetic acid at 140°C in a sealed tube has been used for the purpose [762], but plasmalogens are degraded and acyl migration with formation of a small proportion of 1,3-diacylglycerol acetates takes place.

On the other hand, it is argued that this occurs intra- rather than intermolecularly [369,370]. Therefore, the same fatty acids remain in molecular combination and the diacylglycerol acetates so produced are still suited to molecular species analysis (but not for analysis of fatty acid positional distributions), on GC columns packed with non-polar phases at least. TMS ether derivatives of diacylglycerols have been prepared from glycerophosphatides by subjecting the latter to very high temperatures (approximately 250°C) for brief periods and then silylating [404,405]. Here also, 1,3-derivatives accompany the 1,2-compounds, and the latter do not accurately represent the original composition of the native phospholipid so that their value for structural analysis is limited. Enzymic hydrolysis is perhaps more tedious and time-consuming, but it undoubtedly gives the best results.

A variety of enzymes capable of releasing 1,2-diacylglycerols from phosphoglycerides or of ceramide from sphingomyelin, and termed "phospholipase C" (EC.3.1.4.3), has been isolated from microorganisms, but especially from *Clostridium welchii* and *Bacillus cereus*. Enzymes from both sources are commercially available, or alternatively highly active preparations of that from *B. cereus* can be obtained by ammonium sulphate precipitation from the supernatant fluid used as a growth medium for the microorganisms [684]. If the enzyme is to be used in structural studies, little further purification is necessary. The properties and substrate specificities of corresponding enzyme preparations from different sources can vary greatly (reviewed by Brockerhoff and Jensen [130]) and in structural studies, it is necessary to choose the enzyme from the most appropriate source for a particular phospholipid class. The enzymes do not possess an absolute specificity for a phosphate bond in position 3 of *L*-glycerol, and that of *B. cereus*, for example, will react with synthetic phospholipids with the phosphate bonds in positions *sn*-1, *sn*-2 and *sn*-3 [213]; that of *C. welchii* hydrolyses *D*-phosphatidylcholine, but much more slowly than the normal *L*-isomer [645]. Although there is some evidence that molecular species containing shorter-chain fatty acids are hydrolysed more rapidly than those with longer chain components, the effect need not be troublesome as the reaction can usually be taken to completion with care.

Phospholipase C preparations from *C. welchii* are used most often for the preparation of 1,2-diacyl-*sn*-glycerols from phosphatidylcholine or of ceramides from sphingomyelin. In addition, they can be used to prepare monoacylglycerols from lysophosphatidylcholine [278] and ceramides from ceramide aminoethylphosphonate [601]. The enzyme is utilised in the following manner [762].

"The phospholipase C of *C. welchii* (1 mg) in 0.5 M tris(hydroxymethyl)methylamine (tris) buffer (pH 7.5; 2 ml), 2×10^{-3} M in calcium chloride, is added to phosphatidylcholine (5 mg) in diethyl ether (2 ml). After the mixture is shaken at room temperature for 3 hours, it is extracted three times with diethyl ether (4 ml portions). The ether layer is dried over anhydrous sodium sulphate before it is evaporated in a stream of nitrogen at ambient temperature. Pure 1,2-diacyl-*sn*-glycerols are obtained by preparative TLC on layers of silica gel G impregnated with boric acid (10 %, w/w), with hexane-diethyl ether (50:50, v/v) as solvent system. The appropriate band is located under UV light after spraying with 2′,7′-dichlorofluorescein (see Chapter 2), and is eluted from the adsorbent with diethyl ether. (Boric acid, which is also eluted, does not interfere with subsequent stages)."

No unnecessary delays are allowable at any stage and the compounds should not be heated or permitted to come in contact with polar solvents, otherwise acyl migration may occur. The diacylglycerols should be acetylated at once with acetic anhydride and pyridine (see Chapter 4 for practical details), as diacylglycerol acetates can be stored indefinitely in an inert atmosphere at low temperatures without coming to harm. In addition, it is advisable that the diacylglycerol acetates be purified by HPLC [168], by TLC on silica gel G (hexane-diethyl ether, 7:3 v/v, is a suitable solvent system), or more conveniently on a short column of Florisil™ or silica gel (0.3 to 0.5 g) eluted with the same solvent mixture, as this can prolong the life of the GC column in sustained use. A small portion of the diacylglycerols or diacylglycerol acetates should be transesterified so that its fatty acid composition can be compared with that of the original lipid, to ensure that random hydrolysis of molecular species has occurred. TMS or *t*-butyldimethylsilyl (BDMS) ether derivatives of diacylglycerols are preferred for some purposes, but only the latter are stable during prolonged storage.

Ceramides prepared by this method from sphingomyelin are comparatively stable, but they should be purified on thin layers of silica gel G or by a mini-column procedure with chloroform-methanol (9:1, v/v) as solvent system, before being derivatised and analysed further.

Although *C. welchii* preparations can be used to hydrolyse phosphatidylethanolamine, provided that some lysophosphatidylcholine [903] or sphingomyelin [762] is present to activate the enzyme, better results are obtained with the phospholipase C of *B. cereus*, which can also be used to prepare diacylglycerols from phosphatidylserine, phosphatidylinositol, phosphatidylglycerol and diphosphatidylglycerol. The author has found the following method to be satisfactory for the purpose [991,1002]. (Note that tris buffer is not suitable).

"Phosphatidylethanolamine (5 mg) and sphingomyelin (3 mg) are mixed with 0.2 M phosphate buffer (pH 7.0; 0.5 ml) containing 0.001 M 2-mercaptoethanol and 0.0004 M zinc chloride, and phospholipase C from *B. cereus* (1 mg) in the same buffer (0.5 ml) is added. The mixture is shaken vigorously for 2 hours at 37°C, when the diacylglycerols produced are extracted and purified as in the procedure immediately above."

The phospholipase C of *Clostridium perfringens* is specific for the diacyl forms of phospholipids, yet that of *B. cereus* hydrolyses the 1-0-alkyl forms of phospholipids three times as quickly as the diacyl or alkenyl forms [962]. The latter enzyme will only hydrolyse sphingomyelin under exceptional circumstances. Phospholipase C preparations that are specific for phosphatidylinositol have been obtained from *Staphylococcus aureus* [563] and *B. thuringiensis* [897]. A related enzyme, found in brain, has been used to prepare diacylglycerols from mono-, di- and triphosphoinositides [396,915]. While hydrolysis does not go to completion, representative diacylglycerols appear to be obtained, although some acyl migration occurs with formation of 1,3-diacylglycerols, probably because of the low pH optimum of the enzyme. Phosphatidic acid has been dephosphorylated by the acidic phosphatase (EC.3.1.3.2) from wheat germ [109].

With many phospholipid classes, and phosphatidylethanolamine and phosphatidylserine especially, it is advisable to separate alkenylacyl, alkylacyl and diacyl forms (the "diradyl" forms) as the acetate derivatives before proceding to GC analysis. Until recently, this was a task for TLC. The compounds migrate in the order stated and can be adequately resolved on silica gel layers, with a first development to half way up the plate with hexane-diethyl ether (1:1, v/v) followed by a full development in the same direction in toluene [767]. Good separations were achieved by Cursted [201,202] with column chromatography on lipophilic Sephadex™, but more recently, Nakagawa and Horrocks [660] obtained excellent results with HPLC, and this is likely to set the standard. For example, diradylacetylglycerols prepared from the ethanolamine-containing glycerophospholipids of bovine brain were separated into the three forms by adsorption HPLC on silica gel. They were eluted in the same order as on TLC from a column (3.9 x 300 mm) containing μPorasil™, maintained at 36°C, with cyclopentane-hexane-methyl-*t*-butyl ether-acetic acid (73:24:3:0.03 by volume) as the mobile phase at a flow-rate of 2 ml/min; UV detection at 205 nm was employed. This procedure could also be used with BDMS ether (but not TMS ether) derivatives.

2. *Preparation of Diacylglycerols from Triacylglycerols*

The determination of the stereospecific distribution of fatty acids in triacyl-*sn*-glycerols can be carried out by a number of procedures, all of which are complex and involve sequential hydrolytic and synthetic steps, followed by chromatographic analysis of an array of products [125,167]. One procedure described by Kuksis and coworkers [645] involves partial hydrolysis of triacylglycerols to *rac*-1,2-diacylglycerols, which are converted chemically to phosphatidylcholines; these are in turn hydrolysed with the enzyme phospholipase C, which reacts very rapidly with the natural *L*-form to yield *sn*-1,2-diacylglycerols and then very slowly with the *D*-form to produce the *sn*-2,3-diacylglycerols. Both types of diacylglycerol can be selectively recovered, then derivatised and subjected to GC analysis so that the pairing of the fatty acids in each can be determined. The same methods that are employed to separate diacylglycerols derived from phosphoglycerides are applicable (see below).

3. Separations on Packed Columns

For much of the early work with packed-column GC, non-polar stationary phases were used. The nature of the columns and the precise operating conditions that have been recommended vary somewhat from laboratory to laboratory but, as a rough guide, 1 m x 3 mm o.d. columns packed with 2 % SE-30™ on a silanised support will give good results. Carrier gas flow-rates of 100-200 ml/min and temperature-programming from 220 to 300°C are commonly quoted. Some deviation from these optimum conditions is permissible, however, and the separations illustrated in Figure 8.7 were obtained on 50 x 0.4 cm. (i.d.) glass columns containing 1 % SE-30™, temperature-programmed from 250 to 300°C at 2°C/min, and with the flow-rate of the carrier gas (nitrogen) at 50 ml/min. Because the elution temperatures are in general lower than with triacylglycerols, the resolutions obtained are sufficient to separate components with combinations of fatty acids differing in chain-length by one carbon atom. The separations depend largely on the molecular weight, as with the triacylglycerols in similar circumstances, and the *carbon number* concept is again used for identification purposes. Distearin has a carbon number of 36 and the diacylglycerol acetate prepared from it has a carbon number of 38 (although the two carbon atoms of the acetate moiety are not counted by some authors in calculating carbon numbers). Components are identified by their retention times (or elution temperatures) relative to authentic materials or to standard triacylglycerols having the same total number of carbon atoms. Quantification is generally relatively straight forward, unlike the situation with triacylglycerols.

Some partial separation on the basis of degree of unsaturation may occur but may not be especially desirable. For example, in Figure 8.7 (A), the fraction of carbon number 38 consists of two partially resolved peaks; the first contains 16:0 together with 20:4 and the second consists of two C_{18} acids of varying degrees of unsaturation. On hydrogenation (a suitable procedure is described in Chapter 4), these components merge and the remaining peaks are distinctly sharper with improved resolution (Figure 8.7 (B)); in particular, trace amounts of components containing odd-chain fatty acids become apparent. With these particular columns and operating

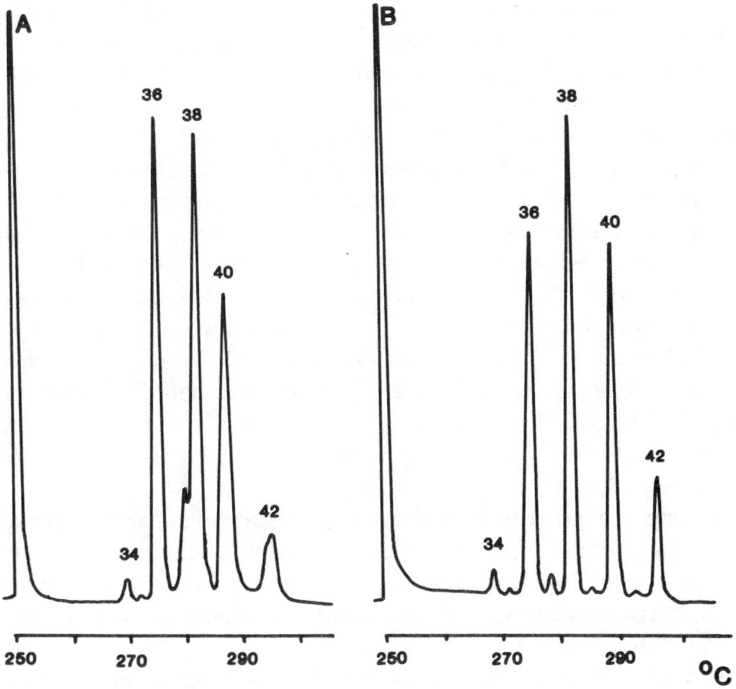

Figure 8.7 GC separation of diacylglycerol acetates prepared from the phosphatidylcholine of pig liver. A 50 x 0.4 cm (i.d.) glass column packed with 1 % SE-30™ on Chromosorb W™ (acid-washed and silanised; 100 to 120 mesh) was used and was temperature-programmed from 250 to 300°C at 2°C/min; nitrogen was the carrier gas at a flow-rate of 50 ml/min. Trace A before and Trace B after hydrogenation. The numbers above each peak refer to the carbon number of the component.

conditions, recoveries of a wide range of standards were essentially complete whether the compounds were hydrogenated or not, and similar results have been reported from many other laboratories for many different phospholipid classes and with acetate, TMS ether and BDMS ether derivatives (c.f. [220,408,477,478,673,764,966,989,991]). The same technique has been employed for the analysis of diacylglycerols generated from triacylglycerols during stereospecific analyses of the latter [585,645,650,651]. As with the triacylglycerols (Section B), silver ion chromatography used to complement GC permits much more extensive separations.

One novel procedure that may repay attention consisted in selectively deuterating the double bonds in the fatty acids of diacylglycerols derived from phospholipids, prior to conversion to the BDMS ether derivatives for GC separation as above [219]. In this instance, mass spectrometry served to identify and quantify unresolved components in specific peaks.

Acetate, TMS ether and BDMS ether derivatives of diacylglycerols are separable to some extent, both by the chain-length and degree of unsaturation of the combined fatty acid constituents, on short columns containing some more polar liquid phases having increased thermal stability. EGSS-X™ was used initially [502], but improved resolution and column durability were obtained with Silar 5CP™ [644] and Silar 10C™ [411] as stationary phases. GC traces obtained by this means are complex, peaks are poorly shaped and the base-line is noisy so that quantification is difficult. In addition, components are not easily identified, although internal standards help and it is possible to use a series of separation factors similar to those used for fatty acid identification as a guide [644]. Most of the important species from rat liver phospholipids were in fact resolved in this work, and only the species 18:0-18:2 and 16:0-20:4 were poorly separated. However, the peak for 18:0-22:6 was not observed, because of its low abundance and long retention time. The high bleed from polar phases in packed columns rules them out for GC-MS. WCOT columns with polar phases are greatly to be preferred (see below).

Mass spectrometry can be of enormous assistance in identifying unresolved components within a fraction, and one reason for the use of BDMS [653] and TMS ether [797] derivatives of diacylglycerols is that they have good mass spectral characteristics. Acetates can also be analysed in this way if need be [369,370]. This is discussed in detail later.

Alkyl- and alkenyl-analogues of diacylglycerols have been prepared and analysed by similar techniques to these [201,203].

The main limitation of all these procedures is that components with two given fatty acids in different positions of the glycerol moiety (e.g. 1-palmito-2-olein and 1-oleo-2-palmitin) are not separated. If this information is required, molecular fractions must be physically isolated by some procedure [168], in order that they may be subjected to enzymatic hydrolysis with pancreatic lipase or a related enzyme [163].

4. Separations on WCOT Columns

The greater resolving power of WCOT columns has been put to good use for the resolution of diacylglycerol species derived from phospholipids, first with apolar and more recently with polar stationary phases. For example, Gaskell and Brooks [278] separated the TMS ether derivatives of diacylglycerols from arterial wall phospholipids on a glass WCOT column, coated with OV-1/Silanox™ and maintained isothermally at 300°C, for identification by MS. Others used similar columns to quantify disaturated phospholipid species in animal tissues [295,560] and to resolve synthetic 1,2- and 1,3-diacylglycerols [772].

Much more impressive separations have come from Kuksis' laboratory. For example, a glass WCOT column (10 m x 0.25 mm) coated with

SP-2330™ (a 68% cyanopropyl-32% phenylsiloxane), with hydrogen as the carrier gas and a maximum temperature of 250°C, was used to separate the TMS ether derivatives of diacylglycerols prepared from many different natural lipids [646]. The phosphatidylcholines from rat liver gave about 30 distinct peaks, ranging from the species 16:0-16:0 to 18:1-22:6, and this is illustrated in Figure 8.8. The existence of di-unsaturated species, such as the the latter,

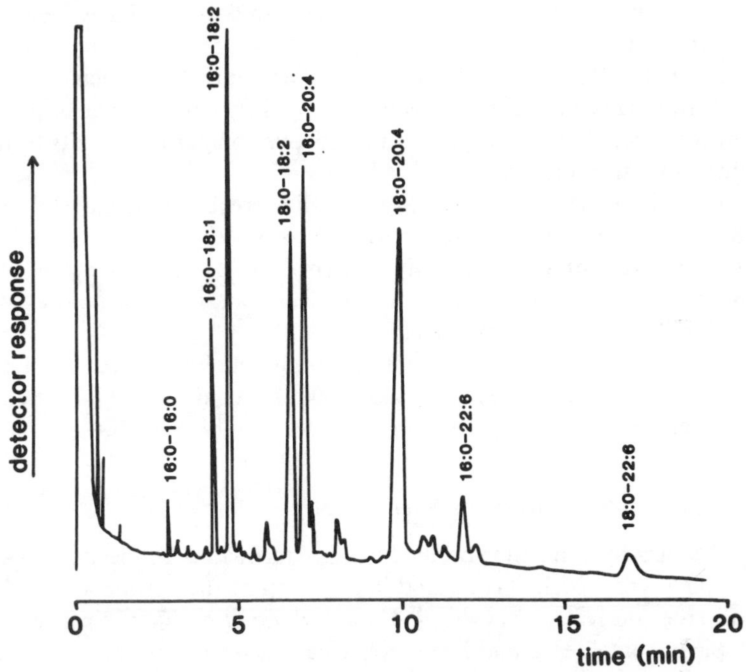

Figure 8.8 Separation by high temperature GLC of the trimethylsilyl ether derivatives of diacylglycerols, prepared by phospholipase C hydrolysis from the phosphatidylcholines of rat liver [646]. The column was a 10 m x 0.25 mm glass capillary coated with SP-2330™, and was temperature-programmed from 190°C to 250°C at 20°C/min, then was held isothermally at 250°C. Splitless injection was used with hydrogen as the carrier gas. A few only of the major peaks are identified here for illustrative purposes. (Reproduced by kind permission of the authors and of the *Canadian Journal of Biochemistry and Cell Biology*, and redrawn from the original paper)

had not previously been demonstrated in animal tissues. Positional isomers and disaturated species within a given carbon number were not separable, but these cannot be resolved by any chromatographic technique. Subsequently, the technique was applied to the determination of the alk-1-enyl-2-acylglycerol moieties of phospholipids [648]. Applications to the analysis of diacylglycerols derived from triacylglycerols have also been described [523,654]. In order to assist others in identifying molecular species separated in this way, comprehensive lists of relative retention times have been published [646,648].

Although the column life was found to be short (about 100 analyses), the newer WCOT columns of fused silica with cross-linked and chemically-bonded stationary phases would be expected to last much longer. Indeed a preliminary report of some outstanding separations with such a column (SE-54^{TM} as the stationary phase) has been published (Myher,J.J., Pind,S. and Kuksis,A., *J. Am. Oil Chem. Soc.,* **65**, 524 (Abst HH7) (1988)). Molecular species with positional isomers in the fatty acid residues were found to be separable.

As the molecular weights of diacylglycerol derivatives are much lower than those of triacylglycerols, there appears to be no difficulties with quantification and uncorrected detector responses should give comparable results to those obtained by other means.

In a novel application of capillary GC, diastereoisomeric diacylglycerols of short-chain fatty acids were resolved on a WCOT column coated with a non-polar phase in the form of the α-methoxy-α-trifluoromethylphenylacetic acid derivatives, and the structures were confirmed by mass spectrometry [608].

GC analysis of 1-O-alkyl-2-acetyl-phosphorylcholine (platelet-activating factor) is a rather specialised application, and for convenience it is discussed in Chapter 10.

5. Gas Chromatography-Mass Spectrometry

GC-MS has for some time been a favoured technique for the identification of diacylglycerol species, separated in the form of the acetate and TMS or BDMS ether derivatives. Myher [642] and Saito *et al.* [784] have reviewed the topic in some depth and have tabulated much valuable data. Normally it is advisable to use the response of the flame ionisation detector for quantification of the main molecular species, and to use GC-MS for identification and quantification of isomers within a single peak.

The mass spectra of acetate derivatives of diacylglycerols are of course those of triacylglycerols (see Section B.3), except that one of the acyl moieties is an acetyl residue [369,370,478,617]. Published mass spectral data are sparse, but it is apparent that the molecular ion tends to be rather small or non-existent, although the ion representing loss of water ($[M\text{-}18]^+$) can usually be seen. Loss of the acetyl group gives ions at $[M\text{-}59]^+$ or $[M\text{-}60]^+$, depending on the degree of unsaturation of the residual ion, and this is probably the best marker for determining the molecular weight. In addition, ions are seen for the loss of one or both of the other acyloxy moieties as expected. The higher the degree of unsaturation, the smaller are the ions in the high mass range, but those ions containing a single fatty acid residue are usually sufficiently abundant for identification purposes. Although it has been suggested that ions equivalent to $[M\text{-}RCOOCH_2]^+$ can serve as an indication of the identity of the fatty acid in position 1 of the glycerol moiety [369,370], Myher [642] has doubted the practicality of this and suggests that further data are necessary. The procedure has been applied to diacylglycerol acetates derived from egg phosphatidylcholine [369,370] and from the phosphatidylglycerol of *Escherichia coli* [478]. No mass spectra obtained by chemical ionisation methods appear to have been published for these compounds.

TMS ether derivatives tend to give much better spectra, which permit differentiation of 1,2- and 1,3-diacylglycerols even [82,200]. When the derivatives are prepared by high-temperature hydrolysis followed by silylation, the main products are the 1,3-isomers [404,405]. The principal difference in the spectrum of the latter is the presence of an abundant ion formed as a result of the loss of an acyloxymethylene radical ($RCOOCH_2$); although this might be expected even with a 1,2-isomer, it is not in fact seen to any significant extent. Most analysts use 1,2-diacylglycerol derivatives prepared by milder methods as described above. With mass spectra from electron-impact ionisation, the molecular ion is rarely seen, but ions equivalent to $[M\text{-}15]^+$ (loss of a methyl group) and $[M\text{-}90]^+$ (loss of the TMS ether moiety) can be used to determine the molecular weight and thence the total carbon number and degree of unsaturation of the acyl moieties. An important diagnostic ion results from the loss of an acyloxy residue, i.e. $[M\text{-}RCOO]^+$ or $[M\text{-}RCOOH]^+$ if the residual acyl group is unsaturated. Other useful ions are equivalent to [RCO + 74], [RCO + 90], [RCO] and [RCO-1]. Characteristic ions at m/z = 145 and 129 contain the TMS group and parts of the glycerol backbone. Molecular species from several glycerophospholipids have been examined by GC-MS in this form [201,203,404,405].

It is now apparent that BDMS ether derivatives of diacylglycerols are especially useful, because their greater chemical stability means that they can be subjected to such techniques as silver ion TLC and HPLC in addition to GC-MS. On GC, corresponding fractions elute about two methylene groups later than the TMS ethers. BDMS ethers also give distinctive fragmentations, similar in many ways to the TMS ethers, both with electron-impact and chemical ionisation procedures. The molecular ion tends to be small, but is often measurable if one of the acyl groups is highly unsaturated, and there is always an abundant ion equivalent to $[M\text{-}57]^+$ [653]. This is seen at m/z = 671 in the mass spectrum of the BDMS ether derivative of 1-palmitoyl-2-eicosapentaenoyl-*sn*-glycerol, illustrated in Figure 8.9. Indeed, the ion at $[M\text{-}57]^+$ is often sufficiently clear for identification of components present at less than 0.5 % of the total. Ions formed by loss of RCOO and RCOOH radicals are of immediate diagnostic value. If the acyl moiety lost is unsaturated, the ion formed by loss of the RCOO radical is much more abundant than that for loss of RCOOH (at m/z = 427 in the figure); if the radical lost is saturated, the two ions are of about the same

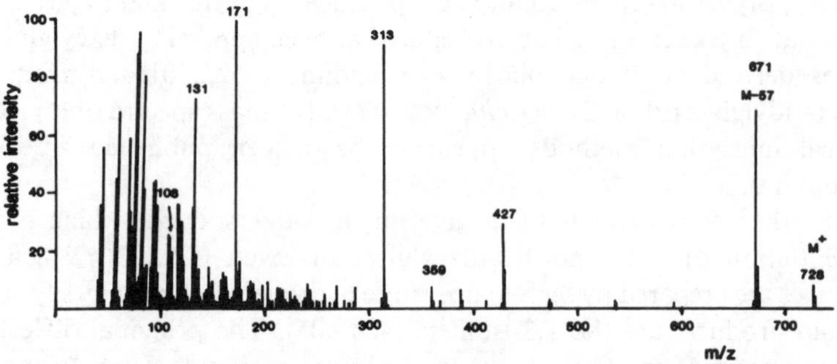

Figure 8.9 The mass spectrum of the *t*-butyldimethylsilyl ether derivative of 1-palmitoyl-2-eicosapentaenoyl-*sn*-glycerol [653]. (Reproduced by kind permission of the authors, and of *Analytical Chemistry*, and redrawn from the original paper)

intensity (at m/z = 472 and 473). The total abundance of ions formed by loss of fragments from position 2 is greater than that from position 1. In addition, there are characteristic ions equivalent to $[RCO]^+$, $[RCO+74]^+$ and $[RCO+148]^+$ for each acyl group. It is possible to use selective ion monitoring of many of these ions for identification and quantification purposes [477]. With chemical ionisation, the intensities of the characteristic ions in the high mass range are enhanced [522-524]. BDMS ether derivatives have been used for the identification of molecular species of phospholipids of animal [477,522-524,653] and microbial origin [220,752]. They have also been employed with diacylglycerols, prepared from phosphoglycerides, and selectively deuterated as an aid to identification [219].

Ether analogues of diacylglycerols are readily identified by GC-MS as the TMS or BDMS ether derivatives in a similar manner (reviewed by Egge [237]). With the former, the molecular weight of a 1-alkyl-2-acylglycerol is given by ions at $[M\text{-}15]^+$ and $[M\text{-}90]^+$ [795]. For saturated species, the relative abundance of these ions is low, but the $[M\text{-}90]^+$ ion especially becomes much more prominent when there is an unsaturated residue. There is an ion of relatively low intensity representing the loss of the alkoxy group, $[M\text{-}RO]^+$, one representing loss of the acyloxy moiety, $[M\text{-}R'COO]^+$, and others for $[R'CO]^+$ and $[R'CO+74]^+$. The base ion at m/z = 130 contains the TMS ether group and the carbons of the glycerol backbone (c.f. diacylglycerol derivatives where this is at m/z = 129). Mass spectra of some TMS ether derivatives of synthetic dialkylglycerols have been described [798]. In addition, mass spectrometry has been invaluable in determining the structures of the complex ether lipids of Archaebacteria (reviewed elsewhere [216]).

The TMS ethers of 1-alk-1-enyl-2-acyl glycerols are similarly identifiable by GC-MS [796]. Some features are analogous to the alkyl ethers, and for example there are ions equivalent to $[M-15]^+$, $[M-90]^+$, $[R'CO]^+$, $[R'CO+74]^+$, $[R'COO+74]^+$, and a base peak at m/z = 129. Also, ions at $[M-RCH=CHO]^+$ and $[R'COO+130]^+$ are abundant. In the spectra of the relatively common 1-hexadec-1-enyl- and 1-octadec-1-enylglycerols, there are characteristic ions at m/z = 311 and 319, representing $[(R'CH=CHO-1)+73]^+$. The TMS and BDMS ether derivatives of ether lipids of this type from several natural sources have been analysed by GC-MS [201,477,648].

D. MONOACYLGLYCEROPHOSPHOLIPIDS AND MONOACYLGLYCEROLS

Lysoglycerophospholipids are converted to monoglycerides for analysis by GC by the same procedures used to generate diacylglycerols from phospholipids (see Section C.1 above). While 2-monoacylglycerols are usually only trace constituents of tissues, they are generated by the action of pancreatic lipase on triacylglycerols during analyses of positional distributions (reviewed elsewhere [163,167]), and they may also be formed in other circumstances. They can be separated with relative ease by GC after conversion to nonpolar volatile derivatives. For convenience, the alkyl and alkenyl ether analogues are not discussed here, but together with other products of the hydrolysis of ether lipids in Chapter 10.

Wood *et al.* [996], for example, were able to separate isomeric 1(3)- and 2-monoacylglycerols from each other as the TMS ether derivatives by GC on packed columns with DEGS as the stationary phase. The 2-isomer eluted first. As monoacylglycerols in the underivatised state isomerise rapidly, the practical value of such separations is doubtful since isomeric compounds differing in degree of unsaturation tend to overlap. By employing more efficient columns and more polar stationary phases, such as Silar 5CP™ or Silar 10C™, much better resolution was obtained, although there were still critical pairs which caused difficulties [411,643,656]. For example, monoacylglycerols such as 1-18:0 and 2-18:1 or 1-18:1 and 2-18:2 overlapped. As a guide to the GC conditions, TMS ether derivatives of monoacylglycerols eluted in reasonable times from a glass column (2 m x 3 mm i.d.) packed with 5% Silar 5C™ on GasChrom Q™ (100-120 mesh) and maintained at 190°C [411]. Carbonate derivatives of 1-monoacylglycerols have been separated under similar conditions [678]. Analysis is greatly simplified if the two isomeric forms of the monoacylglycerols are first resolved by TLC on silica gel impregnated with boric acid, immediately prior to derivatisation [163].

GC of TMS ethers on WCOT columns greatly improves the resolution attainable, although there are still likely to be problems with complex samples

[628,772].

The mass spectrometric fragmentations of monoacylglycerol acetate and TMS ether derivatives have been studied systematically both by direct probe insertion and by GC-MS [200,450,656]. With the acetates, there is no simple means of distinguishing the 1(3)- and 2-isomers, although there are differences in the intensities of ions representing loss of the long-chain acyloxy group with position. More characteristic spectra are obtained from the TMS ethers, which are the best derivatives for GC purposes. The molecular ion is usually detectable, especially when the acyl residue is unsaturated. As with the diacylglycerols discussed above, $[M\text{-}15]^+$ and $[M\text{-}90]^+$ are abundant ions, in addition to those equivalent to $[RCO]^+$ and $[RCO+74]^+$. An ion formed by cleavage between carbons 2 and 3 in the spectra of the 1(3)-monoacylglycerol derivatives at $[M\text{-}103]^+$ appears to be absolutely characteristic for this isomer, while a smaller ion at $m/z = 205$ is of further diagnostic value. Similarly, an ion at $m/z = 218$ is highly favoured in the spectra of 2-isomers. When isomers are incompletely resolved by GC, they can be still be quantified by selective ion monitoring, by making use of the distinctive ions in the mass spectra. BDMS ethers would no doubt be equally suitable for the purpose. Carbonate derivatives of 1- and 3-monoacyl-*sn*-glycerols exhibit particularly good molecular ions in MS [677].

E. STEROL ESTERS

Fatty acid esters of cholesterol are abundant components of plasma and other animal tissues, while other sterol esters are common if minor constituents of plant lipids and are occasionally found in animals. Cholesterol esters were first successfully separated on nonpolar silicone phases in packed columns, similar to those used initially for triacylglycerol separations (see Section B.1 above), and they elute at temperatures intermediate between those of diacylglycerol derivatives and triacylglycerols [498]. Components differing by one carbon atom are separable in this way, but useful resolutions of saturated and unsaturated compounds of the same chain-length cannot be achieved. Cholesteryl alkyl ethers from bovine heart were subjected to GC under similar conditions [271]. Later, cholesterol esters were successfully separated according to degree of unsaturation on short packed columns containing a thermally-stable polar phase, such as Silar 10C™ [902].

With complex samples, the resolution is still far from ideal, and much better results can now be obtained on modern WCOT columns. For example, cholesterol esters from plasma were well separated according to their chain-lengths and partly by degree of unsaturation on WCOT columns of fused silica and coated with a nonpolar phase, OV-1™, temperature-programmed to 330°C; with the polar phase, SP-2330™, better separation by degree of unsaturation was achieved, though peaks were less sharp [861,863]. A report

[862] that hydrogenation can occur on a polar column when hydrogen is the carrier gas has been discounted by others (see Section B.2 above).

The additional resolution attainable with WCOT columns is essential if the nature of the sterol moiety as well as that of the fatty acid varies. Studies of sterol esters from geochemical samples [961], plants [249,566], beeswax [550], dinoflagellates [196], and plasma in patients with phytosterolemia and xanthomatosis [527,528] illustrate the difficulties that can be encountered. Figure 8.10 shows a separation of sterol esters from the plasma of a patient

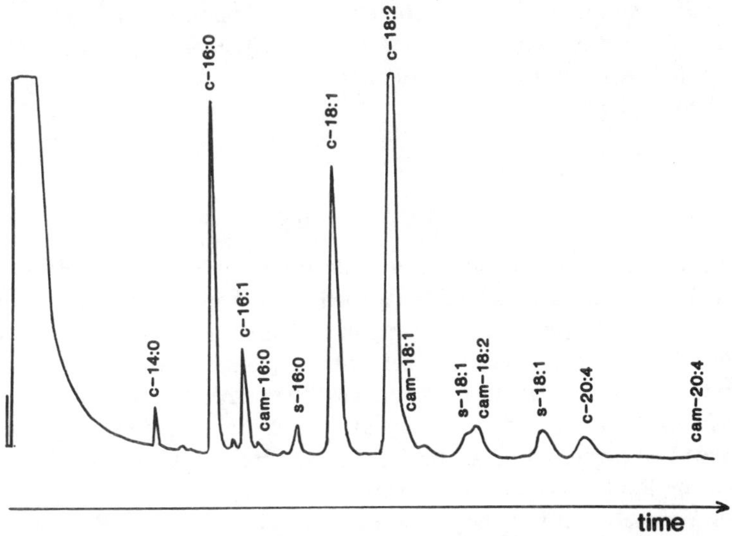

Figure 8.10 GC elution profile of low- and high-density lipoproteins of plasma from a patient with phytosterolemia [527]. A glass WCOT column (10 m x 0.25 mm i.d.), coated with SP-2330™, was maintained isothermally at 250°C with hydrogen as the carrier gas. Abbreviations: c, cholesterol; cam, campesterol; s, β-sitosterol. (Reproduced by kind permission of the authors and of *Lipids*, and redrawn from the original paper).

with phytosterolemia [527]; fatty acid esters of cholesterol, campesterol and β-sitosterol were identified. In most of the work cited, nonpolar phases were used, although a more polar phenylmethyl silicone (SP-2330™) gave the excellent separation of the figure. Double bonds and other substituents in the the sterol moieties contribute to the retention times of the esters, and as these may vary from sample to sample, it is not easy to put forward general rules to predict their behaviour in GC. Although mass spectrometry helps greatly to identify components and to quantify mixtures hidden under a single GC peak, it still appears advisable to pre-fractionate complex samples by reversed-phase HPLC and silver ion chromatography before GC analysis [249,527].

With electron-impact ionisation in MS, sterol esters rarely give detectable molecular ions [566]. Ions representing loss of the fatty acyl moiety are seen with model compounds, but these are of limited value with unknowns. Nonetheless, valuable structural information can be obtained with some samples [196]. Chemical ionisation, however, tends to give base ions representative of the nature of the sterol component [636]. With a suitable choice of reagent gas, and ammonia seems to be the best for the purpose, a good quasimolecular ion ($[M+NH4]^+$) and ions diagnostic for both the sterol and fatty acid moieties are obtained [566,961]. These findings have been confirmed and extended by others in GC-MS studies of sterol esters from various sources; in particular, it was demonstrated that negative chemical ionisation with ammonia as the reagent gas is to be preferred [248-250]. The EI mass spectrum of cholesteryl hexadecyl ether has also been published [271].

F. WAX ESTERS

Natural waxes can consist of a wide range of different lipid classes, including esters of various kinds, hydrocarbons, ketones, hydroxy-ketones, β-diketones, aldehydes, acids and terpenes. With crude mixtures of this kind, it is usually necessary to react them with diazomethane to methylate free carboxyl groups, to acetylate (or to prepare TMS ethers) in order to deactivate free hydroxyl groups and to convert any aldehydes to oximes, prior to GC analysis. The preen glands of birds, for example, contain a wide range of fascinating lipid molecules. However, detailed analyses of these and many other waxy materials is rather a specialised topic, and the reader is referred to reviews that have appeared elsewhere [423,425,928,929]. Esters of long-chain alcohols and fatty acids, the compounds commonly termed "wax esters", are widespread in nature and have some commercial importance, while their analysis presents some general problems that merit attention here. The fatty acids and alcohol constituents from different sources reflect their origins; frequently, both are saturated or monoenoic compounds from 16 to 30 carbon atoms in length, but those of marine origin may be highly unsaturated, for example.

Wax esters tend to have similar molecular weights to diacylglycerol acetates (see Section C) and are eluted from GC columns under comparable conditions. Initially, short packed columns containing nonpolar silicone phases were employed, and these were easily capable of separating species differing in carbon number by one or two units, where the carbon atoms of the alcohol and fatty acid moieties have equal value [422]. Thus, a 16:0 acid with an 18:1 alcohol and an 18:1 acid with a 16:0 alcohol will co-chromatograph, and both will emerge with species such as 14:0-20:1 and so forth. With marine oils, the high degree of unsaturation in some components can lead to peak broadening, so hydrogenation and/or prefractionation by silver ion chromatography are often recommended to improve resolution.

Further complications can arise when branched-chain fatty acid or alcohol constituents are present. Several applications of this methodology to wax esters from marine mammals [17,18,152,556] and to the commercial vegetable oil from jojoba [872,874] have appeared. Subsequently, separation of wax esters according to degree of unsaturation was achieved on short packed columns containing polar phases, such as Silar 10C™ [899,901]. Rather complex chromatographic traces were obtained if the sample was not first fractionated by silver ion chromatography, and the high bleed from such columns precluded the use of mass spectrometry for identification purposes.

As with the other lipid classes discussed above, greatly improved resolution of wax esters can now be achieved on WCOT columns. Indeed, short-chain ($\ll C_{24}$) esters from the bottlenose dolphin were resolved on a stainless steel WCOT column coated with DEGS and other stationary phases by Ackman and colleagues in 1973 [29,554]. Jojoba wax has been fractionated on a glass WCOT column coated with OV-1™ [302]. More recently, a WCOT column (25 m x 0.2 mm) of fused silica, coated with a methylsilicone phase, was utilised with temperature-programming from 250°C to 350°C for the analysis of the wax esters from the alga, *Chlorella kessleri* [769]. The separation obtained is illustrated in Figure 8.11; within each carbon number group,

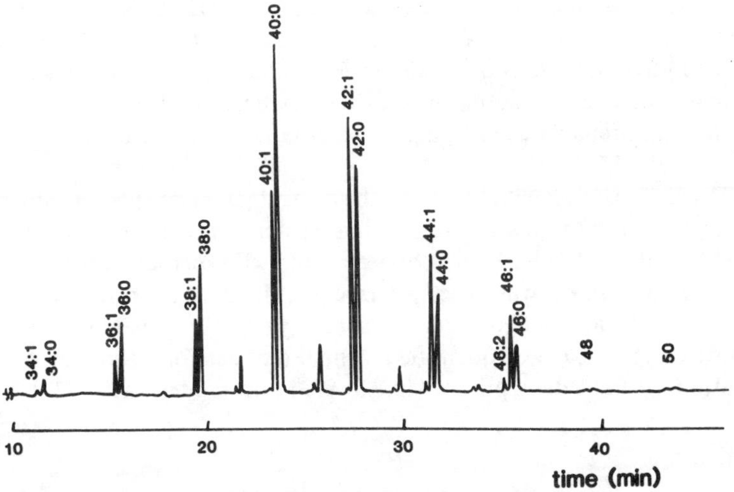

Figure 8.11 GC separation of the wax esters from the alga, *Chlorella kessleri* [769]. A fused silica WCOT column (25 m x 0.2 mm i.d.), coated with a methylsilicone phase, was temperature-programmed from 250 to 350°C at 2°C/min, with helium as the carrier gas. (Reproduced by kind permission of the authors and of the *Journal of Chromatography*, and redrawn from the original paper).

species with zero, one and two double bonds are clearly resolved. With on-column injection, good quantification of individual peaks is possible, while mass spectrometry permits estimation of multiple components within a single peak. Excellent resolution by degree of unsaturation was achieved on a glass WCOT column coated with the polar phase SP-2340™ [412]. Within carbon number groups, some separations according to the chain-lengths of the individual constituents of the esters were seen. Thus, 14:0-14:0 and 10:0-18:0 were separable, but 14:0-14:0 and 12:0-16:0 were not. It was also observed that the elution temperature had an appreciable effect on the resolution of specific critical pairs.

MS is a valuable means of identifying individual wax esters separated by GC. In a systematic study of electron-impact ionisation spectra of a number of model compounds of the form RCOO.R′ [1], it was shown that a good molecular ion is always obtained while the base peak is generally the protonated acid ion ($[RCOOH_2]^+$), except when the alcohol chain is less than 10 carbon atoms long. The alcohol moiety is indicated by a prominent ion equivalent to $[R'-1]^+$. Relative abundances of all ions are dependent on the chain-lengths of the individual aliphatic groups. In order to use MS for quantitation of species within each GC peak, a method was developed in which double bonds were reduced with deuterohydrazine and this was followed by ozonolysis to remove any residual unsaturated species. Relative peak intensities for ions corresponding to $[RCOOH]^+$, $[RCOOH_2]^+$ and $[R'-1]^+$ were then measured, since it was observed that the sum of these for each species was a more reliable indicator of the total amount present than were the individual intensities. This approach has been utilised successfully by others [769,872]. In addition, GC-MS has been used to identify phytyl esters in a dinoflagellate [196], and other wax esters in sediments [961] and psoriatic nail [575].

Chemical ionisation with butane as the reagent gas gives spectra with much less fragmentation and with the quasimolecular ion ($[M+1]^+$) as the base ion [718]. Others [961] obtained better results with methane as the reagent gas, although caution was necessary in using the data quantitatively, since the ions for saturated and unsaturated species varied appreciably in intensity. Intact wax esters have been quantified without the need for a chromatographic separation by field desorption [640,820-822] and tandem MS [873].

G. GLYCOSYLDIACYLGLYCEROLS

Mono- and digalactosyldiacylglycerols are polar molecules of high molecular weight, so they would not be expected to be good candidates for analysis by high-temperature GC. Nonetheless, the TMS ether derivatives of monogalactosyldiacylglycerols from plants were successfully subjected to GC on short packed columns containing methylsilicone phases, under conditions similar to those required for intact triacylglycerols [63]. Sulphoquinovosyldiacylglycerol from plants was separated into molecular species in the same way after methylating the sulphonic acid group with diazomethane and converting the carbohydrate moiety to the TMS ether derivative [938]. In essence, three species were obtained equivalent to the combinations C_{16}-C_{16}, C_{16}-C_{18} and C_{18}-C_{18} (irrespective of degree of unsaturation). Later, both mono- and digalactosyldiacylglycerols and their monoacyl equivalents were chromatographed in the same manner, while the products of deacylation were separated as the *O*-methyl, *O*-acetyl, *O*-TMS ether and *O*-trifluoroacetate derivatives [976].

A procedure for the release of the diacylglycerol moieties from galactosyldiacylglycerols (including sulphoquinovosyldiacylglycerols) has been described, involving periodic acid oxidation in methanol followed by incubation with 1,1-dimethylhydrazine [382]. The diacylglycerols are then converted to UV-absorbing derivatives for separation by means of HPLC [475], but equally it should be possible to prepare TMS or BDMS ethers and use the GC conditions developed for the analysis of the equivalent compounds released from phospholipids (see Section C above).

H. SPHINGOLIPIDS

1. Preliminaries to GC Separation

The basic lipid moiety of a sphingolipid is a ceramide, consisting of a long-chain (sphingoid base) linked via an amide bond to a fatty acid. Though free ceramides can occur in small amounts in tissues, they usually form part of complex lipids containing sugar moieties, as in glycosphingolipids, or a phosphorus group, as in sphingomyelin and certain phosphonolipids. One approach to the analysis of sphingolipids by GC, therefore, consisting in preparation of the ceramides by suitable procedures for conversion to nonpolar derivatives such as the TMS ethers. Intact sphingolipids can be analysed by GC only with difficulty, and this is one area where HPLC has particular advantages; the compounds can be analysed in native form or after conversion to benzoyl derivatives, which can be detected with some sensitivity by UV spectrophotometry [168]. The analysis of long-chain bases is described in Chapter 10.

It is a relatively simple matter to hydrolyse sphingomyelin and ceramide aminoethylphosphonate to ceramides with the enzyme phospholipase C from *Clostridium welchii*, and practical details of the hydrolysis procedure and of methods of purifying the product are given in Section C.1 above. Hydrolysis of sphingomyelin can also be accomplished chemically by reaction with hydrofluoric acid [760].

Unfortunately, it is much less easy to prepare ceramides from glycosphingolipids. Compounds containing dihydroxy bases can be converted to ceramides by a chemical procedure devised by Carter *et al.* in 1961 [150], in which the glycosidic ring is opened with periodate and the resulting product reduced with sodium borohydride before being hydrolysed under mild acidic conditions. Derivatives of trihydroxy bases cannot be converted to ceramides by this procedure, as the bases would be cleaved across the vicinal diol group by the peroxide oxidation step. Although many enzymes have been described that will cleave bonds between the ceramide and carbohydrate residues, none appear to be readily obtainable with a sufficiently high specific activity for structural studies of ceramides.

The nature of the fatty acids and long-chain bases in sphingolipids are described in Chapter 2, but it is perhaps worth reiterating briefly that the fatty acids are commonly saturated and monoenoic components (up to C_{26}), and they may also have a free hydroxyl group in position 2; di- and trihydroxy bases occur, varying in degree of unsaturation, but frequently with a *trans*-double bond in position 4. Ceramides derived from these therefore have variable numbers of free hydroxyl groups in different regions of the molecule, i.e. two to four in total (two or three in the base and zero or one in the fatty acid). Some preliminary separation prior to analysis by GC or other means can then often be helpful. Karlsson and Pascher have published a valuable paper on TLC of ceramides [467], and layers of silica gel or better of silica gel containing diol-complexing agents, such as sodium tetraborate ($Na_2B_4O_7.10H_20$) or sodium meta-arsenite ($NaAsO_2$), can be used to effect separations that depend on the number and configuration of the hydroxyl groups present. Layers are prepared by incorporating 1 % (w/v) of the salt into the water used to prepare the slurry of adsorbent. In addition, ceramides having a *trans*-double bond in position 4 of the long-chain base are separable on silica gel impregnated with sodium borate (saturated compounds migrate ahead of unsaturated), although the reason for the effect is not understood [624]. The long-chain base and fatty acid constituents may each contain up to two double bonds, so TLC with silica gel impregnated with silver nitrate (5 % w/w) can be used to good effect.

From these studies, Karlsson and Pascher were able to recommend a programme for the preliminary fractionation of natural mixtures by TLC, although they recognise that, as the components of natural ceramides vary greatly in chain length, bands may tend to spread more than is the case with pure model compounds. Four groups of ceramide can be separated on the arsenite layers with chloroform-methanol (95:5, v/v) as the solvent system; dihydroxy base-normal acid, dihydroxy base-hydroxy acid and similar species containing trihydroxy bases, as illustrated in Figure 8.12 (plate A). Derivatives of dihydroxy bases isolated in this manner can then be separated on borate-impregnated layers using the same solvent mixture for development, according to whether they contain long-chain bases with *trans*-double bonds in position 4 or not (Figure 8.12, plate B). Finally all the components isolated thus far can be further separated into groups, according to the combined numbers of *cis*-double bonds (*trans*-double bonds have comparatively little effect) in the component fatty acids and in the long-chain bases on silver nitrate-impregnated layers. Better results are obtained in this instance if the free hydroxyl groups are first acetylated with acetic anhydride and pyridine (see

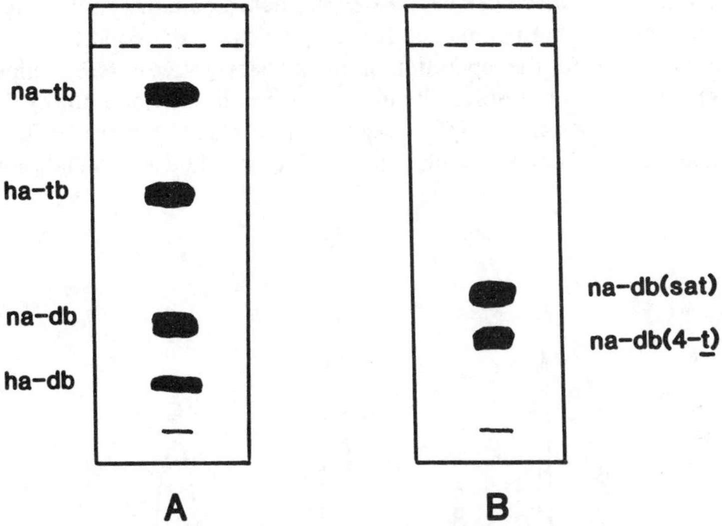

Figure 8.12 Schematic separation of ceramides by TLC on layers of silica gel G, impregnated with 2 % sodium arsenite (plate A) and 2 % sodium borate (plate B), using a solvent system of chloroform-methanol (95:5, v/v) for development [467]. Abbreviations: na, normal fatty acid; ha, hydroxy fatty acid; db, dihydroxy base; tb, trihydroxy base; sat, saturated.

Chapter 4); chloroform-benzene-acetone (80:20:10, by volume) was chosen as a suitable solvent system for the development, but it would now be considered safer to replace the benzene with toluene.

The preliminary separation of ceramides is undoubtedly an area where HPLC could make an effective contribution, assuming that some enterprising analyst is prepared to devote time to the problem.

For GC analysis, ceramides have generally been converted to TMS ether derivatives, but BDMS ethers [649] and cyclic boronates [279] have also been prepared.

2. Gas Chromatography-Mass Spectrometry

Mass spectrometry of intact glycosphingolipids by direct probe insertion and soft ionisation procedures has become one of the major means of structure determination for these compounds [460,896,974]. Information is thereby obtained on the nature of both the lipid and carbohydrate moieties, although the presence of different molecular species hampers the interpretation of the spectra greatly. Detailed discussion of this aspect is beyond the scope of this book.

The molecular weights of most glycosphingolipids are too high to permit fractionation by GC, but monoglycosylceramides (cerebrosides) have been successfully analysed by the technique in the form of the TMS ether derivatives [347,350,405,633,681,788,892]. Packed columns containing non-polar phases have been employed for the purpose under conditions similar to those required for the separation of intact triacylglycerols (see Section B.1 above), so species are resolved by molecular weight only and not by degree of unsaturation. As an example, a separation of glucocerebrosides from the spleen in Gaucher's disease is illustrated in Figure 8.13 [681]. A glass column

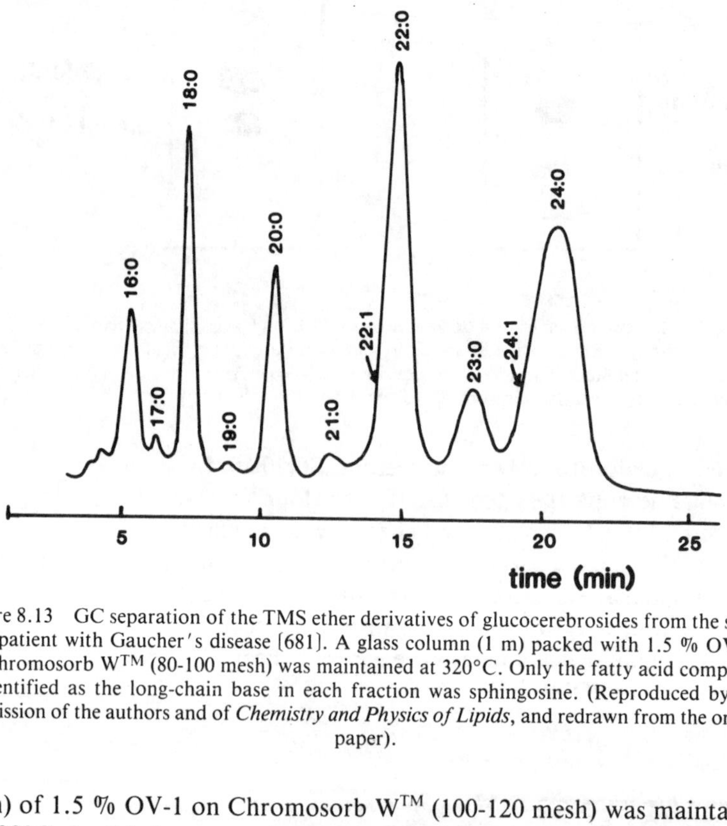

Figure 8.13 GC separation of the TMS ether derivatives of glucocerebrosides from the spleen of a patient with Gaucher's disease [681]. A glass column (1 m) packed with 1.5 % OV-1™ on Chromosorb W™ (80-100 mesh) was maintained at 320°C. Only the fatty acid component is identified as the long-chain base in each fraction was sphingosine. (Reproduced by kind permission of the authors and of *Chemistry and Physics of Lipids*, and redrawn from the original paper).

(1 m) of 1.5 % OV-1 on Chromosorb W™ (100-120 mesh) was maintained at 320°C. In this instance, there was essentially only one long-chain base (sphingosine), so the separation depends simply on the nature of the fatty acid constituents. Components were identified by chemical ionisation MS with isobutane as the reagent gas, as this technique gives a detectable quasimolecular ion, together with other ions indicative of the molecular weight ($[MH-90]^+$ and $[MH+73]^+$), while ions revealing the molecular size of the carbohydrate, ceramide, long-chain base and fatty acid moieties are also prominent. Naturally-occurring ceramides in spleen were similarly

subjected to GC-MS as the TMS ethers in this study. While electron-impact MS can give valuable data, ions in the high mass range are less easily detected [347,405,788].

It is perhaps rather surprising, but intact ceramide aminoethylphosphonate from shell fish has been successfully subjected to GC separation in the form of the TMS ether derivative [601]. Presumably the phosphonate bond is much more stable than a phosphate at elevated temperatures.

Much more work has been done on GC of sphingolipids after conversion to simpler ceramides and thence to appropriate volatile derivatives, and these are eluted from GC columns under similar conditions to diacylglycerols (see Section C above). As ceramide standards are not readily available, and considerable variation in molecular structure can occur, identification of peaks can present enormous difficulty unless the analyst has access to GC-MS. Much of the early work in this area consisted of systematic studies with model ceramides, that provided a solid foundation for subsequent analyses of natural samples. The chromatographic conditions employed in the pioneering studies in 1969 were a glass column (1.2 m x 3 mm i.d.), containing 2 % OV-1™ on GasChrom Q™ (60-80 mesh), maintained at temperatures of 300 to 320°C, with helium as the carrier gas [786]; similar general conditions have been employed by most others who subsequently worked on related topics. As discussed before for triacylglycerols, separation is solely on the basis of molecular weight. The detector response was found to vary with such factors, as column temperature, the flow-rate of the carrier gas and the molecular weight of the ceramide derivative, so careful standardisation of conditions and calibration was necessary for quantitative work [789]. In order to ascribe meaningful relative retention times to ceramide derivatives, they were assigned "triglyceride carbon unit" (TGCU) values according to the retention times of saturated triacylglycerol standards. Some representative values of this kind are listed for reference in Table 8.1.

Table 8.1
Triglyceride carbon units (TGCU) for the TMS ether derivative of sphingosine and sphinganine ceramides [789].

	TGCU			
	sphingosine		sphinganine	
Fatty acid	normal	2-hydroxy	normal	2-hydroxy
16:0	37.4	38.1	37.7	38.0
18:0	39.5	40.0	39.7	40.0
18:1	39.2		39.1	
19:0	40.5		40.7	
20:0	41.6	42.0	41.7	41.9
22:0	43.6	43.9	43.7	43.9
22:1	43.3			
23:0	44.6		44.7	
24:0	45.7	45.8	45.8	45.8
24:1	45.5			
25:0	46.7		46.7	

The electron-impact mass spectrum of the 1,3-di-O-TMS ether derivative of *N*-stearoyl-sphingosine is illustrated in Figure 8.14 [786,789]. A molecular

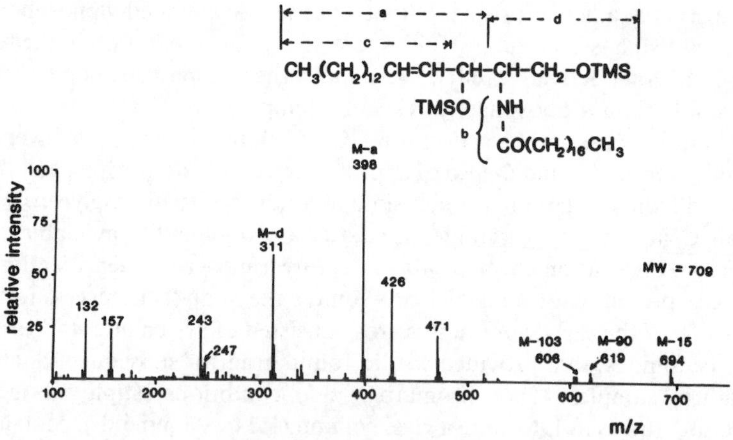

Figure 8.14 The mass spectrum of 1,3-di-*O*-trimethylsilyl-*N*-stearoyl sphingosine [789]. (Reproduced by kind permission of the authors and of *Chemistry and Physics of Lipids*, and redrawn from the original paper).

ion (M = 709) is not apparent, but there are ions at $[M\text{-}15]^+$ (m/z = 694), $[M\text{-}90]^+$ (m/z = 619) and $[M\text{-}103]^+$ (m/z = 606), formed by elimination of a methyl, trimethylsilanol and the terminal methylene plus trimethylsilanol groups respectively, and these serve to indicate the molecular weight. Loss of the fatty acid residue produces characteristic ions for $[M\text{-}(b+1)]^+$ at m/z = 426 and $[M\text{-}(b+1+90)]^+$ at m/z = 336 among others. Cleavage between carbons 2 and 3 of the molecule gives a prominent ion, diagnostic for the long-chain base, at m/z = 311. In addition, ions at m/z = 471, equivalent to $[M\text{-}(a\text{-}73)]^+$, and m/z = 398, equivalent to $[M\text{-}a]^+$, are of great value in structure assignment. Minor differences only are seen when the long-chain base is saturated. Analogous spectra are seen when the ceramide contains a 2-hydroxy fatty acid, although there are also certain distinctive fragmentations [351,789], and this is also true of ceramides containing trihydroxy bases, such as phytosphingosine [345].

Because of the difficulty of preparing ceramides from cerebrosides, relatively little work has been done on GC separations of such compounds. Nonetheless, cerebrosides from bovine [346] and mouse [350] brain were converted to ceramides by the chemical procedure described in the previous section, and these were separated into hydroxy and nonhydroxy acid-containing substituents and then by degree of unsaturation as the acetates, before hydrolysis and conversion to the TMS ethers for analysis by GC-MS. The GC trace obtained from a fraction containing sphingosine in combination with saturated 2-hydroxy fatty acids is illustrated in Figure 8.15. The

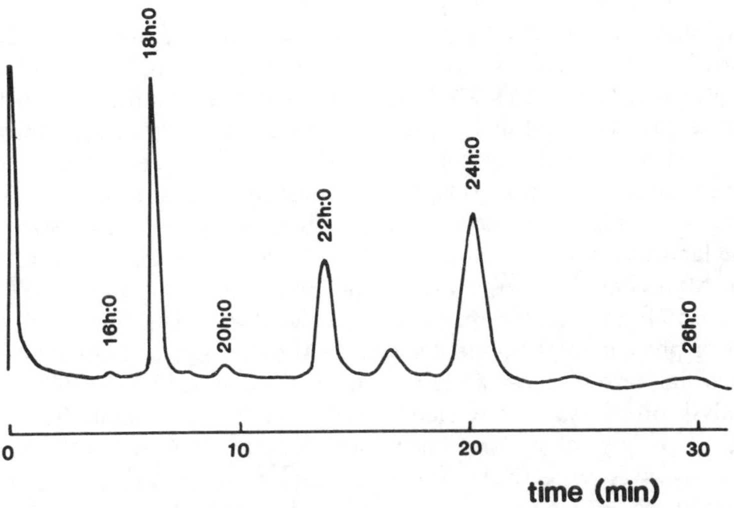

Figure 8.15 GC separation of the TMS ethers of a ceramide fraction containing sphingosine and non-hydroxy fatty acids from brain cerebrosides [346]. A glass column (1.7m x 3.5mm), packed with 2.4% OV-1™ on Gas Chrom Q™ (100-120 mesh), was maintained at 300°C. Only the fatty acid component is identified. (Reproduced by kind permission of the authors and of the *European Journal of Biochemistry*, and redrawn from the original paper).

chromatographic trace is relatively simple enabling mass spectrometric identification of each peak, and similar uncomplicated traces were obtained from the remaining TLC fractions.

More attention has been given to the separation of molecular species of ceramides derived from sphingomyelin by GC-MS. For example, native ceramides and sphingomyelins from human plasma [787,789] and bovine intestines [126], and ceramide 2-*N*-methylaminoethylphosphonate from the shell fish *Turbo cornutus* [601,603] have been analysed as the TMS ethers. In addition, ceramides from sphingomyelins of human plasma [649] and bovine brain [278,279] have been examined as the BDMS and cyclic boronate derivatives respectively, since these have improved mass spectral fragmentation characteristics. In this last work, glass WCOT columns were used resulting in greatly improved resolution. Little work has been done in general in this area in recent years because of the dominance of HPLC methodology, yet it ought to be possible to devise an analytical scheme to rival this but involving separation of molecular fractions, as stable BDMS or boronate derivatives, by adsorption chromatography (TLC or HPLC) followed by GC-MS on modern WCOT columns of fused silica with thermally-stable polar phases.

I. DETERMINATION OF LIPID PROFILES BY GAS CHROMATOGRAPHY

The ultimate objective of all analysts is to obtain as much information as possible on the lipid composition of a particular sample in the shortest possible time. Ideally this would mean obtaining a quantitative profile of the lipid classes present in a tissue together with the fatty acid composition of each in a single chromatographic step. This ideal goal cannot yet be attained, but a great deal has been accomplished by separating natural lipid mixtures according to molecular weight by high-temperature GC, principally in the laboratory of Kuksis in Canada but with a significant contribution from Mares in Czechoslovakia, who have both reviewed the topic [507,516,590]. In brief, the lipids are first digested with phospholipase C which converts phosphatidylcholine, lysophosphatidylcholine and sphingomyelin to diacylglycerols, monoacylglycerols and ceramide respectively. The hydrolysis products are converted to the TMS ether (or related) derivatives, while the cholesterol and free fatty acids also react to form TMS ether and ester derivatives respectively. Tridecanoin is added as an internal standard for quantification purposes, and the mixture is subjected to GC separation over a large temperature range so that as many as possible of the components are separated. The absolute amounts of the various lipid classes are easily determinated, while the proportions of the molecular species give an indication of the chain-length distributions of the fatty acid constituents.

Although such methodology could in theory be applied to any tissue, most work has been done on human plasma lipids and related body fluids such as lymph, as rapid screening procedures here can lead to the diagnosis of disorders of lipid metabolism and can assist in monitoring the effects of clinical therapy. For these purposes, a partially-resolved lipid profile may often contain sufficient information, provided that enough data are available on normal and diseased states to enable significant comparisons to be made. In historical terms, the development of the methodology has followed closely behind that for the separation of intact triacylglycerols since these inevitably must be resolved if the procedure is to be of value. Thus in the pioneering paper in 1967 [513], a short packed column containing a non-polar methylsilicone phase was employed, whereas nowadays WCOT columns of fused silica would be preferred. Improvements in computerised data handling have contributed greatly to applications involving routine screening.

Those aspects of quantification in high temperature GC described in Section B above, including column preparation, injection technique and calibration, are equally apposite here and need not be duplicated. It is, however, worth noting that calibrations should be carried out with standards similar in composition and overall concentration to the corresponding lipids in the tissue under study. For example if this is plasma, the tridecanoin used as an internal standard should be made up to a concentration of 100 μgram per ml of chloroform, and 100-200 μlitres of this is added to 0.25 to 0.5 ml of plasma during extraction of the lipid with chloroform/methanol (see Chapter 2). The lipid extract is then incubated with the phospholipase C of *Clostridium welchii*, as described in Section C above, and the products are isolated for analysis. Alternatively [525] -

> "Whole plasma (0.1 to 1 ml) with ethylenediaminetetraacetate (EDTA) added as an anticoagulant is digested in a stoppered tube with phospholipase C (2 to 4 units) in Tris buffer (17.5 mM; pH 7.3; 4 ml) with 1 % calcium chloride (1.3 ml) and diethyl ether (1 ml) for two hours at 30°C with shaking. The reaction is stopped by adding 0.1 M hydrochloric acid (0.2 ml), and the mixture is extracted with chloroform-methanol (2:1, v/v; 10 ml) containing tridecanoin (0.1 to 0.25 mg). After brief centrifugation to separate the layers, the chloroform layer is removed from the bottom of the tube by pasteur pipette and is taken to dryness in a stream of nitrogen."

The lipids are immediately converted to TMS or BDMS ether derivatives (see Chapter 4 for details), the latter now often being preferred because of their greater chemical stability and good MS fragmentation properties. In some of the early work [529], acetate derivatives were employed, but free carboxylic acid groups are not rendered inert by this means.

With the short packed columns used initially for this work, the resolution obtainable was limited, although adequate for many purposes, as shown in Figure 8.16 [520]. Derivatives of the free acids (two peaks) are the first significant components to emerge, followed in turn by cholesterol, the internal standard tridecanoin, and the diacylglycerol and ceramide derivatives, which run together (six peaks); cholesterol esters (three peaks) are then eluted before the triacylglycerols (five peaks). The relative proportions of ceramides and diacylglycerols were determined using a correction factor, based on the area of the first peak in the group which contains palmitoylsphingosine only, and this gave a reasonable approximation to the true result. The method was found to give accurate results over a range of lipid concentrations, and was capable of a high degree of automation in terms of sample injection and computerised data handling [525]. In a direct comparison of data obtained in this manner with that from automated colorimetric (enzymatic) methods, the GC technique gave results that were more accurate and reproducible, i.e. with a within-day standard deviation of 2.3 mg % for cholesterol and 3.5 mg % for triacylglycerols [520]. The colorimetric methods tended to overestimate these lipids because of interfering chromogens in the plasma extracts. Similarly in comparisons of total phospholipids, phosphatidylcholine, lysophosphatidylcholine and sphingomyelin between results obtained by the GC method and by an alternative TLC separation

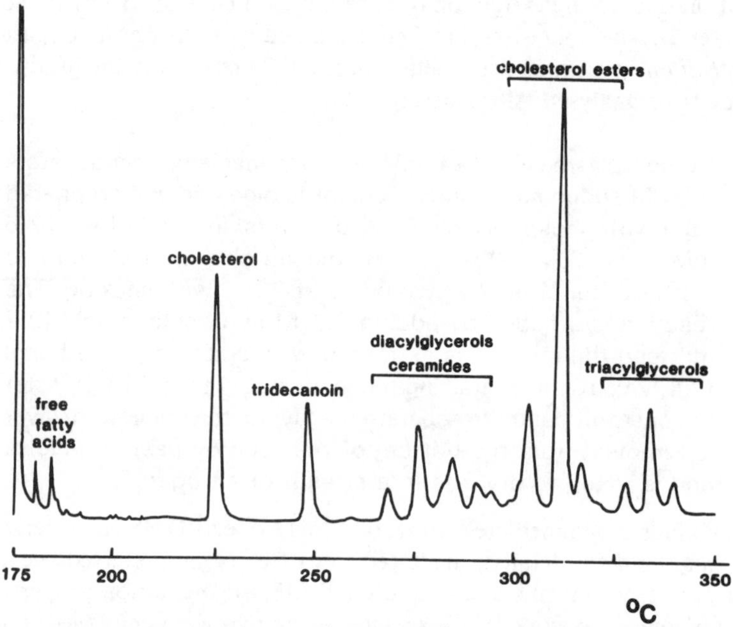

Figure 8.16 GC trace of the total lipid profile, following phospholipase C digestion and preparation of the TMS ether derivatives, of plasma from a normolipemic subject [520]. A stainless-steel column (50 cm x 2 mm i.d.), packed with 3 % OV-1™ on Gas Chrom Q™ (100-120 mesh), was temperature-programmed from 175 to 350°C at 4°C/min with nitrogen as the carrier gas at a flow-rate of 40 ml/min. Tridecanoin is the internal standard. Peak "A" is the TMS ether derivative of palmitoylsphingosine. (Reproduced by kind permission of the authors and of the *Journal of Chromatography*, and redrawn from the original paper).

and phosphorus assay, agreement was excellent for all parameters except for the lysophosphatidylcholine, where variable amounts of endogenous monoacylglycerols in plasma caused errors [521]. The value of this GC procedure in clinical diagnosis has been confirmed by a number of studies of which those cited are representative only [518-520,528,650,651,908].

Apart from the limited resolution of this procedure, it was not possible to assay small amounts of endogenous diacylglycerols and monoacylglycerols in plasma as these were masked by similar lipids derived from phospholipase C digestion of the phospholipids. A comparable automated procedure was therefore developed by the Czechoslovakian group in which the phospholipase C digestion step was omitted and only the simple lipids were assayed [595,596]. Here also the methodology has proved its worth in clinical trials [848,849] and in comparisons with alternative methods [593].

WCOT columns of glass or fused silica have so far only been used to obtain plasma lipid profiles to a limited degree [518,528,544,647]. but greatly enhanced resolution is obtained and they will surely be employed more often in the future. Figure 8.17 illustrates the separation that can be achieved with TMS ether derivatives of plasma lipids (digested with phospholipase C) on a WCOT column of fused silica coated with an apolar stationary phase,

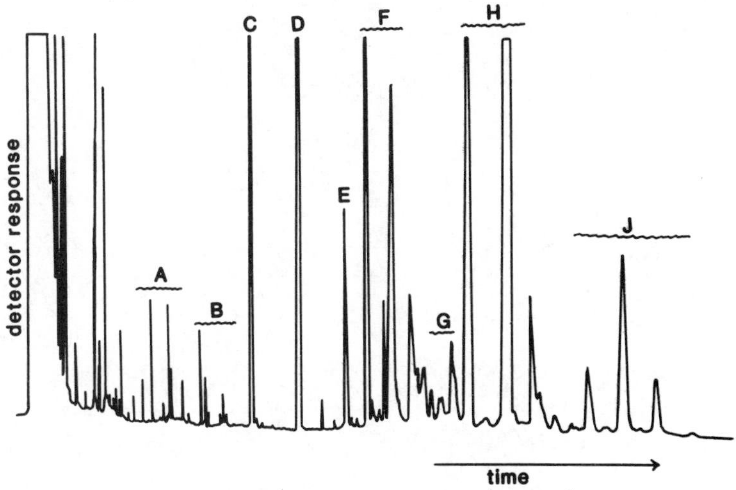

Figure 8.17 High-temperature GC of derivatised plasma lipids on a WCOT column (8 m x 0.3 mm) of fused silica, coated with SE-54™, and temperature-programmed to a maximum of 340°C [647]. Abbreviations: A, TMS ester derivatives of free acids; B, TMS ethers of monoacylglycerols (derived from lysophosphatidylcholine); C, TMS ether of cholesterol; D, tridecanoin (internal standard); E, TMS ether of 16:0-sphingosine (derived from sphingomyelin); F, TMS ethers of diacylglycerols (derived from phosphatidylcholine) and ceramides; G, more TMS ethers of ceramides; H, cholesterol esters; J, triacylglycerols. (Reproduced by kind permission of the authors and of the *Journal of Biochemical and Biophysical Methods*, and redrawn from the original paper).

temperature-programmed to 340°C, and with on-column injection [647]. While there is some separation of lower molecular weight components by degree of unsaturation, the chief feature is improved resolution among lipid classes; many more fractions are obtained and for example, free acid and monoacylglycerol derivatives are clearly seen, better resolution of ceramide and diacylglycerols is apparent and there is no overlap of the peaks for cholesterol esters and triacylglycerols. Quantification was comparable to that with packed columns, and indeed the accuracy of the determination of the ratio of phosphatidylcholine to sphingomyelin, an important parameter in clinical research, was certainly greatly superior. In addition, identification of the peaks by mass spectrometry is greatly facilitated when WCOT columns of fused silica are utilised. Unfortunately, on-column injection cannot yet be automated, but hopefully this will come in time.

Among other applications of methodology of this kind, the products of hydrolysis of triacylglycerols with pancreatic lipase [628] and of re-esterification of glycerol with long-chain fatty acids [218] have been analysed.

CHAPTER 9

ALTERNATIVE OR COMPLEMENTARY METHODS FOR THE ANALYSIS OF MOLECULAR SPECIES OF LIPIDS

A. INTRODUCTION

The complexity of natural lipid samples is such that it is unrealistic to expect to be able to separate all lipid classes into molecular species in a comprehensive way by GC alone. A preliminary separation by some other chromatographic method is often essential if sense is to be made of complicated GC traces. In addition, it must be recognised that there are many circumstances in which GC may not be the best method to resolve particular molecular fractions, although sometimes it may be the only one available to a particular analyst. The three most important alternative or complementary techniques are adsorption, reversed-phase partition and silver ion chromatography. They may be used in conjunction with either TLC or HPLC, although the latter is greatly to be preferred when the required equipment is available. Adsorption and silver ion chromatography can be regarded as valuable adjuncts to GC, since they provide a form of separation that may not always be attainable by the last. HPLC in the reversed-phase mode is probably used more as an alternative to GC, especially with lipids of high molecular weight and with polar complex lipids in intact form. All of these procedures have one distinct advantage over GC in that they operate at ambient temperature, so there is less opportunity for thermal degradation. They can also be utilised for small-scale preparative purposes as well as for analysis *per se*. On the other hand, HPLC has a higher initial capital cost and higher running costs than GC. The author has reviewed these methods elsewhere in some detail [163,168], and the following is intended merely as an introduction to and brief summary of the subject so that the reader will be aware of the broader analytical strategies available. Whatever the analytical approach, it is frequently necessary to have recourse to GC at some stage, if only to analyse the alkyl and acyl constituents of fractions separated by the alternative means.

B. ADSORPTION CHROMATOGRAPHY

Adsorption chromatography is of most value when some of the molecular species in a lipid class contain a polar moiety, such as a free hydroxyl group, or when the fatty acids have a much wider spread of chain-lengths than normal. For example, the triacylglycerols of the Beluga whale contain isovaleric acid, and species with zero, one and two molecules of this acid have been isolated by TLC on silica gel G with a solvent system of hexane-diethyl ether-acetic acid (87:12:1 by volume) as illustrated in Figure 9.1(A);

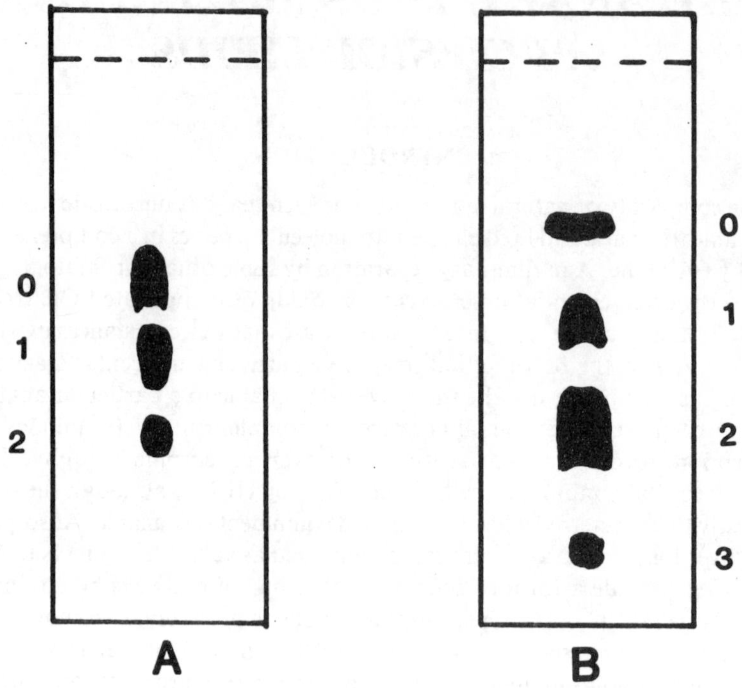

Figure 9.1 Schematic TLC separation of triacylglycerols containing short-chain or polar fatty acids on layers of silica gel. Plate A: triacylglycerols from the Beluga whale; the numbers refer to the isovaleroyl residues in each molecule [555]. Plate B: seed oil containing vernolic acid; the numbers refer to the epoxy fatty acid residues in each molecule [192]. The compositions of the mobile phases are given in the text.

the presence of a short chain acid retards the migration of the molecule relative to that of triacylglycerols with only long-chain fatty acids [555]. Similarly, the triacylglycerols of ruminant milks can be separated by related TLC systems into two rough fractions, one containing three fatty acids of normal chain-length and one containing two long-chain fatty acids and one butyrate or hexanoate residue. It is not uncommon to see individual glycosphingolipids and sphingomyelin migrating as double bands on TLC or HPLC in the adsorption mode, because of the presence of fatty acids of longer than usual chain-length in some molecular species.

Fatty acids with polar substituents also retard the migration of triacylglycerols on TLC adsorbents, and those from seed oils that contain epoxy fatty acids can be separated into simpler species that contain from zero to three epoxide residue per mole. For example, silica gel layers and a solvent system of hexane-diethyl ether (3:1, v/v) were used to isolate such fractions from seed oils (Figure 9.1(B)) [192]. Similar separations have been achieved with triacylglycerols that contain hydroxyl, estolide and other functional groups in the fatty acids.

As HPLC is a relatively new technique in lipid methodology, fewer separations of this kind have been described, but sufficient experience is available to leave little doubt as to the potential of the technique [168]. HPLC in the adsorption mode is also of great value for the separation of peroxidised lipids, including both triacylglycerols and glycerophospholipids, from unoxidised material (c.f. [690,761,910]). As an example, a separation of a monohydroperoxide fraction prepared from autoxidised trilinolein is illustrated in Figure 9.2 [690]. An HPLC column containing silica gel was

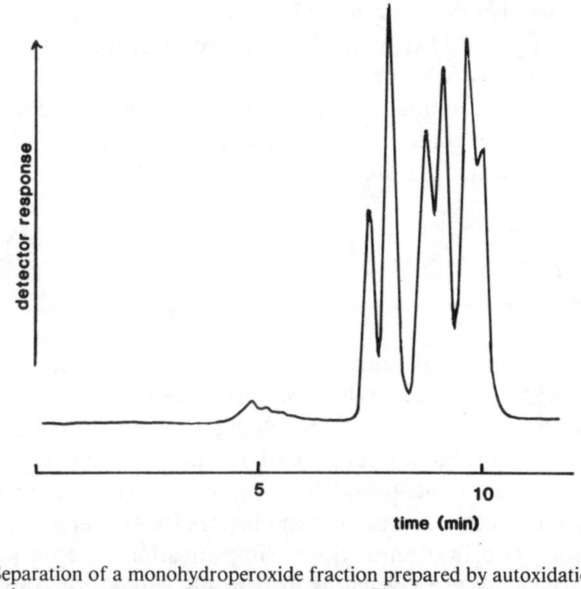

Figure 9.2 Separation of a monohydroperoxide fraction prepared by autoxidation of trilinolein by HPLC [690]. A column (250 x 4 mm) of Zorbax SIL™ was eluted with hexane-isopropanol (99:1, v/v) at a flow-rate of 1 ml/min with spectrophotometric detection at 235 nm. The peaks represent different positional and configurational isomers of the hydroperoxide group and the associated conjugated double bond systems. (Reproduced by kind permission of the authors and of *Agricultural and Biological Chemistry (Japan)*, and redrawn from the original paper).

eluted with 1 % isopropanol in hexane, and the six fractions obtained represent different positional and configurational isomers of the hydroperoxy group within the fatty acids of the triacylglycerols molecules.

C. SILVER ION CHROMATOGRAPHY

1. Triacylglycerols

The principles of silver ion chromatography and its value for the simplification of complex fatty acid mixtures have already been discussed in Chapter 6. As the nature of the separation involves resolution according to degree of unsaturation of the combined fatty acid residues, the technique is perhaps of even greater value when applied to the fractionation of molecular species of intact lipids. High-temperature GC is at its best for separation simply by chain-length, in spite of the recent developments discussed in the previous chapter. When fractions obtained by silver ion chromatography are subsequently resolved further by GC, there is an enormous gain in the amount of information acquired. For example, eight or more fractions can be obtained by silver ion chromatography from many natural triacylglycerols; if these each give four peaks when separated by high-temperature GC, 32 different molecular species are thereby seen.

Because of the historical development of the technique most research has been carried out using TLC with silver nitrate incorporated into the silica gel layer (10 to 15 %, w/w). For example, the normal range of triacylglycerols found in seed oils or animal fats contains zero to three double bonds per fatty acid, so they can contain species with up to nine double bonds in each molecule. Components migrate in the order:

SSS > SSM > SMM > SSD > MMM > MMD > SMD > MMD > SDD > SST > MDD > SMT > MMT > DDD > SDT > MDT > DDT > STT > MTT > DTT > TTT

- where S, M, D and T denote saturated, mono-, di- and trienoic acids respectively (they do not indicate the positions of the fatty acids on the glycerol moiety), although there may be some changes in this order depending on the nature of the solvent mixtures used for development [330,775]. It is noteworthy that one linoleate moiety is more strongly retained than two oleates, and that species containing one linolenate are retarded more than those containing two linoleates. Some simplification of even more highly unsaturated triacylglycerols, such as in fish oils, has also been attained by silver nitrate TLC [121], but HPLC systems give much better results (see below). The solvent systems generally employed for development consist of hexane-diethyl ether, toluene-diethyl ether or chloroform-methanol mixtures. (Note that when chloroform is utilised in a mobile phase in this way, it is necessary to wash it with water to remove any ethanol present as stabiliser, then to dry it, to ensure that the composition of the mobile phase is that required [163]).

As all the fractions listed above cannot be separated on one plate, it is common practice to separate the least polar fractions first with hexane-diethyl ether (80:20,v/v) or chloroform-methanol (197:3, v/v) as the mobile phase, as illustrated in Figure 9.3, plate A, and then to re-chromatograph the

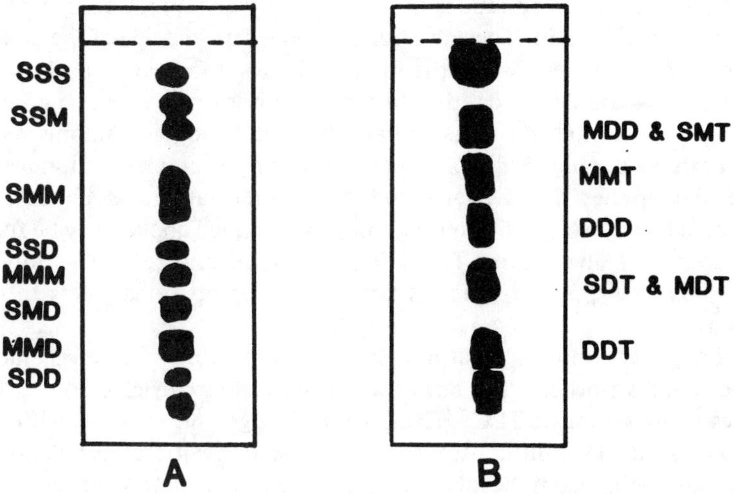

Figure 9.3 Schematic TLC separation of maize oil triacylglycerols on layers of silica gel G, impregnated with 10 % silver nitrate. Mobile phases: Plate A, chloroform-methanol (99:1, v/v); Plate B, chloroform-methanol (96:4, v/v). Abbreviations: S, M, D and T denote saturated, mono-, di- and trienoic fatty acyl residues respectively, esterified to glycerol.

remaining fractions with more polar solvents such as diethyl ether alone or chloroform-methanol (96:4, v/v) as illustrated in the same figure but plate B. Bands are detected under UV light after spraying with 2′,7′-dichlorofluorescein solution, the components are recovered (see Chapter 6, B.2) and they are identified and determined by GC of the fatty acid constituents with an added internal standard after trans-methylation (Chapter 4). As an alternative, tridecanoin can be added as the internal standard for the analysis of intact fractions by high-temperature GC. For analytical purposes only, it is possible to estimate fractions by densitometry following bromination and charring [155]. Approximately 10 mg of triacylglycerols can be fractionated on a 20 x 20 cm plate (0.5 mm thick layer), and excellent separations of large numbers of components have been obtained with 20 x 40 cm plates.

In addition, some separation of isomeric compounds is possible. For example, triacylglycerols of the type S_2M, in which the monoenoic component is in position 2, are separable with care from the related compounds in which the monoenoic component is in position 1(3). Presumably, the presence of long-chain fatty acids on either side of the monoenoic acid weaken the *pi*-complex with the silver ion, permitting the component with the monoenoic fatty acid in position 2 to migrate ahead of isomers in which this residue is in position 1, since separation of this kind cannot be obtained with diacylglycerol acetates where one side of the molecule is comparatively open. When the fatty acid constituents of triacylglycerols contain *trans*-double bonds, the elution pattern is much more complicated as components with *trans*-acids migrate ahead of analogous compounds with fatty acids containing *cis*-double bonds, producing a complex elution pattern. Molecular species of tracylglycerols with polar fatty acid constituents, separated by adsorption chromatography as described above, may be further fractionated by silver nitrate TLC [192]. One disadvantage of silver ion TLC is that it is messy; purple-stained fingers are an occupational hazard for the analyst.

Although column chromatography (low pressure) with silver nitrate-impregnated adsorbents has been used to separate triacylglycerols, it lacks the resolving power of TLC. HPLC with silica gel impregnated with silver salts has so far been little used for the purpose because of problems with silver ions eluting from the adsorbent, although some impressive separations have been recorded (reviewed elsewhere [168]). On the other hand, the column prepared by binding silver salts to an adsorbent containing chemically-bonded sulphonic acid residues, and discussed already in Chapter 6, has shown considerable promise for triacylglycerol separations [169]. It is necessary to employ mobile phases that do not contain alcohols, otherwise residual free sulphonic acid groups on the stationary phase catalyse transesterification of the solute. As an example, a separation of the triacylglycerols palm oil was illustrated in the original paper, in which a linear gradient of acetone into 1,2-dichloroethane was employed as the mobile phase with mass (or light-scattering) detection, and fractions with up to three double bonds were resolved, i.e. up to SMD. More recently, the author (W.W. Christie, *J. Chromatogr.*, **454**, 273-284 (1988)) has shown that species of a higher degree of unsaturation may be eluted by incorporating acetonitrile into the mobile phase at a later stage of the gradient. With the same palm oil sample, species up to SDD can then be analysed, as illustrated in Figure 9.4, and indeed trilinolenin elutes as a sharp single peak from linseed oil if the gradient is extended. The order of elution differs slightly from that obtained with the TLC system, one linoleate residue having an effect on retention time equivalent to just over two monoenes while one α-linolenate is equivalent to two linoleates, i.e.

SSS > SSM > SMM > SSD > MMM > SMD > MMD > SDD = SST > SMT = MDD > MMT > SDT = DDD > MDT > STT > = DDT > MTT > DTT > TTT.

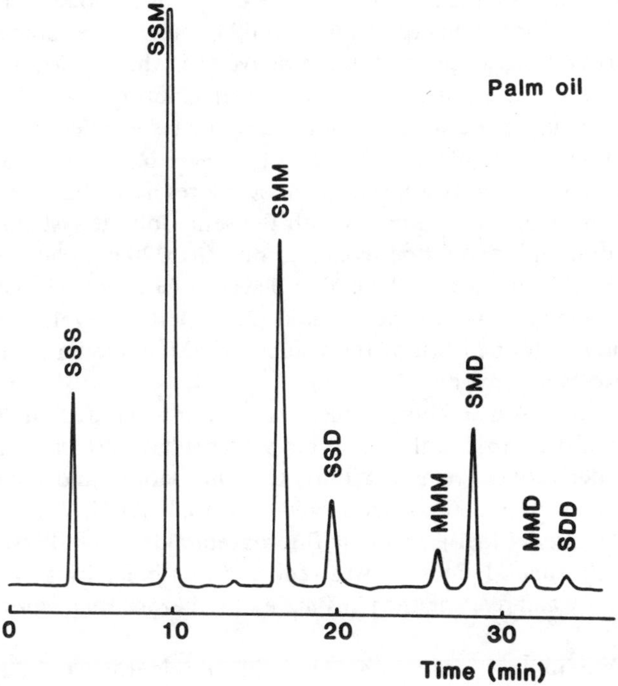

Figure 9.4 Silver ion HPLC separation of molecular species of triacylglycerols from palm oil on a column of Nucleosil 5SA™ (250 x 4.6 mm) loaded with silver ions. The mobile phase was a linear gradient of 1,2-dichloroethane-dichoromethane (1:1, v/v) to this solvent with 50 % acetone over 15 min, when acetonitrile was introduced to give a final mixture of acetone-acetonitrile (9:1, v/v) after a further 30 min, at a flow-rate of 0.75 ml/min and with mass detection (see W.W. Christie, *J. Chromatogr.*, **454**, 273-284 (1988)). Abbreviations - as in the legend to Figure 9.3.

A major advantage of using HPLC as opposed to TLC in this way is that silver ions are not eluted with the mobile phase, so that fractions collected for analysis are particularly clean. In addition, the column is long lived. It will undoubtedly be used more often in future.

2. Phospholipids and Diacylglycerol Derivatives

It is not easy to subject intact phospholipids to silver ion TLC, because the polar head group masks the comparatively small changes in polarity produced by the formation of *pi*-complexes between silver ions and the double bonds of the unsaturated fatty acids. Nonetheless, Arvidson [60,61] has achieved some valuable separations of phosphatidylcholines by using highly-active TLC plates. TLC layers (0.35 mm thick) were prepared with silica gel H (without a binder) and silver nitrate in the proportions 10:3 (w/w), and were air-dried at room temperature (in the dark) for 24 hours initially, then either at 175°C for 5 hours or at 180°C for 24 hours; plates prepared under the latter conditions were much more active than those dried at the lower temperature. By utilising a mobile phase of chloroform-methanol-water (60:35:4 by volume) and with the more active plates, species with one to six double bonds per molecule were separable from each other. Only the saturated and monoenoic fractions were not completely resolved, but they could be separated on the less active plates with the same solvent system. Up to 20 mg of phospholipids could be separated on a 20 x 20 cm plate, and bands were recovered from the adsorbent after detection (see above) by elution with chloroform-methanol-acetic acid-water (50:39:1:10 by volume), washed subsequently with one third of the volume of 4M ammonia to remove the dye and excess silver ions.

Native phosphatidylethanolamines have been separated under similar conditions to these [61], although other workers have preferred to prepare non-polar derivatives (retaining both the phosphorus and ethanolamine moieties), such as the *N*-dinitrophenyl-*O*-methyl- [765], *N*-acetyl- or *N*-benzoyl-*O*-methyl- [890], and trifluoroacetamide [1009] compounds. Similarly, phosphatidylinositol was rendered non-polar by acetylating the inositol moiety and reacting the phosphate with diazomethane, prior to silver ion TLC [567].

Chromatography of intact phospholipids in this way can be invaluable in biochemical studies, but the number of fractions obtainable is limited and subsequent analysis by high-temperature GC is not possible. If a more detailed analysis is required, it is therefore necessary to hydrolyse phospholipids with phospholipase C, as described in the previous Chapter, for analysis as the diacylglycerol acetate or BDMS ether derivatives. The former have been used most for separation by silver ion TLC, and they migrate in the order -

SS > SM > MM > SD > MD > DD > ST > MT > STe > MTe > DTe > SP > SH

- where S, M, D, T, Te, P and H denote saturated, mono-, di-, tri-, tetra-, penta- and hexaenoic fatty acid residues respectively (they do not indicate the relative positions of the fatty acids on glycerol as positional isomers are eluted together) [512,739]. Figure 9.5 illustrates the separation of diacylglycerol acetates prepared from the phosphatidylcholines of pig liver on layers of silica gel G impregnated with 10 % silver nitrate, developed in either chloroform-methanol (99:1, v/v) or hexane-diethyl ether (95:15, v/v).

When the sample contains high proportions of polyunsaturated fatty acids, it may be necessary to repeat the the elution step with a solvent system

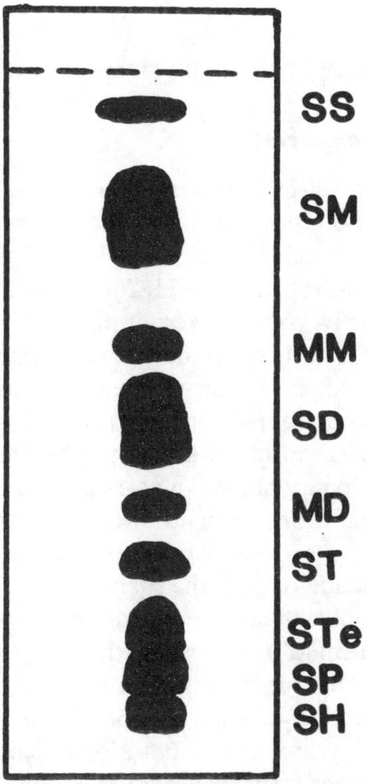

Figure 9.5 Schematic TLC separation of diacylglycerol acetates prepared from the phosphatidylcholine of pig liver, on layers of silica gel G impregnated with 10 % (w/w) silver nitrate. Mobile phase: chloroform-methanol (99:1, v/v). Abbreviations: S, M, D, T, Te, P and H denote saturated, mono-, di-, tri-, tetra-, penta- and hexaenoic fatty acyl residues respectively esterified to glycerol.

containing an increased concentration of the more polar component. Fractions are eluted from the adsorbent for further analysis as described above for triacylglycerols. Diacylglycerol acetates with up to twelve double bonds in the combined fatty acid moieties have been separated on similar plates with chloroform-methanol-water (65:25:4 by volume) as the solvent system [766]. With care, it is possible to separate molecular species of positional isomers of polyunsaturated fatty acids, and fractions containing 18:3(n-6) migrate ahead of those containing 18:3(n-3), while those containing 20:4(n-6) migrate ahead of 20:4(n-3) [161]. Alkylacyl and alk-1-enylacyl derivatives can be analysed in a similar manner [762].

Although HPLC in the silver ion mode does not appear to have been used with diacylglycerol acetates, there is no reason why the column described above and utilised in triacylglycerol separations should not be employed for the purpose with very little modification to the elution scheme.

D. REVERSED-PHASE PARTITION CHROMATOGRAPHY

1. Some Practical Considerations

While many practical TLC systems have been devised that utilise the principle of reversed-phase partition in the separation process, they have generally proved inconvenient and of doubtful reproducibility in use. The newer pre-coated high-performance TLC plates of this kind are much better but they are very costly. When reversed-phase chromatography is used in conjunction with HPLC, there are no such drawbacks and this approach is now generally favoured. It has been reviewed in detail by the author [168] and in relation to phospholipids specifically by others [661,696,779]. The technique is a form of partition chromatography, the term "reversed-phase" implying that the stationary phase is a non-polar liquid and the mobile phase is more polar. Separations depend on differences in the equilibrium distribution coefficients of solutes between the two phases, and in the case of lipids, this is a function of both the chain-lengths of the fatty acid residues and the number of double bonds, one *cis*-double bond reducing the retention volume by the equivalent of about two methylene groups. The principles and applications of the technique to analyses of simple fatty acids are described briefly in Chapter 6 above. When it is applied to molecular species of intact lipids, such as triacylglycerols, the separation is dependent on the combined chain-lengths of the three fatty acyl residues and the total number of double bonds.

One of the major problems facing lipid analysts who wish to make use of HPLC is the choice of a suitable detection system, since lipids in general lack chromophores which can be detected spectrophotometrically. This problem has been discussed in relation to the separation of specific lipid classes in Chapter 2. It is frequently the availability of a particular detection system that is the principal factor in determining the strategy to be adopted for particular separations of molecular species. There is little doubt that detectors operating on the transport-flame ionisation principle are ideal for the purpose, although the commercial instruments are relatively costly and are not yet in wide-spread use. In essence, there are no restrictions on the range of solvents that can be employed in the mobile phase, and the response is quantitative. Similarly, mass detection permits most solvent combinations to be employed, but quantification is more problematical. UV detection at wavelengths around 205 nm, where isolated double bonds absorb, has been used more often, but the range of solvents transparent in the required range is limited and the detector response is highly dependent on the nature of each lipid fraction.

Some better results in triacylglycerol separations have been obtained by detection at 220 nm by a careful choice of mobile phase (see below). Good linearity of response is possible with differential refractometry with the better quality instruments, but it is not possible to use gradient elution. Of course, mass spectrometry coupled to HPLC is a splendid research tool, but it is available to a relatively few analysts only (reviewed elsewhere [168,515]). The discussion that follows therefore treats the subject in terms of specific detection systems.

2. Triacylglycerols

The relative retention time of a given component has been defined in terms of an "equivalent carbon number" (ECN) or "partition number" value, defined as the actual number of carbon atoms in the aliphatic residues (CN) less twice the number of double bonds (n) per molecule (the carbons of the glycerol moiety are not counted for this purpose), i.e.

$$\text{ECN} = \text{CN} - 2n$$

Two components having the same ECN value are said to be "critical pairs". For example, triacylglycerol species containing the fatty acid combinations 16:0-16:0-16:0, 16:0-16:0-18:1, 16:0-18:1-18:1 and 18:1-18:1-18:1 have the same ECN value and tend to elute close together. (The positions of the fatty acids within the triacylglycerol molecules again have no effect on the nature of the separation). The ECN concept was useful in the early days of the technique, when the resolving power was relatively limited. On the other hand, the formula is now only of utility as a rough rule of thumb, by way of a guide to what may elute in a given area of a chromatogram, since the greatly increased resolving power of modern HPLC phases means that the factor for each double bond has to be defined much more precisely. Also, this factor can no longer be treated as a constant, as a second double bond in a molecule has a slightly different effect from the first. Accordingly, more complex formulae are necessary to define the order of elution of triacylglycerols from modern reversed-phase columns, which in essence means from octadecylsilyl (ODS) stationary phases, as these have been used almost exclusively for the purpose.

Under steady state conditions, i.e. with a given column and isocratic elution at a constant temperature and flow-rate, it would be expected from theoretical considerations that there should be a rectilinear relationship between the logarithm of the capacity factor (k), or retention time or volume, and the carbon number of each member of a homologous series of triacylglycerols containing saturated fatty acids only. This was first shown to be true in practice by Plattner *et al.* [720], and it also holds for homologous series of triacylglycerols containing particular unsaturated fatty acids.

From such evidence, it was suggested that triacylglycerol species might be identified from a *theoretical carbon number* (TCN), defined as

$$TCN = ECN - \sum U_i$$

- where U_i is a factor determined experimentally for different fatty acids (zero for saturated fatty acids, and roughly 0.2 for elaidic, 0.6 to 0.65 for oleic, and 0.7 to 0.8 for linoleic acid residues) [242]. Thus, the TCN value for a triacylglycerol containing 18:1-18:1-18:1 (triolein) is calculated as -

$$TCN = (3 \times 16) - (3 \times 0.6) = 46.2$$

- while for 18:1-18:1-16:0, 18:1-16:0-16:0 and 16:0-16:0-16:0, which have the same ECN value, the TCN values are 46.8, 47.4 and 48.0 respectively. The values of U_i will vary with the elution conditions and have to be determined independently by each analyst for his own system. Although alternative identification methods have been suggested by others, the concept of the TCN has been found to be useful by many workers.

Virtually all the work on the reversed-phase separation of triacylglycerols has been carried out with ODS phases. Most commercial brands of ODS phase have been used by one analyst or another for the purpose, and while it is evident from direct comparisons that some are better than others, the reasons for this are not clear.

Perhaps the single most important factor in the separation of triacylglycerols is the choice of the mobile phase. Early in the development of the methodology, it became apparent that solvent combinations based on acetonitrile gave much better resolution than any others tested. These are now used almost universally, although propionitrile is preferred by some analysts. In order to obtain the optimum separations of triacylglycerols, it is necessary to add some other solvent to the acetonitrile to increase the solubility of the solute, to change the polarity of the mixture and to modify the selectivity. It is well established that the relative proportion of acetonitrile to the modifier solvent has a marked effect on the elution time of a given triacylglycerol species, i.e. the lower the polarity of the mobile phase, the lower the retention volume. No concensus has emerged as to which combination is best, partly because there appears to be no objective criterion that can be used to assess relative merits, and partly because the nature of the columns and other equipment used in different laboratories may impose constraints.

The separations that are described below are examples selected from the wide body of literature reviewed elsewhere [168] to illustrate the use of particular detection systems. Acetonitrile-acetone mixtures have been used more often than any other solvent combination as the mobile phase in the reversed-phase separation of triacylglycerol species, in an isocratic manner with refractive index detection and in gradients with the mass detector. It has certainly permitted some fine separations, especially with seed oils, and an application to palm oil is illustrated in Figure 9.6 [212]. Here, the mobile phase was acetone-acetonitrile (62.5:37.5 by volume) at 30°C. Five main groups of peaks, distinguishable by their ECN values, were recognized. Group

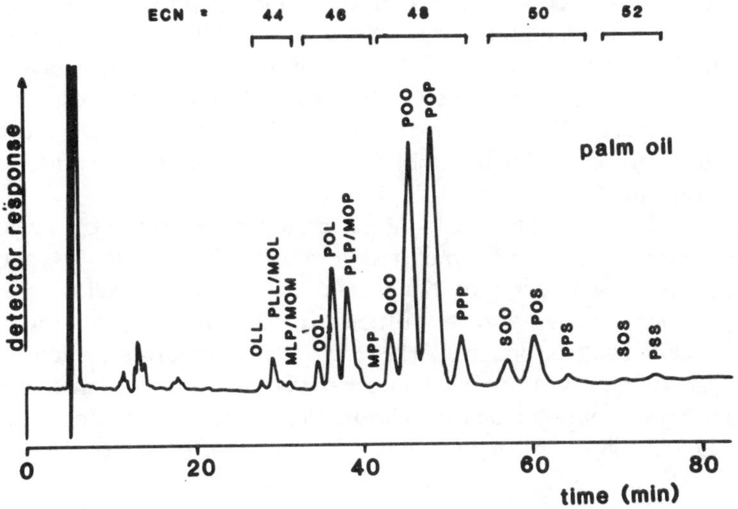

Figure 9.6 HPLC (reversed-phase) separation of triacylglycerols from palm oil [212]. Two ODS columns in series were maintained at 30°C, with acetone-acetonitrile (62.5:37.5, v/v) as the mobile phase at a flow-rate of 1.1 ml/min, and with refractive index detection. The abbreviations refer to fatty acyl residues: M, 14:0; P, 16:0; S, 18:0; A, 20:0; O, 18:1; L, 18:2. The ECN values are listed above the appropriate groups of peaks. (Reproduced by kind permission of the author, and of *Revue Francaise des Corps Gras*, and redrawn from the original paper).

1 (ECN = 44) consisted of the combinations 18:1-18:2-18:2, 16:0-18:2-18:2 with 14:0-18:1-18:2, and 14:0-18:2-16:0 with 14:0-18:1-14:0. Group 2 (ECN = 46) consisted of 18:1-18:1-18:2, 16:0-18:1-18:2, 16:0-18:2-16:0 with 14:0-18:1-16:0, and 14:0-16:0-16:0. Group 3 (ECN = 48) is a particularly important one in confectionery fats and comprised predominantly 18:1-18:1-18:1, 16:0-18:1-18:1, 16:0-18:1-16:0 and 16:0-16:0-16:0. In group 4 (ECN = 50), only three main components could be recognized, i.e. 18:0-18:1-18:1, 16:0-18:1-18:0, and 16:0-16:0-18:0, while the fifth group (ECN = 52) contained 18:0-18:1-18:0 and 16:0-18:0-18:0. Although gradient elution is not possible with differential refractometry, temperature-programming can be utilised with a detector of sufficient quality, in order to speed up separations and improve resolution with complex mixtures, and some excellent analyses of triacylglycerols from seed oils and from milk fat have been obtained in this way [264,265].

The main difficulty associated with acetonitrile-acetone as the mobile phase, other than its unsuitability with UV detection, is that trisaturated species of high molecular weight, i.e. above C_{48}, tend to be insoluble and can crystallize out in the column. Other solvent combinations have therefore been sought. Dichloromethane-acetonitrile has been used to good effect as a mobile phase, especially by Privett's group, with a transport-flame ionization detector of their own design [711,712]. The eluent was tested first with model mixtures, and so good was the resolution that it was possible to distinguish molecular species containing petroselinoyl (i.e. with the *cis*-double bond in position 6) from those with oleoyl (position 9) residues. It also gave good results with natural samples such as vegetable oils. Only one application of the commercial Tracor™ transport-flame ionisation detector to triacylglycerols has so far been published, and it confirms the potential of the instrument [676].

By a careful choice of solvents, it has been possible to use UV detection in the reversed-phase separation of triacylglycerols. Shukla *et al.* [844] showed that by using a wavelength of 220 nm, i.e. away from the region in which isolated double bonds absorb, sufficiently sensitive detection and good quantification could be obtained. They employed tetrahydrofuran-acetonitrile (73:27 by volume) as the mobile phase with two columns of a 3 micron ODS phase in series, and obtained the chromatogram of cocoa butter shown in Figure 9.7. Each of the main groups with the same ECN value were well

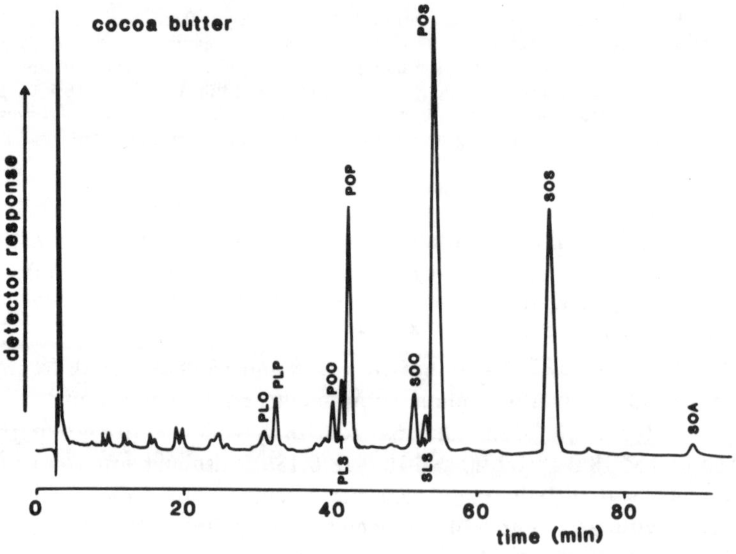

Figure 9.7 Separation of triacylglycerols from cocoa butter by reversed-phase HPLC using two columns of a 3 micron ODS phase in series [844]. The mobile phase was acetonitrile-tetrahydrofuran (73:27, v/v) at a flow-rate of 1 ml/min, with spectrophotometric detection at 220 nm. See Fig. 9.6 for a list of abbreviations. (Reproduced by kind permission of the authors and of *Fette Seifen Anstrichmittel*, and redrawn from the original paper).

resolved from the others, as were molecular species within each group. Indeed the resolution obtained is at least as good as any published to date. It may, however, be necessary to purify the tetrahydrofuran component of the mobile phase immediately prior to use by distillation from lithium aluminium hydride to remove UV-absorbing oxidised material.

Several research groups have found propionitrile on its own to be an excellent mobile phase for the elution of triacylglycerol molecular species from reversed-phase columns, and indeed some would claim that it may be the best available, although it is costly and highly toxic. Schulte [817] first made use of it for the fractionation of cocoa butter, but it has been since been employed for many different samples and with most detection systems. In Kuksis' laboratory, a gradient of 30 to 90% propionitrile in acetonitrile has been used to effect separations of triacylglycerols from vegetable oils and animal fats [514,516,522-524]. Chemical ionisation MS, with the solvent as the ionising agent, was utilized to detect and identify components in this research (and incidentally to correct some misidentifications made in earlier work from other laboratories).

3. Phospholipids and Diacylglycerol Derivatives

Methods for the separation of intact glycerophospholipids by means of HPLC in the reversed-phase mode have evolved rapidly in recent years. Although it has always been considered that non-polar derivatives of phospholipids are capable of being resolved more cleanly than are the intact compounds, the difference is now much less marked than it was formerly. With reversed-phase HPLC of molecular species of phospholipids, the nature of the separation is similar to that discussed above for triacylglycerols, in that it is dependent on the combined chain-lengths and degree of unsaturation of the fatty acyl or alkyl chains. Similarly, the detector limitations are the same as for triacylglycerol analyses. It has become apparent that some ODS phases are much better than others for phospholipid separations, although the reason for this is not known, and a number of analysts have obtained excellent resolution of phospholipids with Ultrasphere™ ODS.

A particularly important paper on the subject of separations of phospholipid molecular species was published by Patton, Fasulo and Robins in 1982 [694]. They fractionated molecular species from different phospholipid classes on a column (4.6 x 250 mm) of Ultrasphere™ ODS, with methanol-water-acetonitrile (90.5:7:2.5 by volume) containing 20 mM choline chloride as the mobile phase. UV spectrophotometric detection at 205 nm was used to locate the fractions, and these were collected for phosphorus assay and fatty acid analysis. In Figure 9.8, the separation obtained for phosphatidylethanolamine from rat liver is illustrated. (Note that with UV detection at 205 nm, the response is highly dependent on the degree of unsaturation, and peak heights do not immediately reflect the

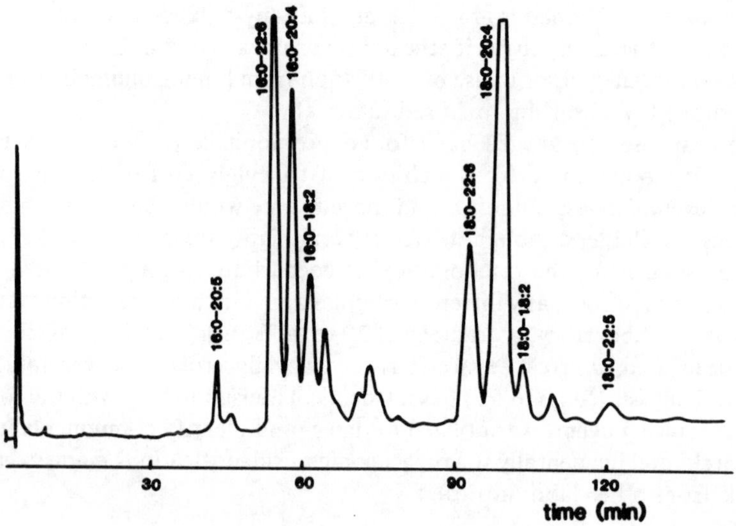

Figure 9.8 HPLC separation (reversed-phase) of molecular species of phosphatidylethanolamine from rat liver [694]. An ODS column was used with methanol-acetonitrile-water (90.5:7:2.5 by volume) containing 20 mM choline chloride as the mobile phase at a flow-rate of 2 ml/min, and with UV detection at 205 nm. Only a few of the peaks are identified here for illustrative purposes. (Reproduced by kind permission of the authors and of the *Journal of Lipid Research*, and redrawn from the original paper).

relative abundances of the components). The separation can be considered as bimodal, with in essence those molecular species containing a 16:0 fatty acyl group eluting before those containing 18:0. As with triacylglycerols, the position of the acyl group within the molecule has no effect on separation in the reversed-phase mode, although the saturated components are known to be located predominantly in position *sn*-1 in this particular sample. All of the major components are clearly resolved, and a few only of the minor fractions contain two distinct species. The problems of identification of molecular species are somewhat less than with triacylglycerols, since only two fatty acids are present.

Phosphatidylcholine and phosphatidylinositol from rat liver were separated into molecular species under exactly the same conditions. Indeed, fractions identical in composition were obtained, although the relative proportions were rather different because of variations in fatty acid profiles, as expected. After modifying the mobile phase to 30 mM choline chloride in methanol-25 mM KH_2PO_4-acetonitrile-acetic acid (90.5:7:2.5:0.8 by volume), molecular species of phosphatidylserine were resolved. Closely-related elution schemes have been used by many other workers in separating molecular species of phospholipids.

Molecular species of phosphatidylglycerol from plant chloroplasts were fractionated by HPLC on a column (4.6 x 250 mm) containing Rainin Microsorb™ reversed-phase packing material, and with 1-ethylpropylamine-acetic acid-methanol-acetonitrile (0.3:0.5:34.7:64.5 by volume) as the mobile phase at a flow-rate of 0.8 ml/min [674,860]. Of particular interest here was what appears to have been the first published application of the Tracor transport-flame ionization detector to lipid analysis. Excellent base-line stability was recorded in spite of the presence of ionic species in the mobile phase, and minor components present at a level of as little as 1.2 nmol could be determined. It was confirmed that direct quantification of components by integration of peaks from the detector gave results which were comparable to those obtained by alternative methods. A stream splitter ahead of the detector enabled fractions to be collected for determination of radioactivity or for GLC analysis of the fatty acid constituents. In the sample studied, species eluted in the order 16:1*t*-18:3, 16:1*t*-18:2 and 16:0-18:2.

The alternative approach is to prepare diacylglycerol derivatives by phospholipase C digestion as described in Section C.2 above and in the previous Chapter. Some improvement in resolution may be obtained, with no requirement for inorganic ions in the mobile phase, while UV-absorbing or fluorescent derivatives of diacylglycerols may be employed, simplifying detection and quantification. In addition, complementary chromatographic techniques can more easily be brought to bear for the further resolution of fractions.

This is probably the only appropriate procedure for the separation of diradylglycerols. For example, the alkenylacyl-, alkylacyl- and diacylglycerol acetates, isolated by adsorption HPLC as decribed in the previous Chapter, were each separated into molecular species by HPLC in the reversed-phase mode [660]. For the alkenylacyl and alkylacyl derivatives, a column (4.6 x 250 mm) containing Zorbax™ ODS and maintained at 33°C was eluted with acetonitrile-isopropanol-methyl-*t*-butyl ether-water (63:28:7:2 by volume) at 0.5 ml/min; for the diacyl form, the same solvents were used but in the proportions 72:18:8:2 respectively, as illustrated in Figure 9.9. UV detection at 205 nm was again employed. Each of the diradyl forms was separated into as many as 22 fractions, and the recorder traces resembled those published by others for intact phospholipids, although the resolution was perhaps slightly better here with the acetate derivatives. This sample also contains a higher proportion of polyunsaturated fatty acids than did that in the previous figure (9.8). Fractions were collected once more for identification and determination of the alkyl and acyl moieties by GLC methods. The resolution was still far from complete, and the most abundant peak for example contained 18:0-22:6(n-3), 18:1-20:3(n-6) and 18:1-18:2(n-6), three fractions that would readily be resolvable by high-temperature GC.

Because disaturated molecular species of phosphatidylcholine, such as those predominating in lung, could not easily be detected and quantifed

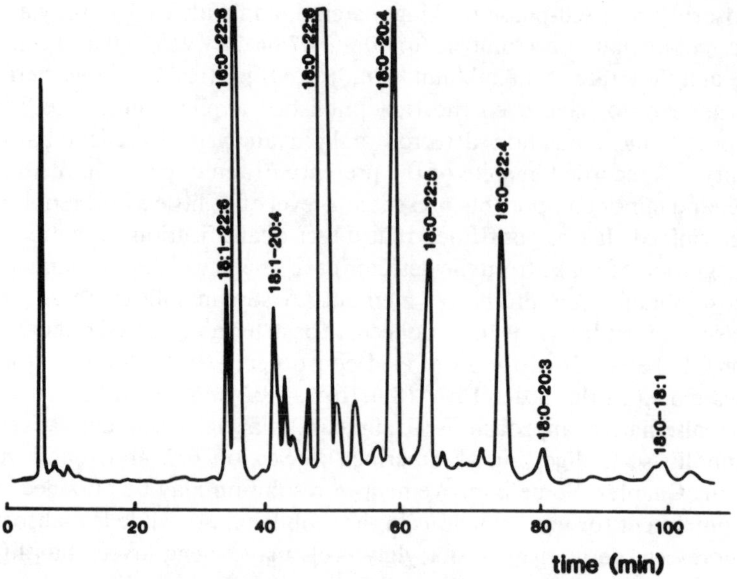

Figure 9.9 Reversed-phase HPLC separation of diacylglycerol acetates prepared from the phosphatidylethanolamine of bovine brain on an ODS column, with acetonitrile-isopropanol-methyl-*t*-butyl ether-water (72:18:8:2 by volume) at a flow-rate of 0.5 ml/min [660]. Detection was at 205 nm and the column was maintained at 33°C. Only a few of the peaks are identified here for illustrative purposes. (Reproduced by kind permission of the authors, and of the *Journal of Lipid Research*, and redrawn from the original paper.

spectrophotometrically at 205 nm, the use of refractive index detection for the purpose was explored and was found to give satisfactory results [413]. BDMS ether derivatives can be analysed in the same way. Methodology of this kind could be used to obtain fractions for further resolution and identification by GC-MS, although there is little to suggest that the two techniques are being used in concert at the moment.

By converting diacylglycerols prepared from phospholipids to UV-absorbing derivatives, it has proved possible to use the high sensitivity and specificity of spectrophotometric detection in the analysis of molecular species. Dinitrobenzoate derivatives of diacylglycerols were employed to obtain some impressive separations of molecular species of phosphatidylcholine from rat tissues, for example [479,905]. With a column (4.6 x 250 mm) of Ultrasphere™ ODS and elution with acetonitrile-isopropanol (4:1 by volume) as the mobile phase, 29 distinct fractions were detected, identified and quantified. 12:0-12:0 or 18:0-18:0 species could be added as an internal standard if required. When methanol-isopropanol (19:1 by volume) was the mobile phase, only 17 fractions were seen but some not separated by the previous system were in fact resolved. Thus by collecting fractions containing more than one component from the first eluent, a more comprehensive analysis could be obtained by re-running with the second eluent. In this way, as many as 36 distinct molecular species were obtained from each lipid class. This method used in sequence with silver ion chromatography might prove to be an even more profitable approach. As methods for the separation of the various diradyl forms of the benzoate derivatives of diacylglycerols involving HPLC in the adsorption mode have recently been described [104,260], a fairly comprehensive HPLC methodology is now potentially available.

CHAPTER 10

SOME MISCELLANEOUS SEPARATIONS OF LIPIDS BY GAS CHROMATOGRAPHY

A. FATTY ALCOHOLS

1. Isolation and GC Analysis

Fatty primary alcohols with similar aliphatic chains to fatty acids tend to occur in the free state in tissues at low concentrations only, but they may be of some metabolic importance as precursors of alkyl lipids, as plant growth regulators and as insect pheromones, for example. In addition, they are found in esterified form in wax esters, which are substantial components of many natural materials. Secondary alcohols may be present in plant surface waxes, together with aliphatic diols which are common constituents of skin lipids. In mammalian tissues, the primary alcohols are saturated or monoenoic, but never di- or polyunsaturated; in wax esters of marine origin, the alcohol constituents are often closely related in structure to the fatty acids from which they may derive biosynthetically. The occurrence, chemistry and metabolism of fatty alcohols [577] and their chromatographic properties [577,579] have been reviewed.

Alcohols may be released from the esterified form by any of the hydrolytic or transesterification procedures described in Chapter 4. If a pure wax ester fraction is hydrolysed, the alcohols are obtained simply by solvent extraction of the alkaline solution. On the other hand, when other lipids are present, it is advisable to isolate them as a class by adsorption chromatography. TLC on layers of silica gel G with the elution system described for simple lipid separations in Chapter 2, i.e. with hexane-diethyl ether-formic acid (80:20:2 by volume) as the mobile phase, is usually used. With such a system, any secondary alcohols migrate ahead of primary alcohols, which in turn are slightly less polar than cholesterol; diols migrate just in front of monoacylglycerols. If cholesterol is present in an extract, it may be necessary to re-run the plate in the same direction to obtain additional resolution and ensure that primary alcohols and cholesterol are fully separated. Procedures of this kind were utilised to isolate trace levels of fatty alcohols from animal tissues, for example [108,662,904]. When wax esters are transesterified, the methyl esters and free alcohols can be separated on a mini-column of Florisil™ or silica gel; hexane-diethyl ether (9:1, v/v) elutes methyl esters, while diethyl ether is utilised to recover the free alcohols. Methods have been described for the simultaneous analysis of alcohol and fatty acid derivatives from specific samples by GC, without such a preliminary separation (see below), but it remains to be determined whether they have wider applicability.

As fatty alcohols are comparable in structure and molecular weight to fatty acid methyl esters, they are usually subjected to GC on the same stationary phases and under near-identical conditions. It is certainly possible to separate alcohols in the free form by GC, especially on modern WCOT columns of fused silica, but sharper peaks are obtained if less polar derivatives such as the acetates, trifluoroacetates or TMS ethers are prepared. Suitable preparation procedures are described in detail in Chapter 4. Jamieson and Reid [438] studied the relative retention times of many different saturated and unsaturated fatty alcohols in the free form and as the acetates on packed GC columns containing polar polyester phases, and concluded that very similar separation factors applied as with the equivalent fatty acid methyl esters. The order of elution was - methyl ester $<$ alcohol acetate $<$ free alcohol. A TMS ether derivative would be expected to have a lower retention time than an acetate, but the separation factors for double bonds in the alkyl chain in this instance were found to be lower than with the acetates and resolution in general was poorer; some changes in retention sequence for specific isomers was noted, depending on the type of derivative [439]. In contrast, the free alcohol eluted before derivatised forms on non-polar phases [944]. It is therefore possible to use equivalent chain-length data for the provisional identification of fatty alcohols in the same way as with methyl ester derivatives of fatty acids (see Chapter 5).

In practice, acetate derivatives appear to have been preferred by most analysts. They have been employed in GC analyses of alcohols from animal tissues [108,662,904], from wax esters of marine origin [17,29,422,757], from microorganisms [657], from avian uropygial glands [424,426] and from vegetable oils such as jojoba [344]. (These are representative analyses from many that could have been cited in which packed columns or the older stainless steel WCOT columns were utilised). Acetates have also been used in more recent work with WCOT columns of fused silica or glass. For example, the resolution of positional and configurational isomers of some insect pheromones (C_{14} and C_{16} olefinic primary alcohols) was studied with four different stationary phases in this way [377].

Jojoba fatty acids and alcohols were analysed simultaneously as the ethyl ester and acetate derivatives on a glass WCOT column coated with OV-1™, from which the ester eluted before the structurally-related acetate [302]. As an alternative, a Grignard reagent (ethyl magnesium bromide) was utilised to hydrolyse the fatty acids and alcohols of jojoba wax to tertiary alcohols and free primary alcohols respectively, and these were analysed simultanously without further derivatisation on a WCOT column of fused silica coated with a phase of the carbowax type [714]. The nature of the separation is illustrated in Figure 10.1, from which it can be seen that the primary alcohols give sharp

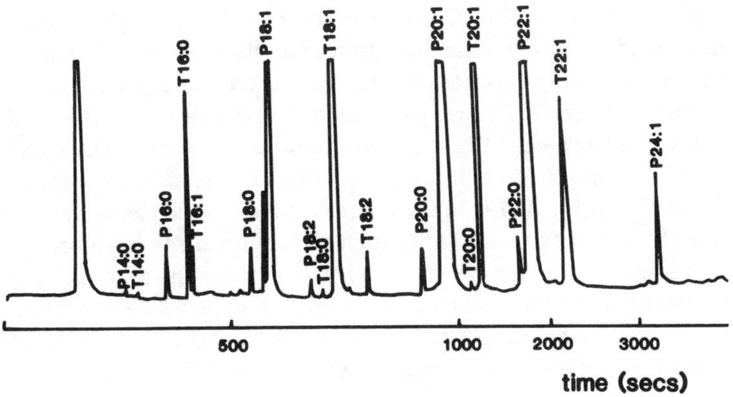

Figure 10.1 GC separation of primary (P) and tertiary (T) alcohols derived from Grignard reduction of the alcohols and fatty acids respectively from jojoba wax [714]. A fused silica WCOT column (20 m x 0.317 mm. i.d.), coated with Carbowax 20M^{TM}, was maintained at 220°C. (Reproduced by kind permission of the authors and of *Lipids*, and redrawn from the original paper).

peaks which emerge well before the tertiary alcohols of related alkyl structure. This methodology may be suitable for waxes of mammalian origin also, but wax esters from fish might be too complex. Nonetheless, free alcohols and methyl ester derivatives of fatty acids from wax esters of marine copepods have been analysed simultaneously on a glass WCOT column coated with the polar phase Silar 10^{TM} [470]. Alcohols eluted before esters of analogous structure, and certainly the main components appeared to be well resolved. None of the alcohols were di- or polyunsaturated, however.

Alkane-2,3-diols from the uropygial glands of birds have been separated on packed columns (SE-30^{TM}) in the form of the isopropylidene derivatives [492], and on glass WCOT columns coated with the same phase in the form of the TMS ethers (both *threo* and *erythro* isomers) [3]. In addition, a number of synthetic 1,2- and 1,3-diols have been subjected to GC analysis in the form of cyclic di-*tert*-butylsilene derivatives [132]. Racemic 2-alkanols have been resolved in the form of *D*-phenylpropionate [349], *N*-(1-phenylethyl) urethane [340] and *R*-(+)-2-phenylselenopropionate [608] derivatives by GC.

2. Other Separatory Procedures and Structure Determination

Fatty alcohols may be separated and analysed by many of the procedures described in detail in Chapter 6 for fatty acid derivatives. For example, silver ion chromatography has been much used as an aid to the isolation and identification of specific components. Saturated and monoenoic fractions were isolated as the acetates from mullet roe by a column procedure [422], but TLC has been used more often; hexane-diethyl ether (9:1, v/v) as the mobile phase in silver ion TLC will separate these fractions [344]. The effect of double bond position on the migration of all the isomeric *cis*-octadecenols and the corresponding acetates on TLC with layers impregnated with silver nitrate was examined [327]. Like the methyl ester derivatives of the analogous fatty acids, the R_f values of isomeric acetates fall on a sinusoidal curve with a minimum around the 5- and 6-isomers, but the free alcohols hardly show this effect. Similarly, HPLC in the reversed-phase mode has been employed for the isolation of fatty alcohols, such as the insect sex hormones, under conditions similar to methyl esters. Thus, various unsaturated aliphatic acetates were resolved on an ODS phase with acetonitrile-water (7:3 or 93:7, v/v) as the mobile phase and with UV detection at 205 nm (or at 235 nm for isomers with conjugated double bond systems) [53].

Determination of double bond position in fatty alcohols or acetates, isolated by silver ion chromatography and preparative GC in combination, has been carried out by means of periodate-permanganate oxidation [344] and by ozonolysis [422,757]. Chain-lengths can be established by hydrogenating and comparing GC retention times with those of authentic standards.

^{13}C NMR spectroscopy has been utilised to assign double bond configurations to unsaturated alkanols [776].

3. *Gas Chromatography-Mass Spectrometry*

Free alcohols are not ideal substrates for GC-MS with electron-impact ionisation as they do not in general give a molecular ion, the first significant ion in the high mass range to be observed usually being equivalent to $[M\text{-}18]^+$. This is followed by a characteristic ion at $[M\text{-}46]^+$, equivalent to $(M\text{-}H_2O\text{-}CH_2{=}CH_2)$. Little other structural information is gleanable from the spectra, and in general they resemble those given by mono-alkenes. Nevertheless, triacontanol, a naturally-occurring plant growth regulator, was identified by MS in the free form [771]. Its spectrum is illustrated in Figure 10.2. There is a barely detectable ion at $[M\text{-}1]^+$, but the ions for $[M\text{-}18]^+$ and $[M\text{-}46]^+$ are reasonably abundant. With negative-ion MS, a good molecular ion is obtained, but there is little other structural information [754]. Secondary alcohols in which the hydroxyl group is located approximately centrally in the alkyl chain are found in some insect waxes. On electron impact, these give characteristic cleavages between the carbon containing the hydroxyl group and the adjacent carbon atom [113]. Collisional activation decomposition spectra may yield further information, but this technique is not compatible with GC [31].

In contrast, acetate derivatives can give adequate molecular ions, although they are not always detected. There is usually a distinctive ion at m/z = 61,

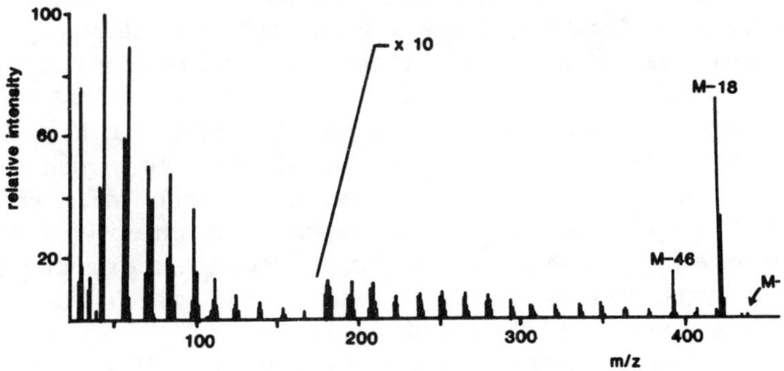

Figure 10.2 The mass spectrum of triacontanol [771]. (Reproduced by kind permission of the authors and of *Science*, and redrawn from the original paper).

equivalent to $[CH_3COOH_2]^+$, together with the corresponding fragment at $[M\text{-}60]^+$ [837,868]. Thus in the mass spectrum of hexadecyl acetate, illustrated in Figure 10.3, the molecular ion is recognisable (m/z = 284), and there are ions at m/z = 224 ($[M\text{-}60]^+$), equivalent to loss of the acetyl

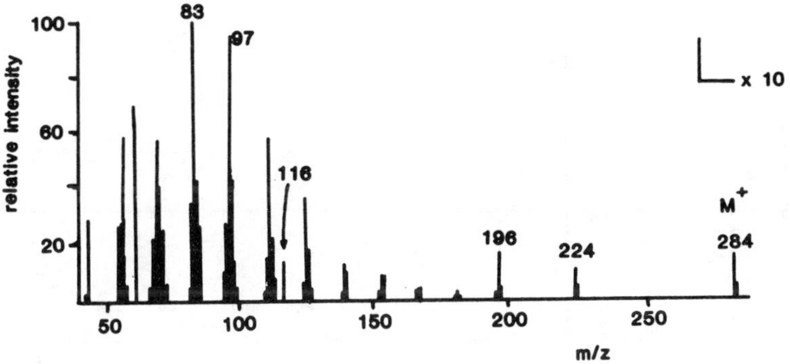

Figure 10.3 The mass spectrum of the acetate derivative of hexadecanol.

group, and at m/z = 196, representing the further loss of an ethylene group. Saturated 1- and 2-alkoxyacetates have spectra that are similar in the high mass range, but small ions at m/z = 73 and 116 are only seen with the former, while ions at m/z = 87 and 102 are found only with the latter [657]. *iso*- and *anteiso*-Methyl branch points in the aliphatic chain are differentiated by the presence of an ion equivalent to $[M\text{-}60\text{-}29]^+$ in the spectrum of the latter, while the base ions are at m/z = 56 and 70 respectively [364]. It is claimed that differences in double bond position in alkenyl acetates give rise to small but significant variations in the intensities of specific ions [402,541]; whether these data can be readily translated to other laboratories and instruments is problematical, however. Tandem mass spectrometry in the negative-ion mode certainly does give information on double bond position in alkenyl acetates, but again the technique may not be used in conjunction with GC [906].

TMS ethers, as might be expected, give good EI mass spectra; the molecular ion may be small, but the ion representing loss of a methyl group is often the base peak [364,836]. Spectra of deuterated decanol TMS ethers gave information on the mechanisms of the fragmentation processes [935]. TMS ethers were utilised with GC-MS to identify fatty alcohols (including diols) in cutin monomers [964]. In addition, related derivatives suited to electron capture detection have been described [134].

N-Alkyl-2-pyrrolidone derivatives, analogous to fatty acid pyrrolidides (see Chapter 6), have given useful spectra with substituted alkanols [95]. However, it soon became evident that alkyl nicotinates, analogous to fatty acid picolinyl esters, give much more informative spectra in which methyl-branch points and double bond positions are readily determined [96,364,950]. These derivatives were used in conjunction with GC-MS for the identification of the aliphatic alcohols from meibomian glands [363,365]. The mass spectrum of the nicotinate derivative of *anteiso*-nonadecanol, obtained at an ionisation potential of 25 eV, is illustrated in Figure 10.4 [364]. The molecular ion (m/z

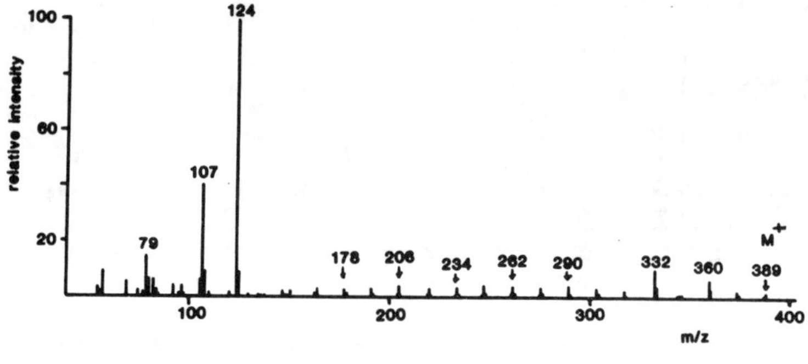

Figure 10.4 The mass spectrum (25 eV) of the nicotinate derivative of *anteiso*-nonadecanol [364]. (Reproduced by kind permission of the authors and of *Biomedical Mass Spectrometry*, and redrawn from the original paper).

= 389) is small but measurable, and the position of the methyl branch is indicated by a gap of 28 amu between m/z = 360 and 332 in the high mass range. The base ion (m/z = 124) represents protonated nicotinic acid. Equally informative spectra are given by nicotinate derivatives of diols.

Regrettably, these are not the easiest of compounds to prepare and this led Harvey [362] to investigate picolinyldimethylsilyl ether derivatives, which also proved to give excellent spectra. The required reagent is not yet available commercially, but can be synthesised without too much difficulty.

Other approaches to locating functional groups in the aliphatic moieties of alcohols have involved more extensive chemical reaction or derivatisation prior to GC-MS. For example, branched-chain primary alcohols have been oxidised to the corresponding acids and methylated for analysis, since the mass spectra of methyl esters are well documented [424,426]. Others prepared pyrrolidides, after oxidation to the acids, as these give spectra which are more readily interpreted [46]. Similarly, secondary alcohols have been oxidised to ketones as an aid to identification [113]. Double bonds in alkyl chains of alcohols have been located by MS after the preparation of suitable chemical adducts, similar to those described for fatty acids in Chapter 7. Oxidation to diols and conversion to the TMS ethers is one method [638], but synthesis of dimethyldisulphide adducts from alcohol acetates is a one-step reaction (see Chapter 4) and is now preferred [143,540]. On the other hand, it may be too much to expect that a single method will provide all the information desired on a given sample; it required partial hydrogenation, coupled with GC-MS and GC/fourier-transform IR (to identify *trans*-double bonds), to determine the structure of a trienoic insect trail pheromone [1006].

B. THE HYDROLYSIS PRODUCTS OF ETHER LIPIDS

1. Isolation and identification of ether lipids

Ether lipids are wide-spread in nature, and only in the higher plants are they rarely encountered. In any complete analysis of lipid samples, the ether forms must be quantified and the alkyl and alkenyl moieties must be identified. Suitable procedures have been reviewed briefly by the author [163,168] and more comprehensively by others [202,588,801,865,875,957]. One important distinguishing feature of alkyl as opposed to acyl lipids is that the ether bonds are stable to alkali.

Alkyl- and alk-1-enyldiacylglycerols sometimes accompany triacylglycerols in tissues, and they can be isolated by a preparative TLC procedure, although some care is required. They tend to migrate close to each other in the order stated and a double development with mobile phases such as toluene-methanol (199:1, v/v) [763] or hexane-diethyl ether (95:5, v/v) [803] can give satisfactory results. The various ether forms of phospholipids are less easily resolved, especially in the native form, but this is discussed in Chapters 8 (C.1) and 9 (D.3) above. When they have been separated by TLC, triacylglycerols and the ether analogues can be quantified by procedures such as charring followed by densitometry, or by GLC of the methyl esters of the fatty acid components of the compounds with a suitable internal standard. Basic transesterification procedures should be used to avoid disruption of vinyl ether bonds, which would result in contamination of the methyl esters by free aldehydes or dimethyl acetals; any other hydrolysis products can be removed by a minicolumn procedure (see Chapter 4). In calculating the results, the differing molar proportions of fatty acids in each lipid class must be taken into account.

Plasmalogens may be detected by spraying TLC plates with 2,4-dinitrophenylhydrazine (0.4 %) in 2 M hydrochloric acid; aldehydes are released which show up as yellow-orange spots on warming the plates [804]. Alternatively, aldehydes released by exposure to fumes of hydrochloric acid appear as purple spots on spraying with a 2% aqueous solution of 4-amino-5-hydrazino-1,2,4-triazole-3-thiol in 1 M NaOH [745]. The total amount of plasmalogenic material in a lipid sample can be determined by preparing the *p*-nitrophenylhydrazone derivatives of the aldehydes released under acidic conditions and estimating these spectrophotometrically at 395 nm relative to a suitable blank [255,755]. They are prepared from the aldehydes or directly from the natural plasmalogens by the following procedure.

> "To a solution of the lipids (1-5 mg) in 95 % ethanol (1.6 ml) is added freshly-prepared 0.02 M *p*-nitrophenylhydrazine in the same solvent (0.2 ml), followed by 0.5 M sulphuric acid (0.2 ml). After heating at 70°C for 20 min, the solution is cooled, water (1 ml) and hexane (2 ml) are added, and the mixture is thoroughly shaken. The hexane layer is washed with water (2 x 2 ml) and dried over anhydrous sodium sulphate, before the solvent is evaporated to yield the required product."

The plasmalogen content can also be measured by a method involving specific binding of mercury salts to the vinyl ether bond [148].

Unfortunately, there is no simple spot test for the identification of alkyl lipids and the ether linkage is not easily disrupted. They must be identified by their chromatographic behaviour on TLC adsorbents relative to authentic standards or by the chromatographic behaviour of their hydrolysis products, i.e. free fatty acids and 1-alkylglycerols. The latter tend to migrate with or just ahead of monoacylglycerols on TLC adsorbents, but they cannot be hydrolysed further, and they react with periodate-Schiff reagent [163].

With plasmalogens, spectroscopic aids to identification are useful; the ether-linked double bond exhibits a characteristic band in the IR spectrum at 6.1 μm, while the olefinic protons adacent to the ether bond produce a doublet centred at 5.89 δ in the NMR spectrum [802]. The ether bond in alkyldiacylglycerols gives a characteristic sharp band at 9 μm in their IR spectrum [90]. In the NMR spectra of the compounds, the protons on the carbon atom adjacent to the ether group give rise to a distinctive signal in the form of a triplet centred on 3.4δ [832]. Mass spectrometric methods of identification are discussed below.

Lipid samples containing plasmalogens should not be stored for long periods in solvents containing acetone, methanol or glacial acetic acid, as some rearrangement or other degradation may occur.

2. GC Analysis of Alkyl- and Alk-1-enylglycerols and Fatty Aldehyde Derivatives

As an alternative to or to complement an estimation by fatty acid analysis, ether lipids can be quantified by determination of the alkyl or alk-1-enyl moieties and indeed in a complete analysis, the individual components should be characterised. Methods are available that are suited to pure lipids with one type of ether moiety, isolated as described in the previous section, or that can be applied to more complex samples containing all of the radyl forms. 1-Alkylglycerols are released from alkyldiacylglycerols both by saponification and transesterification, but better recoveries are obtained if hydrogenolysis with lithium aluminium hydride is used, and this also brings about the release of alk-l-enylglycerols from plasmalogens [914,1001]. Although the technique has been used to remove the phosphorus group from phosphoglycerides, better results are reportedly obtained by hydrogenolysis with Vitride reagent (70% sodium bis-(2-methoxyethoxy) aluminium hydride in benzene) [867, 886]. The following method is recommended -

> "Vitride reagent (0.5 ml) is added to the lipid (1 to 10 mg) in diethyl ether-benzene (2.5 ml; 4:1, v/v) in a test-tube, and the solution is heated with occasional shaking for 30 min at 37°C. On cooling, water-ethanol (10 ml; 5:1,v/v) is added cautiously, then the products are extracted with diethyl ether (3 x 6 ml portions); hexane (10 ml) is added and the solution is dried over anhydrous sodium sulphate. After filtering, the solvent is removed from the combined extracts in a rotary evaporator. The samples are dissolved in a little chloroform and applied to a silica gel G TLC plate, which is developed in diethyl ether-hexane (4:1, v/v). The products are identified by their R_f values relative to standards, alkyl ethers migrating just ahead of the alk-1-enyl analogues, and they are eluted from the adsorbents with several volumes of diethyl ether."

Once the alkyl- and alk-l-enylglycerols are recovered from the adsorbent, they may be analysed separately by appropriate procedures. The GC methods used to identify individual components can also yield information on the total amount of sample if suitable internal standards are added. Various alternative procedures for determining the total amount of each ether form have been described, and the simplest appears to be acetylation with ^{14}C-acetic anhydride, followed by separation of the acetate derivatives for liquid scintillation counting [921]. Alk-l-enylglycerols are not normally analysed as such, but are converted to aldehyde derivatives as described below. The fatty acid components of the sample are reduced to fatty alcohols during the reaction, but they can be analysed independently from a further sample of the lipid.

Alkylglycerols must be converted to less-polar volatile derivatives such as isopropylidene compounds, TMS ethers or acetates for GLC analysis (see Chapter 4 for details of preparation) and of these, isopropylidene derivatives appear to be generally favoured. In this form, they may be separated both according to chain-length and to the number of double bonds in the alkyl chain on GLC columns packed with similar polyester liquid phases as are used to separated methyl esters, e.g. EGS, EGSS-X™, EGSS-Y™ and related polyesters (c.f. [643,656,999]). Because of the higher molecular weights, slightly higher column temperatures are necessary. It is a common practice to add an internal standard, such as the heptadecylglycerol derivative so that the amount of alkyl ether lipid in the sample can be estimated, when the nature of the aliphatic moieties is determined [688,886]. Non-polar phases can also be used as, for example, when the ethers have branched and multibranched alkyl moieties. Figure 10.5 illustrates a separation of the

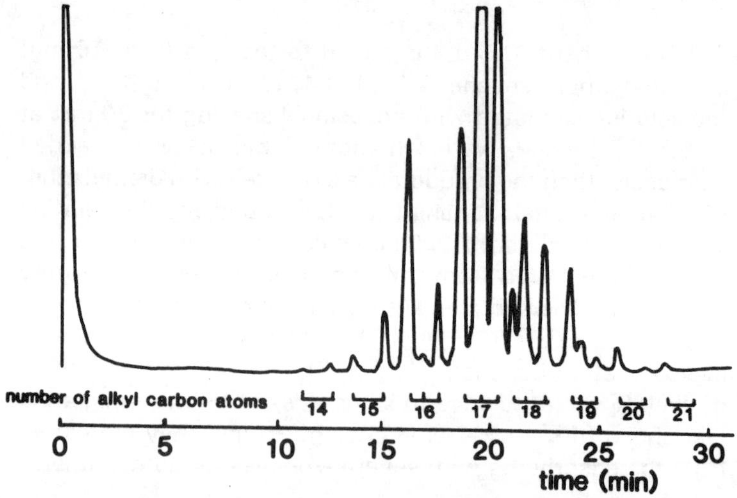

Figure 10.5 GC separation of the isopropylidene derivatives of 1-O-alkylglycerols from the glycerol ether diesters of skin surface lipids from the guinea pig [228]. A glass column (2 m x 4 mm), packed with 3% SE-30™, was temperature-programmed from 150 to 300°C at 3°C/min. In groups of two peaks, the first component is multibranched and the second is straight-chain; a third central component has a single methyl branch. (Reproduced by kind permission of the authors and of *Biochimica Biophysica Acta*, and redrawn from the original paper).

isopropylidene derivatives of 1-*O*-alkylglycerols from the glycerol ether diesters of the skin surface lipid of the guinea pig on a column packed with

3 % SE-30™ and temperature-programmed from 150 to 300°C [228]; the straight-chain compounds are separated from those with a single methyl branch, and these are in turn separable from isomers with several methyl branches.

Although it does not appear to have been taken up to any significant degree, one other of the many methods of analysing alkylglycerols that have been described deserves further consideration, i.e. the preparation under basic conditions of thionocarbonate derivatives [749]. These may be estimated by their absorbance at 235 nm, and they are also suited to GLC analysis (this may be also true for alk-l-enylglycerol derivatives).

From preparations of alkylglycerols or their derivatives from natural sources, individual isomers or homologues can be isolated by procedures related to those described for fatty acids in Chapter 6, for example, for determination of the positions of double bonds by oxidation with permanganate-periodate [352] or by ozonolysis [750] with GLC identification of the fragments. The IR and NMR spectra of alkylglycerols and their derivatives are similar to those of the lipids from which they are derived, except that the free hydroxyl residues or specific functional groups in the derivatives introduce additional features. Mass spectrometry of these compounds is discussed below. Racemic short-chain alkylglycerols, as the isopropylidene derivatives, have been resolved on a WCOT column of fused silica coated with a chiral phase [807].

In order to determine the composition of the alk-1-enyl moieties, it is necessary to liberate the aldehydes quantitatively from the plasmalogens and this is usually achieved simply by treatment with acid. The perfect method for the release of aldehydes has yet to be devised, and Anderson et al. [44] have critically examined some of the acidic hydrolysis procedures that have been described and recommend the following:

> "The plasmalogens (0.2 to 2 mg) in diethyl ether (1.5 ml) are shaken vigorously for 2 min with conc. hydrochloric acid (1 ml). The ether layer is removed and the aqueous phase is extracted once more with ether (2 ml) and once with hexane (2 ml). The combined extracts are washed with distilled water before the solvent is evaporated in a stream of nitrogen. The free aldehydes are obtained by preparative TLC on silica gel G layers, with hexane-diethyl ether (90:10, v/v) as the mobile phase. Aldehydes migrate to just below the solvent front, and they can be recovered from the adsorbent for further analysis by elution with diethyl ether. The other products of the reaction are found much further down the plate."

It is now known that, although complete hydrolysis of the vinyl ether bond occurs with this method, only 80 % recovery of aldehydes is likely to be attained, although these are probably representative in composition of the alk-1-enyl moieties originally present in the natural compound [992]. Addition of an odd-chain aldehyde to the reaction medium for use as an internal standard will correct for a proportion of these losses. Generation of aldehydes can also be effected as part of a more comprehensive scheme for the analysis of ether lipids (see below).

The properties of fatty aldehydes have been reviewed [303,576]. In the native form, they can be analysed by GLC on similar columns to those used for fatty acid analysis, and they can be identified and estimated by analogous procedures. Standard aldehyde mixtures are available from commercial sources, or they can be prepared from the corresponding fatty acids by a number of simple methods. Aldehydes have been reported to be stable for long periods, if stored at -20°C in solution in carbon disulphide or other inert solvents such as pentane or diethyl ether [990], but they should not be kept in contact with other lipids, especially those containing ethanolamine, which catalyses a condensation reaction in which 2,3-dialkylacroleins are formed [805].

On the other hand, because free aldehydes have some tendency to polymerise on standing, especially in the presence of traces of alkali, it is more usual to convert them to more stable derivatives. Of these, acetals are the most popular, especially dimethyl acetals which are easy to prepare, although cyclic acetals (of 1,3-propanediol in particular) are also used because of their greater stability. Dimethyl acetals are prepared by heating the aldehydes or alk-1-enylglycerols under reflux with 5 % methanolic hydrogen chloride in the same manner as was described earlier for the preparation of methyl esters (Chapter 4). They can also be prepared directly from plasmalogens, but prior to GC analysis it may be desirable to separate them from the methyl ester derivatives of the fatty acids, which are formed at the same time, by means of adsorption chromatography (see Chapter 4 also) or by saponification of the esters. On the other hand, the resolving power of modern WCOT columns with polar stationary phases is such that the common range of aldehydes found in animal tissues is well-resolved from the methyl esters. The C_{16} and C_{18} dimethyl acetals emerge clearly ahead of the corresponding esters in a region of the chromatographic trace that tends to be comparatively empty (c.f. [91]).

Cyclic acetals or 1,2-dioxolanes are prepared by condensing 1,3-propanediol with aldehydes in the presence of an acidic catalyst. They have greater thermal stability and are sometimes favoured for GC analysis [889]. They are prepared as follows:

"The aldehydes or plasmalogens (up to 10 mg) are heated with *p*-toluenesulphonic acid (0.5 mg) and 1,3-propanediol (50 mg) in chloroform (5 ml) in a sealed tube at 80°C for 2 hr. On cooling, chloroform (3 ml), methanol (4 ml) and water (3 ml) are added, and the mixture is shaken. The lower solvent layer, which contains the required derivatives, is evaporated."

Hydrazone derivatives of aldehydes can be converted directly to acetals by heating them with the required alcohol and an acid catalyst in the presence of acetone, which serves as an exchanger [580]. All acetal derivatives are stable to alkaline conditions, but they are hydrolysed by aqueous acid. Although decomposition of dimethyl acetals to methyl enol ethers tended to occur on packed columns in the early days of the technique [583,881], this should not be a problem with modern packing materials. On the other hand, there is a report of some decomposition of dimethyl acetals on a modern WCOT column of fused silica, when a "hot needle" injection technique was employed [135]; this should not happen with on-column injection. Like the free aldehydes, dimethyl acetals can be separated according to chain-length and the number of double bonds, under comparable GLC conditions as are used with the analogous methyl esters, and similarly they can be identified by their retention times relative to authentic standards or by using equivalent chain-length (ECL) values, as illustrated earlier (Chapter 5) for methyl esters. In addition, individual components can be isolated and characterised by the same procedures as those used to determine the structures of fatty acids [689].

The IR spectra of free aldehydes are similar to those of the related fatty acid esters, except that the distinctive frequency for the carbonyl functions is at 5.9 μm with an additional band at 3.7 μm [581]. In the NMR spectra of aldehydes, a triplet at 9.7δ is characteristic of the proton on the carbonyl group [576]. Mass spectra of aldehydes have been recorded and these are described in the next section. The elution properties of isomeric monounsaturated aldehydes on silver ion chromatography have been investigated [327].

An integrated method for the simultaneous determination of both the alkyl and alk-1-enyl moieties of lipids is obviously desirable. In one such [886], the plasmalogens are converted to cyclic acetals in the presence of heptadecanal as an internal standard, before the products are submitted to hydrogenolysis with Vitride reagent to release the alkylglycerols in the presence of heptadecylglycerol as a further internal standard. The alkylglycerols are converted to the isopropylidene derivatives, and each type of product is isolated by TLC prior to identification and quantification by GC. As an alternative, the products of the Vitride reaction are acetylated, and acidified to release the aldehydes; the two types of compound are then analysed by GC simultaneously, the aldehydes emerging well ahead of the alkylglycerol acetates [920]. The vitride reaction is described above in detail, a procedure for acetylation with acetic anhydride and pyridine is given in Chapter 4, and the method for the release of aldehydes with acid is also given above. The nature of the GC separation is illustrated in Figure 10.6. A packed column, with Silar 5CP™ as the stationary phase, was used and was maintained

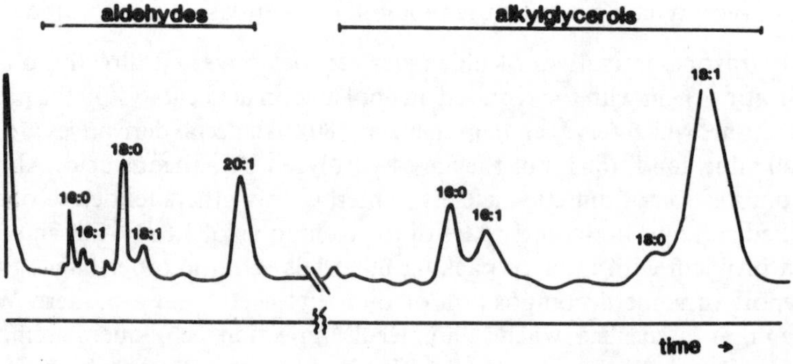

Figure 10.6 "Typical" GC separation of alkyldiacetylglycerols and aldehydes derived from plamalogens [920]. A glass column (200 x 0.5 cm), packed with 10% Silar 5CP™ on GasChrom Q™, was maintained at 220°C. The chart speed was reduced from 20 to 5 mm/min after elution of the aldehydes. (Reproduced by kind permission of the author and of *Fette Seifen Anstrichmittel*, and redrawn from the original paper).

isothermally at 220°C. If temperature-programming and a modern WCOT column were employed, a much tidier chromatogram would be anticipated.

3. Gas Chromatography-Mass Spectrometry

Mass spectrometry of ether lipids has been reviewed by Egge [237] and Myher [642]. There have been a number of electron-impact spectra published for isopropylidene derivatives of alkylglycerols, and that for the C_{16} compound is illustrated in Figure 10.7 [106,337,339,758,868]. The molecular

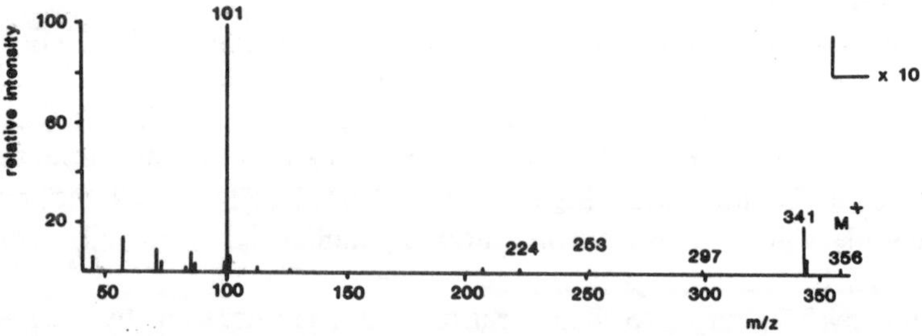

Figure 10.7 The mass spectrum of the isopropylidene derivative of hexadecylglycerol.

ion is rarely detectable, but there is usually a sufficiently abundant ion equivalent to $[M-15]^+$ (i.e. at m/z = 341, loss of methyl from the isopropylidene group) for determination of molecular weight. The base peak is at m/z = 101 and represents cleavage between carbons 1 and 2 of the glycerol moiety, the charge being retained by the fragment containing the isopropylidene group. Among other fragments obviously derived from the molecular ion are $[M\text{-}31]^+$, $[M\text{-}59]^+$ (m/z = 297), $[M\text{-}76]^+$, $[M\text{-}103]^+$ (m/z = 253) and $[M\text{-}132]^+$ (m/z = 224). The diagnostic ions are more easily seen in spectra obtained at an ionisation potential of 25 eV [237]. Mass spectra of 2-methoxy- [337,339] and 2-hydroxyalkylglycerols [629] in the form of isopropylidene derivatives have also been published. In many respects, the mass spectra of isopropylidene derivatives of thioglycerol ethers are similar to those of the related *O*-alkyl compounds, but they differ in that the former tend to have a somewhat greater molecular ion while the ion at $[M\text{-}15]^+$ is smaller [106,254].

With TMS ether derivatives of 1-*O*-alkylglycerols, the molecular ion is again of low intensity but ions equivalent to $[M\text{-}15]^+$, $[M\text{-}90]^+$, $[M\text{-}103]^+$, $[M\text{-}147]^+$ and $[M\text{-}180]^+$ are significant [643,656]. The base ion is usually at m/z = 205 for cleavage between carbons 1 and 2 of the glycerol moiety. Acetate derivatives tend to give a small but significant molecular ion, especially with higher homologues, and there are characteristic ions at $[M\text{-}43]^+$, $[M\text{-}60]^+$ (loss of an acetic acid moiety), $[M\text{-}73]^+$, $[M\text{-}103]^+$ (often the base peak), $[M\text{-}120]^+$ and $[M\text{-}145]^+$ [237]. A 2-hydroxyalkylglycerol in the Harderian gland of the rabbit was identified as the acetate derivative; its mass spectrum resembled that of an unsaturated compound because the 2-acetyl group was lost so readily [774].

With none of these derivatives can the positions of double bonds and methyl branches in the alkyl chain be located. Nicotinates have been employed successfully with other diols to fix the positions of such substituents [364], and there would appear to be no reason why they should not be used here. This would certainly appear to be simpler than an alternative approach in which aldehydes were cleaved to form iodides, which were in turn converted to nitriles, thence to acids and finally to pyrrolidide derivatives for identification by GC-MS [859].

Alk-1-enylglycerols have been subjected to mass spectrometry in the form of the TMS ether derivatives, and they give distinctive spectra with a good molecular ion [656]. However, most analysts have preferred to study aldehydes and their derivatives prepared by acid hydrolysis of plasmalogens. An important paper on the subject was published by Christiansen *et al.* [156]. Saturated aldehydes give a small but significant molecular ion, and there is a series of peaks of the form 68 + 14n, where n = 0, 1, 2 and so on. In the high mass range, there are characteristic peaks at $[M\text{-}18]^+$, $[M\text{-}28]^+$, $[M\text{-}44]^+$ and $[M\text{-}46]^+$, which are not seen in the spectra of unsaturated isomers, although these have somewhat greater molecular ions. With dimethyl acetals, molecular ions are not observed in the electron-impact mass spectra, but an ion equivalent to $[M\text{-}32]^+$ (loss of methanol) permits the determination of molecular weight. An ion at $[M\text{-}64]^+$ represents loss of a vinyl methyl ether moiety. In homologues at the higher end of the molecular weight range, the base peak is at m/z = 71. *O*-Methyl- and *O*-*t*-butyldimethylsilyloximes tend to give much better molecular ions, especially when chemical ionisation is employed [710], and reduction with lithium aluminium hydride to alcohol derivatives has also been recommended [624]. Similarly, negative-ion MS tends to give an excellent molecular ion with aldehydes [754]. On the other hand, none of these methods give any information on the positions of double bonds or methyl branches in the alkyl chain, so a more informative approach might be to oxidise the aldehyde to an acid for conversion to a picolinyl ester derivative for mass spectrometric analysis. Indeed, methodology of this kind has been employed in the analysis of long-chain bases (see Section E.1 below).

4. Platelet-Activating Factor

1-Alkyl-2-acetyl-*sn*-glycerophosphorylcholine or "platelet-activating factor" (commonly abbreviated to PAF) is present in minute concentrations in platelets and certain other cells, yet exerts profound physiological effects. Its chemistry and biochemistry have been reviewed [353,866]. It is such a polar molecule that HPLC in the adsorption mode is required for its isolation. As the author and others [107,168] have reviewed these methods in some detail, it need only be mentioned here as an example that it elutes between sphingomyelin and lysophosphatidylcholine from a column of silica gel with a gradient of hexane-isopropanol-water in which the water content is increased [110]. No existing method is sufficiently sensitive to demonstrate the separation at natural concentrations, however, other than by incorporating a radioactive label by biosynthetic means.

The most satisfactory method of quantification and of identifying isomers of PAF in which the nature of the alkyl-chain varies is probably GC or GC-MS, after collection of the appropriate fraction from an HPLC column. Interestingly, intact ether-linked equivalents of lysophosphatidic acid can be subjected to GC separation after conversion to the TMS ether derivative [917], but this does not appear to be a practical approach to the analysis of PAF. The preferred method is to hydrolyse PAF to 1-*O*-alkyl-2-acetylglycerol with the phospholipase C of *B. cereus* or *C. welchii* (see Chapter 8 for detailed procedures), then to convert to a suitable derivative for GC analysis. With synthetic samples, GC conditions similar to those described above for alkyl ether derivatives can be employed. With PAF at the concentrations existing in natural tissues, the problem of detection is technically demanding. GC-MS with selective ion monitoring of BDMS [793,794] and TMS ether [672] derivatives of 1-alkyl-2-acetylglycerols has given good results with some samples. What is probably the most sensitive method involves preparation of the pentafluorobenzoyl derivative of 1-alkyl-2-acetylglycerols, which gave 92 % of the total ion current as the molecular ion when subjected to GC-MS with negative-ion chemical ionisation [753]. It was necessary to add deuterium-labelled PAF as an internal standard for quantification purposes, and as little as 100 fg could be measured. Others prepared a heptafluorobutyrate derivative from the 1-alkyl-2-acetylglycerols derived from PAF, then employed GC on a glass WCOT column with highly-sensitive electron-capture detection; amounts as low as 20 pg were determined [120]. PAF [916] and lyso-PAF [358] have been analysed by GC-MS with somewhat lower sensitivity, after chemical hydrolysis, as propionyl and isopropylidene derivatives respectively.

C. CHOLESTEROL

Free and esterified cholesterol can be measured during the determination of a total lipid profile by the methods described in Chapter 8 (Section I). On the other hand, so important is the absolute concentration of cholesterol in plasma as a diagnostic marker in disease states believed to be that a number of methods have been developed for the rapid determination of cholesterol alone by various means. For routine clinical applications, all such methods, including GC, should be capable of a high degree of automation. The procedures available have been reviewed and compared elsewhere [336,659,1011]. Enzymatic and colorimetric methods appear to be favoured in most routine clinical applications (commercial kits are available for the purpose), but GC procedures have certain advantages in terms of precision and specificity, and may also permit detection of sterols other than cholesterol.

All such methods require that the lipids be extracted from the plasma or other tissue, and an appropriate internal standard is added at this stage. If the total cholesterol concentration is required, as well as that of the free cholesterol, a hydrolysis step is inserted. Finally, the products are derivatised for GC analysis. To simplify the first steps, it is possible to hydrolyse a plasma sample directly. In most of the published papers, the internal standard has been cholestane, octadodecane, desmosterol or epicoprostanol (3α -hydroxy -5β -cholestane), while the GC separation has been effected on a column (50 to 100 cm long) packed with 3 % SE-30™ and operated isothermally in the range 200 to 240°C; preparation of TMS ethers gives sharper peaks and improved quantification. Innumerable procedures, varying in minor details, have been described for the purpose. The following candidate reference method is recommended [214].

> "The serum sample (about 0.2 ml) is weighed into a screw-capped vial, and a solution of epicoprostanol, in an amount equivalent to the concentration of cholesterol expected, in ethanol (0.6 ml) is added, followed by ethanol (1 ml) and 0.4 ml of ethanolic potassium hydroxide solution (4.6 mol of KOH in 0.3 L of water and 1 L of ethanol). The mixture is left at

37°C for 3 hours, it is cooled and water (2 ml) and hexane (4 ml) are added. The whole is shaken thoroughly for 15 min, centrifuged and the organic layer recovered. An aliquot of this is taken to dryness and the residue is converted to the TMS ether derivatives (see Chapter 4 for a method).

While many different GC columns could be used, a glass WCOT column (25 m x 0.3 mm), coated with OV-1™ and operated isothermally at 240°C, gave satisfactory results in the work cited; the TMS ether derivatives of epicoprostanol and cholesterol eluted in 17.7 and 20 min respectively [214]. Of course, the procedure must be calibrated carefully with suitable standards, and the original publication should be consulted for the fine detail of the protocol. Other workers have obtained results adequate for many purposes with packed columns (1.5 to 2 m x 4 mm i.d.) and similar non-polar silicone phases (e.g. 3 % SE-30™ relative to the support). In routine use, a coefficient of variation of 0.35 to 0.5 % was obtained. Even higher accuracy is claimed for "definitive" procedures in which [3,4-^{13}C]-cholesterol [273,703] and deuterated cholesterol [187,983] are employed as internal standards and GC-MS is used for quantification.

There is an enormous body of work on the analysis of steroids other than cholesterol by chromatographic means (reviewed elsewhere [381]), and detailed discussion is outwith the scope of this book. It may be worth noting, however, that GC on WCOT columns has been utilised to study the products of oxidation of cholesterol [691] and to identify phytosterols accumulating in the plasma of patients suffering from phytosterolaemia [527].

D. GLYCEROL

Free glycerol or that released by hydrolysis of triacylglycerols is readily estimated by enzymatic or chemical means using one of the many kits that are available commercially for the purpose (reviewed elsewhere [659]). GC methods may also be used, and the following is suited to the determination of the glycerol content of simple lipids [390].

"The glycerolipid (1 to 10 mg) and a known amount of a suitable methyl ester as an internal standard, say methyl pentadecanoate, are dissolved in dry diethyl ether (2 ml), and an ethereal solution of lithium aluminium hydride (20 mg in 3 ml) is added in portions of 0.1 ml until the boiling stops. After addition of a one volume excess of the lithium aluminium hydride solution, the mixture is refluxed for 1 hour. Acetic anhydride is added dropwise with cooling to destroy the excess reagent, followed by additional acetic anhydride (2.5

ml) and xylene (3 ml). The ether is removed by evaporation and the residue is refluxed for 6 hours; the temperature should reach 110°C if no ether remains. Finally, the reagents are removed on a rotary evaporator and the products are taken up in dry diethyl ether for analysis by GC."

The procedure may be scaled down appreciably if this is required. A somewhat different method is recommended for determining the glycerol in phosphoglycerides [389], because lithium aluminium hydride is insufficiently vigorous, but replacement of this reagent with Vitride (see Section B.1 above) should suffice. On a packed column of EGS, triacetin elutes just before octadecanyl acetate (fatty alcohol acetates are also produced in the reaction), but it elutes nearer octanyl acetate on a non-polar stationary phase, such as Apiezon L™ or SE-30™. An alternative method for simple glycerides, involving alkaline hydrolysis followed by determination of the glycerol as the TMS ether derivative on a WCOT column, has recently been described [138].

E. LONG-CHAIN BASES

1. Isolation, Derivatisation and GC Separation

The long-chain or sphingoid bases are the characteristic structural components of sphingolipids and very many different compounds, including homologues and isomers, can exist in a single natural source (see Chapter 2 for a brief description) [460,974,981]. Before these constituents of sphingolipids can be analysed, it is first necessary to hydrolyse any glycosidic linkage or phosphate bond as well as the amide bond to the fatty acyl group. Ideally, this should be accomplished by a procedure in which no degradation or rearrangement of the bases occurs, but the perfect method has not yet been devised. Base-catalysed hydrolysis has been advocated by many analysts, and the following method appears to give much less degradation than others to have been described [625].

> "The sphingolipids (up to 5 mg) are dissolved in warm dioxane (2.5 ml), 10 % aqueous barium hydroxide solution (2.5 ml) is added and the mixture is heated in a sealed tube at 110°C for 24 hours. On cooling, water (10 ml) is added, and the solution is extracted with chloroform (2 x 15 ml). After drying over anhydrous sodium sulphate, the solvent is evaporated to yield the required long-chain bases".

Some degradation of trihydroxy bases especially may be caused by even this procedure, but it is troublesome only if these are present in small amounts.

In other laboratories, acid-catalysed hydrolysis has been used, although rearrangement and substitution at C-3 and C-5 inevitably occurs to a certain extent, thereby altering the configuration of the bases from the *erythro* to the *threo* form. In addition, *O*-methoxy artefacts are formed in the presence of methanol, and compounds containing a tetrahydrofuran ring may be formed from trihydroxy bases. A recently-described procedure [454], in which aqueous hydrochloric acid in acetonitrile (an aprotic solvent) is employed for hydrolysis, reportedly produces fewer artefacts than earlier methods, in which methanol is the solvent. It gives particularly good yields of long-chain bases from gangliosides. The detailed method is -

> "The hydrolysis reagent consists of 0.5 M HCl and 4 M water in acetonitrile (0.3 ml), and is added to the glycolipids (up to 200 μ grams) in a teflon-lined screw-capped tube, which is flushed with nitrogen, sealed and heated at 75°C for 2 hours. The solvents are evaporated in a stream of nitrogen, chloroform (5 ml) is added followed by 0.05 M sodium hydroxide in methanol-0.9% saline solution-chloroform (48:47:3 by volume) (1 ml), and the mixture is shaken thoroughly before being centrifuged at 3000 *g*. The lower phase is washed with three further portions of the NaOH solution, and then with two portions of the same solvents but without NaOH. Finally, the lower phase is evaporated in a stream of nitrogen to recover the required bases."

With sphingolipids other than gangliosides, somewhat milder hydrolysis conditions were preferred, i.e. sphingomyelins were reacted for 1 hour in 1 M HCl in water-methanol (2:1 by volume) at 70°C for 16 hours, while cerebrosides required 3 M HCl in water-methanol (1:1 by volume) at 60°C for 1.5 hours [573]. (A related procedure, but for the isolation of the fatty acid constituents of sphingolipids for further analysis, has also been described [65] (see Chapter 4)). An objective experimental comparison of the above base- and acid-catalysed methods with a range of different substrates would now appear to be desirable.

HPLC procedures for the analysis of long-chain bases have been described and are reviewed elsewhere [168]. These appear to be sensitive, and some of the separations are impressive, but the methods have as yet been applied to a limited range of relatively simple samples only. Much more experimental work with GC methods has been published. As with other lipids with polar functional moieties, it is necessary to prepare volatile non-polar derivatives, and most analysts have made use of *O*-TMS or *N*-acetyl-*O*-TMS ether derivatives. Acetylation is carried out as follows [280]:

> "The long-chain bases (0.1 to 0.2 mg) are reacted with

freshly-prepared acetic anhydride in methanol (1:4, v/v; 50 μ litres) at room temperature overnight. *n*-Butanol (2 ml) is added to facilitate the removal of the excess acetic anhydride during evaporation in a stream of nitrogen".

A silylation reagent consisting of hexamethyldisilazane (2.6 ml), dry pyridine (2 ml) and trimethylchlorosilane (1.6 ml) has been recommended for long-chain bases [149], but any of the more powerful silylating reagents described in recent years (see Chapter 4) should give good results.

The sphingoid bases from plasma sphingomyelin, for example, were separated as *N*-acetyl-*O*-TMS ether derivatives on a packed column (2 m x 3 mm) packed with 3 % SE-30™, maintained isothermally at 230°C [724]. Saturated and unsaturated isomers are only partly resolved by this means. More recently, other workers used a WCOT column (25 m), coated with OV-101 and operated isothermally at 260°C, for similar separations [356]. In this instance, base-line resolution of sphingosine and dihydrosphingosine derivatives was possible. Simple *O*-TMS ethers (i.e. not *N*-acetylated) tend to elute at slightly lower temperatures, but peaks tail somewhat, and these derivatives may not be quite so useful for mass spectrometric identification purposes (see below) [461]. As an example, a separation of the straight- and branched-chain dihydroxy bases from the cerebrosides of the Harderian gland of the guinea pig is illustrated in Figure 10.8 [1008]. Again a packed column containing a non-polar phase, OV-101™, was employed to effect the separation.

Such procedures may be suitable for the analysis of samples containing a relatively simple range of long chain bases, but some natural lipid extracts are very complex. Alternative complementary techniques must then be employed to obtain the resolution needed. The most widely-used approach is to stabilise the amino group by conversion to the dinitrophenyl (DNP) derivative, for separation into different classes by TLC procedures; these compounds are yellow in colour so are easily seen on a TLC plate. DNP derivatives can later be degraded to aldehydes by periodate oxidation for GC analysis in this form [464,465]. The procedure for the preparation of *N*-DNP derivatives of sphingoid bases is:

"The sphingoid bases (up to 5 mg) are reacted with 1-fluoro-2,4-dinitrobenzene (5 mg) in methanol (1ml), with addition of potassium borate buffer (pH 10.5; 4 ml) dropwise followed by heating at 60°C for 30 min. After cooling, the mixture is partitioned between chloroform, methanol and water in the ratio of 8:4:3 by volume, and the lower phase is collected and evaporated. The products are purified by chromatography on a short column of silicic acid (1g)., from which non-polar impurities are eluted first with hexane-diethyl

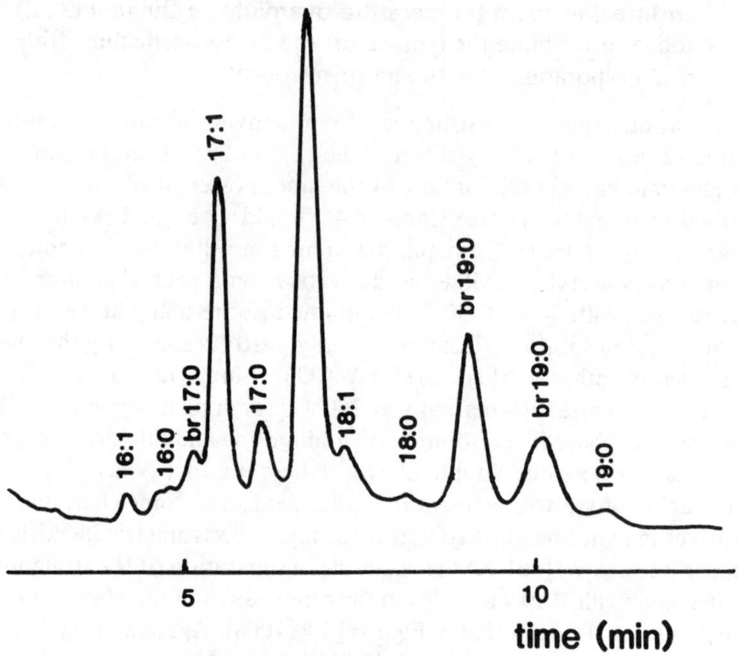

Figure 10.8 GC separation of the TMS ether derivatives of long-chain bases from the cerebrosides of the Harderian gland of the guinea pig [1008]. A glass column (2 m x 3 mm i.d.), packed with 1 % OV-1™, was maintained at 220°C with nitrogen at 30 ml/min as the carrier gas. (Reproduced by kind permission of the authors and of the *Journal of Biochemistry (Tokyo)*, and redrawn from the original paper).

> ether (7:3; 20 ml), while the required DNP derivatives are recovered with the same solvents in the ratio 1:1."

They can be separated into three groups on layers of silica gel G impregnated with 2 % boric acid, i.e. saturated dihydroxy-, unsaturated dihydroxy (with a *trans* double bond in position 4) and trihydroxy bases, with chloroform-hexane-methanol (5:5:2 by volume) as the mobile phase [465]. When acidic hydrolysis procedures are utilised in the preparation of the bases, unnatural *threo*-isomers of the unsaturated dihydroxy bases are found just below the natural *erythro* compounds on the TLC plate. Each of the fractions separated by TLC can be recovered from the adsorbent by elution with chloroform-methanol (2:1, v/v), but the eluate should be washed with one quarter the volume of water to remove boric acid which is also eluted. If need be, the bases can be further resolved by silver ion TLC [462] or by HPLC in the reversed-phase mode (reviewed elsewhere [168]). It is possible that HPLC in the adsorption or silver ion modes could also contribute to the problem of analysis, but these do not appear to have been tried.

To simplify the resolution of saturated, unsaturated and branched-chain isomers, Karlsson [464,465] and others have oxidised the DNP derivatives of each group of bases, separated by TLC, to aldehydes. As similar compounds are produced from both di- and trihydroxy bases of the *threo* or *erythro* configurations, it is essential to carry out the TLC separation prior to oxidation. Aldehydes are readily separated by chain-length and degree of unsaturation on GC columns similar to those used for the separation of methyl esters of fatty acids (see also Section B above). Either periodate or lead tetraacetate may be used for the oxidation step, and the following method can be recommended [466].

> "The long-chain bases (2 mg) are oxidised by reaction with lead tetraacetate (30 mg) in benzene (0.5 ml) (caution!) at 50oC for 1 hour. Water (5 ml) and hexane (5 ml) are added, the mixture is shaken thoroughly and the solvent layers are dried over anhydrous sodium sulphate before the solvent is evaporated. The aldehydes are analysed immediately by GC."

Aldehydes are more easily identified than are the parent compounds, since a wide range of standards is available from commercial sources or can be prepared synthetically from other lipids. As an example of the full application of this methodology, more than 30 different bases were detected in the sphingolipids of bovine kidney [469]. Mass spectrometry can be utilised as an aid to identification of aldehydes (see also Section B above), although some workers have preferred to reduce them to fatty alcohols and then to prepare acetate or TMS ether derivatives for this purpose [624]. In addition, all the methods for the location of double bonds in fatty acids, such as ozonolysis or hydroxylation with osmium tetroxide and preparation of TMS ethers for MS, have been utilised with aldehydes prepared from sphingoid bases [464,465].

One further approach to identification has been to oxidise the aldehydes to fatty acids by the following procedure [887,888,971].

> "The aldehydes (up to 1 mg) are oxidised in tetrahydrofuran-water (9:1, v/v; 2 ml) to which silver oxide (30 mg) is added. The mixture is shaken gently for 24 hours at room temperature, then the solvent is removed in a stream of nitrogen, 6 M nitric acid (1 ml) is added, and the products are extracted with diethyl ether (8 ml). The extraction is repeated and the combined ether layers are washed with water, dried over anhydrous sodium sulphate and evaporated under nitrogen."

If methyl ester derivatives are prepared for GC analysis, it may be necessary to remove some residual aldehydes (in the form of dimethyl acetals) by preparative TLC (see Chapter 4). These will not be troublesome if picolinyl ester derivatives are prepared for identification by GC-MS (see Chapter 7).

Note that aldehyde derivatives prepared from trihydroxy bases will be one carbon shorter than those from equivalent dihydroxy bases, and the number of hydroxyl groups must be determined from the TLC behaviour of the base or its DNP derivative. The proportions of the various isomers within each class of base can be determined with reasonable accuracy by GC, but artefact formation during the hydrolysis stage may distort the apparent relative proportions of the various classes of base to each other.

Recently, it has been demonstrated that carbon-13 NMR spectroscopy may be used to determine the configuration of long-chain bases while they still form part of intact natural lipids [792]. For the first time, it was shown unequivocally that they were exclusively of the *erythro* configuration.

2. Gas Chromatography-Mass Spectrometry

It is not necessary to consider the degraded forms of sphingoid bases, such as aldehydes, here as equivalent compounds are dealt with above (Section B.3). Mass spectrometry is certainly the most powerful tool available to the analyst for identifying long-chain bases, although appropriate derivatives must be prepared for the purpose. Karlsson [464,465] favours the preparation of TMS ether or methyl ethers of the DNP derivatives, with hydroxylation of double bonds and similar derivatisation of the resulting hydroxyl groups. Certainly, this approach gives definitive spectra, but the molecular weights are then frequently too high for GC-MS. Pure individual compounds are then required, and direct insertion probes into the instrument must be utilised. Similar types of derivative have been employed with HPLC coupled to mass spectrometry, and this has been discussed elsewhere [168].

O-TMS ether and *N*-acetyl-*O*-TMS derivatives are better suited to analysis by GC-MS. Of these, the latter have been most used and the details of the fragmentation processes involved with electron-impact ionisation have been well worked out, principally in Sweeley's laboratory [280,495,724]. The mass spectrum of bis-*O*-trimethylsilyl-*N*-acetylsphinganine is illustrated in Figure 10.9 [280,495]. There is no detectable molecular ion (expected at m/z = 487), but the molecular weight is clearly indicated by an ion equivalent to $[M-15]^+$ at m/z = 472. There are small peaks at $[M-59]^+$ (m/z = 428), representing loss of the acetamido group, and at $[M-90]^+$ (m/z = 397), for the loss of a trimethylsilanol moiety. The ion at m/z = 384 is the part of the molecule remaining after the loss of the terminal methylene and its TMS ether group. That at m/z = 313 represents cleavage between carbons 2 and 3 of the molecule, but the corresponding fragment from the remainder of the molecule at m/z = 174 is small (c.f. the spectra of unsaturated isomers). In contrast, an ion at m/z = 157 is particularly abundant, while the base peak is at m/z = 73.

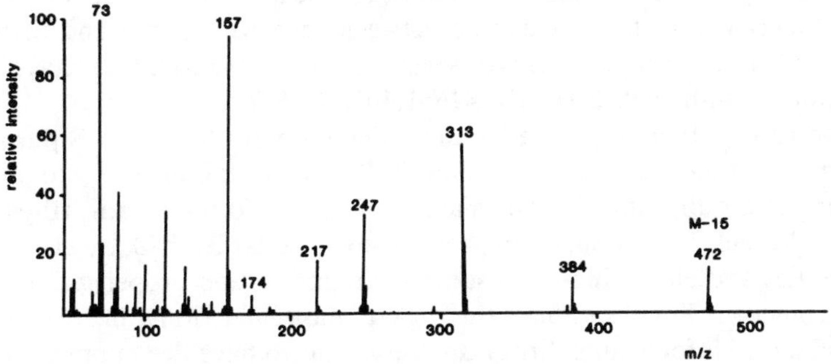

Figure 10.9 The mass spectrum of bis-*O*-trimethylsilyl-*N*-acetyl-sphinganine [495]. (Reproduced by kind permission of the authors and of the *Chemistry and Physics of Lipids*, and redrawn from the original paper).

Many features in the mass spectrum of the *N*-acetyl-*O*-TMS ether derivative of sphinga-4,14-dienine, shown in Figure 10.10 [724], are similar to this, but

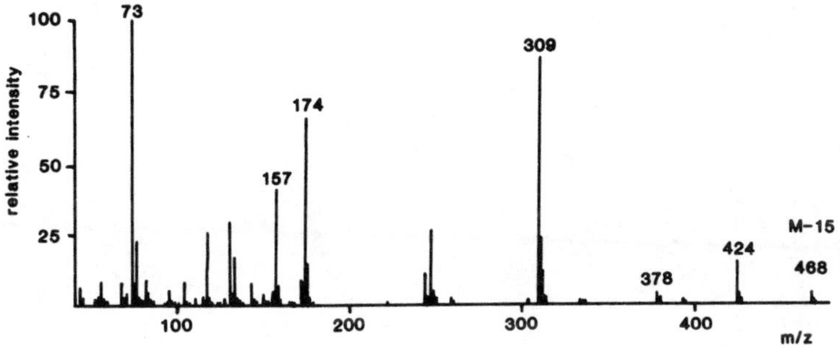

Figure 10.10 The mass spectrum of the *N*-acetyl-*O*-trimethylsilyl ether derivative of sphinga-4,14-dienine, isolated form plasma sphingomyelin [724]. (Reproduced by kind permission of the authors and of *Biochemistry*, and redrawn from the original paper).

the ion for $[M\text{-}59]^+$ at m/z = 424 is more abundant, and there are major peaks for both fragments formed by cleavage between carbons 2 and 3, i.e. at m/z = 309 and 174. Derivatives of this kind have been employed with many natural samples [374,375,600,888,971]. The positions of double bonds or of methyl branches cannot be deduced from such spectra, and this is one reason for approaching the problem of structure determination via the degradative route discussed in the previous section. On the other hand, these compounds can be subjected to hydroxylation of the double bonds and formation of TMS ethers for identification, although the molecular weights are thereby increased very substantially [375,724].

The TMS ethers of the base with a free amine group have also been employed for GC-MS [461]. Again there is no molecular ion, although the molecular weight is indicated by an ion equivalent to $[M-15]^+$, and there are prominent fragments for cleavage between carbons 1 and 2 and between carbons 2 and 3. The structures of several natural sphingoid bases have been determined with such derivatives [601,602,811,887].

One further type of derivative of sphingoid bases for GC-MS purposes is worthy of note, i.e. cyclic boronates [277]. Usually, it is necessary to protect the amine group, but a bis-boronate derivative is formed from trihydroxy bases. Model compounds, prepared from standards, had excellent GC properties and also exhibited distinctive fragmentation properties on mass spectrometry. For example, they gave molecular ions in reasonable abundance. Unfortunately, they do not appear to have been applied to the analysis of natural samples.

REFERENCES

1. Aasen,A.J., Hofstetter,H.H., Iyengar,B.T.R. and Holman,R.T., *Lipids*, **6**, 502-507 (1971).
2. Aasen,A.J., Lauer,W.M. and Holman,R.T., *Lipids*, **5**, 869-877 (1970).
3. Abalain,J.H., Picart,D., Berthou,F., Ollivier,R., Amet,Y., Daniel,J.Y. and Floch,H.H., *J. Chromatogr.*, **274**, 305-312 (1983).
4. Abbot,G.G., Gunstone,F.D. and Hoyes,S.D., *Chem. Phys. Lipids*, **4**, 351-366 (1970).
5. Abe,K. and Tamai,Y., *J. Chromatogr.*, **232**, 400-405 (1982).
6. Abrahamsson,S., Stallberg-Stenhagen,S. and Stenhagen,E., *Prog. Chem. Fats other Lipids*, **7**, 1-164 (1964).
7. Ackman,R.G., *J. Am. Oil Chem. Soc.*, **40**, 558-564 (1963).
8. Ackman,R.G., *J. Gas Chromatogr.*, **2**, 173-179 (1964).
9. Ackman,R.G., *J. Chromatogr.*, **34**, 165-173 (1968).
10. Ackman,R.G., *Prog. Chem. Fats other Lipids*, **12**, 165-284 (1972).
11. Ackman,R.G., *J. Chromatogr. Sci.*, **10**, 243-246 (1972).
12. Ackman,R.G., *Lipids*, **12**, 293-296 (1977).
13. Ackman,R.G., in *Handbook of Chromatography. Vol. I. Lipids*, pp. 95-240 (1984) (edited by H.K.Mangold, CRC Press, Boca Raton).
14. Ackman,R.G., in *Analysis of Oils and Fats*, pp. 137-206 (1986) (edited R.J. Hamilton & J.B. Rossell, Elsevier Applied Science, London).
15. Ackman,R.G. and Burgher,R.D., *J. Am. Oil Chem. Soc.*, **42**, 38-42 (1965).
16. Ackman,R.G. and Eaton,C.A., *Fette Seifen Anstrichm.*, **80**, 21-37 (1978).
17. Ackman,R.G., Eaton,C.A., Kinneman,J. and Litchfield,C., *Lipids*, **10**, 44-49 (1975).
18. Ackman,R.G., Eaton,C.A. and Litchfield,C., *Lipids*, **6**, 69-77 (1971).
19. Ackman,R.G., Eaton,C.A., Sipos,J.C. and Crewe,N.F., *Can. Inst. Food Sci.*, **14**, 103-107 (1981).
20. Ackman,R.G. and Hansen,R.P., *Lipids*, **2**, 357-362 (1967).
21. Ackman,R.G. and Hooper,S.N., *J. Chromatogr.*, **86**, 73-81 (1973).
22. Ackman,R.G. and Hooper,S.N., *J. Chromatogr.*, **86**, 83-88 (1973).
23. Ackman,R.G. and Hooper,S.N., *J. Chromatogr. Sci.*, **12**, 131-138 (1974).
24. Ackman,R.G., Hooper,S.N. and Hooper,D.L., *J. Am. Oil Chem. Soc.*, **51**, 42-49 (1974).
25. Ackman,R.G., Manzer,A. and Joseph,J., *Chromatographia*, **7**, 107-114 (1974).
26. Ackman,R.G., Sebedio,J-L. and Ratnayake,W.M.N., *Methods Enzymol.*, **72**, 253-275 (1981).
27. Ackman,R.G. and Sipos,J.C., *J. Chromatogr.*, **16**, 298-305 (1964).
28. Ackman,R.G. and Sipos,J.C., *J. Am. Oil Chem. Soc.*, **41**, 377-378 (1964).
29. Ackman,R.G., Sipos,J.C., Eaton,C.A., Hilaman,B.L. and Litchfield,C., *Lipids*, **8**, 661-667 (1974).
30. Adams,J., Deterding,L.J. and Gross,M.L., *Spectroscopy Int. J. (Ottawa)*, **5**, 199-228 (1987).
31. Adams,J. and Gross,M.L., *J. Am. Chem. Soc.*, **108**, 6915-6921 (1986).
32. Adams,J. and Gross,M.L., *Anal. Chem.*, **59**, 1576-1582 (1987).
33. Adlof,R.O. and Emkem,E.A., *J. Am. Oil Chem. Soc.*, **57**, 276-278 (1980).
34. Adlof,R.O. and Emken,E.A., *J. Am. Oil Chem. Soc.*, **58**, 99-101 (1981).
35. Adlof,R.O. and Emken,E.A., *J. Am. Oil Chem. Soc.*, **62**, 1592-1595 (1985).
36. Adlof,R.O., Rakoff,H. and Emken,E.A., *J. Am. Oil Chem. Soc.*, **57**, 273-275 (1980).
37. Ahmad,M.S., Ahmad,M.U., Osman,S.M. and Ballantine,J.A., *Chem. Phys. Lipids*, **25**, 29-38 (1979).

38. Albertyn,D.E., Bannon,C.D., Craske,J.D., Hai,N.T., O'Rourke,K.L. and Szonyi,C., *J. Chromatogr.*, **247**, 47-62 (1982).
39. Alexander,L.R., Justice,J.B. and Madden,J., *J. Chromatogr.*, **342**, 1-12 (1985).
40. Allan,D. and Cockcroft,S., *J. Lipid Res.*, **23**, 1373-1374 (1982).
41. Allen,K.G., MacGee,J., Fellows,M.E., Tornheim,P.A. and Wagner,K.R., *J. Chromatogr.*, **309**, 33-42 (1984).
42. Amer,M.A., Kupranycz,D.B. and Baker,B.E., *J. Am. Oil Chem. Soc.*, **62**, 1551-1557 (1985).
43. Anderson,B.A., Miller.R. and Pallansch,M.J., *J. Dairy Sci.*, **57**, 156-159 (1974).
44. Anderson,R.E., Garrett,R.D., Blank,M.L. and Snyder,F., *Lipids*, **4**, 327-330 (1969).
45. Andersson,B.A., *Prog. Chem. Fats other Lipids*, **16**, 279-308 (1978).
46. Andersson,B.A. and Bertelsen,O., *Chem. Scripta*, **8**, 135-139 (1975).
47. Andersson,B.A., Christie,W.W. and Holman,R.T., *Lipids*, **10**, 215-219 (1975).
48. Andersson,B.A., Dinger,F. and Dinh-Nguyen,N., *Chem. Scripta*, **8**, 200-203 (1975).
49. Andersson,B.A., Dinger.F. and Dinh-Nguyen,N., *Chem. Scripta*, **9**, 155-157 (1976).
50. Andersson,B.A., Dinger,F. and Dinh-Nguyen,N., *Chem. Scripta*, **19**, 118-121 (1982).
51. Andersson,B.A. and Holman,R.T., *Lipids*, **9**, 185-190 (1974).
52. Andersson,B.A. and Holman,R.T., *Lipids*, **10**, 716-718 (1975).
53. Ando,T., Hasegawa,Y. and Uchiyama,M., *Agric. Biol. Chem.*, **50**, 2935-2937 (1986).
54. Aneja,R., Bhati,A., Hamilton,R.J., Padley,F.B. and Steven,D.A., *J. Chromatogr.*, **173**, 392-397 (1979).
55. Anthony,G.M., Brooks,C.J.W., MacLean,I. and Sangster,I., *J. Chromatogr.*, **7**, 623-631 (1969).
56. Apon,J.M.B. and Nicolaides,N., *J. Chromatogr. Sci.*, **13**, 467-473 (1975).
57. Argoudelis,C.J. and Perkins,E.G., *Lipids*, **3**, 379-381 (1968).
58. Ariga,T., Araki,E. and Murata,T., *Chem. Phys. Lipids*, **19**, 14-19 (1977).
59. Arnold,R.G. and Hartung,T.E., *J. Food Sci.*, **36**, 166-168 (1971).
60. Arvidson,G.A.E., *J. Lipid Res.*, **6**, 574-577 (1965).
61. Arvidson,G.A.E., *Eur. J. Biochem.*, **4**, 478-486 (1968).
62. Aubourg,P., Bougneres,P.F. and Rocchiccioli,F., *J. Lipid Res.*, **26**, 263-269 (1985).
63. Auling,G., Heinz,E. and Tulloch,A.P., *Hoppe-Seylers's Z. Physiol. Chem.*, **352**, 905-912 (1971).
64. Aveldano,M.I., *J. Biol. Chem.*, **262**, 1172-1179 (1987).
65. Aveldano,M.I. and Horrocks,L.A., *J. Lipid Res.*, **24**, 1101-1105 (1982).
66. Aveldano,M.I. and Sprecher,H., *J. Biol. Chem.*, **262**, 1180-1186 (1987).
67. Aveldano,M.I., Van Rollins,M. and Horrocks,L.A., *J. Lipid Res.*, 24, 83-93 (1983).
68. Ayanoglu,E., Kurtz,K., Kornprobst,J.M. and Djerassi,C., *Lipids*, **20**, 141-144 (1985).
69. Ayanoglu,E., Popov,S., Kornprobst,J.M., Aboud-Bichara,A. and Djerassi,C., *Lipids*, **18**, 830-836 (1983).
70. Ayanoglu,E., Walkup,R.D., Sica,D. and Djerassi,C., *Lipids*, **17**, 617-625 (1982).
71. Badami,R.C. and Patil,K.B., *Prog. Lipid Res.*, **19**, 119-153 (1982).
72. Badings,H.T. and De Jong,C., *J. Chromatogr.*, **279**, 493-506 (1983).
73. Badings,H.T. and De Jong,C., *J. Am. Oil Chem. Soc.*, **65**, 659 (1988).
74. Bagby,M.O., Smith,C.R. and Wolff,I.A., *J. Org. Chem.*, **30**, 4227-4229 (1965).
75. Bandi,Z.L. and Mangold,H.K., *Separ. Sci.*, **4**, 83-88 (1969).
76. Bandi,Z.L., Moslen,M.T. and Reynolds,E.S., *J. Chromatogr.*, **269**, 93-101 (1982).
77. Bannon,C.D., Craske,J.D., Felder,D.L., Garland,I.J. and Norman,L.M., *J. Chromatogr.*, **407**, 231-241 (1987).
78. Bannon,C.D., Craske,J.D. and Hilliker,A.E., *J. Am. Oil Chem. Soc.*, 62, 1501-1507 (1985).
79. Bannon,C.D., Craske,J.D. and Hilliker,A.E., *J. Am. Oil Chem. Soc.*, **63**, 105-110 (1986).
80. Banschbach,M.W., Geison,R.L. and O'Brien,J.F., *Anal. Biochem.*, **59**, 617-627 (1974).
81. Baranska,J., *Adv. Lipid Res.*, **19**, 163-184 (1982).
82. Barber,M., Chapman,J.R. and Wolstenholme,W.A., *Int. J. Mass Spectrom. Ion Phys.*, **1**, 98-101 (1968).
83. Barber,M., Merren,T.O. and Kelly,W., *Tetrahedron Letts.*, **18**, 1063-1067 (1964).
84. Barton,F.E., Himmelsbach,D.S. and Walters,D.B., *J. Am. Oil Chem. Soc.*, **55**, 574-576 (1978).

85. Barve,J.A., Gunstone,F.D., Jacobsberg,F.R. and Winlow,P., *Chem. Phys. Lipids*, **8**, 117-126 (1972).
86. Batrakov,S.G., Sadovskaya,V.L., Rozynov,B.V. and Bergelson,L.D., *Chem. Phys. Lipids*, **33**, 331-353 (1983).
87. Batt,R.D., Hodges,R. and Robertson,J.G., *Biochim. Biophys. Acta*, **239**, 368-373 (1971).
88. Battaglia,R. and Froelich,D., *Chromatographia*, **13**, 428-431 (1980).
89. Bauer,S., Neupert,M. and Spiteller,G., *J. Chromatogr.*, **309**, 243-260 (1984).
90. Baumann,W.J. and Ulshofer,H.W., *Chem. Phys. Lipids,* **2**, 114-128 (1968).
91. Beaumelle,B.D. and Vial,H.J., *J. Chromatogr.*, **356**, 187-194 (1986).
92. Beaumelle,B.D. and Vial,H.J., *J. Chromatogr.*, **356**, 409-412 (1986).
93. Beneytout,J.L., Tixier,M. and Rigaud,M., *J. Chromatogr.*, **351**, 363-365 (1986).
94. Berson,J.A., Poonian,M.S. and Libbey,W.J., *J. Am. Chem. Soc.*, **91**, 5567-5579 (1969).
95. Bertelsen,O. and Dinh-Nguyen,N., *Chem. Scripta*, **19**, 172-175 (1982).
96. Bertelsen,O. and Dinh-Nguyen,N., *Fette Seifen Anstrichm.*, **87**, 336-342 (1985).
97. Bezard,J.A., *Lipids*, **6**, 630-634 (1971).
98. Bezard,J.A. and Bugaut,M., *J. Chromatogr. Sci.*, **7**, 639-634 (1969).
99. Bianchini,J.P., Ralaimanarivo,A. and Gaydou,E.M., *Anal. Chem.*, **53**, 2194-2201 (1981).
100. Bierl,B.A., Beroza,M. and Aldridge,M.H., *Anal. Chem.*, **43**, 636-641 (1971).
101. Bitman,J., Wood,D.L., Hamosh,M., Hamosh,P. and Mehta,N.R., *Am. J. Clin. Nutr.*, **38**, 300-312 (1983).
102. Bitman,J., Wood,D.L., Mehta,N.R., Hamosh,P. and Hamosh,M., *Am. J. Clin. Nutr.*, **40**, 1103-1119 (1984).
103. Bjerve,K.S., Daae,L.N.W. and Bremer,J., *Anal. Biochem.*, **58**, 238-245 (1974).
104. Blank,M.L., Cress,E.A. and Snyder,F., *J. Chromatogr.*, **392**, 421-425 (1987).
105. Blank,M.L., Kasawa,K. and Snyder,F., *J. Lipid Res.*, **13**, 390-395 (1972).
106. Blank,M.L., Rainey,W.T., Christie,W.H., Piantodosi,C. and Snyder,F., *Chem. Phys. Lipids*, **17**, 201-206 (1976).
107. Blank,M.L., Robinson,M. and Snyder,F., in *Platelet-Activating Factor and related Lipid Mediators*, pp. 33-52 (1987) (edited by F.Snyder, Plenum Press, New York).
108. Blank,M.L. and Snyder,F., *Lipids*, **5**, 337-341 (1970).
109. Blank,M.L. and Snyder,F., *Biochemistry*, **9**, 5034-5036 (1970).
110. Blank,M.L. and Snyder,F., *J. Chromatogr.*, **273**, 415-420 (1983).
111. Blau,K. and King,G.S. (editors), *Handbook of Derivatives for Chromatography* (1978) (Heyden & Son, London).
112. Blomberg,J., *Lipids*, **9**, 461-470 (1974).
113. Blomquist,G.J., Soliday,C.L., Byers,B.A., Brakke,J.W. and Jackson,L.J., *Lipids*, **7**, 356-362 (1972).
114. Body,D.R., in *Handbook of Chromatography. Vol. I. Lipids*, pp. 241-275 (1984) (edited by H.K.Mangold, CRC Press, Boca Raton).
115. Bohov,P., Balaz,V. and Hrivnak,J., *J. Chromatogr.*, **286**, 247-252 (1984).
116. Boniforti,L., Lorusso,S., Chiaccherini,E., Mariani,C. and Fedeli,E., *Riv. Ital. Sostanze Grasse*, **67**, 455-457 (1985).
117. Boon,J.J., De Leeuw,J.W., van der Hoek,G.J. and Vosjan,J.H., *J. Bact.*, **129**, 183-1191 (1977).
118. Boon,J.J., Van de Graaf,B., Schuyl,P.J.W., De Lange,F. and De Leeuw,J.W., *Lipids*, **12**, 717-721 (1977).
119. Borch,R.F., *Anal. Chem.*, **47**, 2437-2439 (1975).
120. Bossant,M.J., Farinotti,R., Mencia-Huerta,J.M., Benveniste,J. and Mahuzier,G., *J. Chromatogr.*, **423**, 23-32 (1987).
121. Bottino,N.R., *J. Lipid Res.*, **12**, 24-30 (1971).
122. Bouhours,J-F., *J. Chromatogr.*, **169**, 462-465 (1979).
123. Brandt,A.E. and Lands,W.E.M., *Lipids*, **3**, 178-181 (1968).
124. Brash,A.R., Ingram,C.D. and Harris,T.M., *Biochemistry*, **26**, 5465-5471 (1987).
125. Breckenridge,W.C., in *Handbook of Lipid Research. Vol. 1. Fatty Acids and Glycerides*, pp. 197-232 (1978) (edited by A.Kuksis, Plenum Press, New York).
126. Breimer,M.E., *J. Lipid Res.*, **16**, 189-194 (1975).
127. Breuer,B., Stuhlfauth,T. and Fock,H.P., *J. Chromatogr. Sci.*, **25**, 302-306 (1987).
128. Brian,B.L. and Gardner,E.W., *Appl. Microbiol.*, **16**, 549-552 (1968).

129. Brian,B.L., Gracy,R.W. and Scholes,V.E., *J. Chromatogr.*, **66**, 138-140 (1972).
130. Brockerhoff,H. and Jensen,R.G., *Lipolytic Enzymes* (1974) (Academic Press, New York).
131. Brondz,I., Olsen,I. and Greibrokk,T., *J. Chromatogr.*, **274**, 299-304 (1983).
132. Brooks,C.W.J., Cole,W.J. and Barrett,G.M., *J. Chromatogr.*, **315**, 119-133 (1984).
133. Brooks,C.J.W. and MacLean,I., *J. Chromatogr. Sci.*, **9**, 18-24 (1971).
134. Brooks,J.B., Liddle,J.A. and Alley,C.C., *Anal. Chem.*, **47**, 1960-1965 (1975).
135. Brosche,T., Platt,D. and Vostrowsky,O., *J. Chromatogr.*, **345**, 219-227 (1985).
136. Brown,A.S., Cho,K.Y., Cheung,H.T.A., Hemmens,V. and Vine,J., *J. Chromatogr.*, **341**, 139-145 (1985).
137. Brown,J.B. and Kolb,D.K., *Prog. Chem. Fats other Lipids*, **3**, 57-94 (1955).
138. Brunnekreeft,J.W.I. and Leijnse,B., *J. Clin. Chem. Clin. Biochem.*, **24**, 445-449 (1986).
139. Budzikiewicz,H., *Fres. Z. Anal. Chem.*, **321**, 150-158 (1985).
140. Bu'Lock,J.D. and Smith,G.N., *J. Chem. Soc. (C)*, 332-336 (1967).
141. Bus,J., Sies,I. and Lie Ken Jie,M.S.F., *Chem. Phys. Lipids*, **17**, 501-518 (1976).
142. Bus,J., Sies,I. and Lie Ken Jie,M.S.F., *Chem. Phys. Lipids*, **18**, 130-144 (1977).
143. Buser,H-R., Arn,H., Geurin,P. and Rauscher,S., *Anal. Chem.*, **55**, 818-822 (1983).
144. Butte,W., *J. Chromatogr.*, **261**, 142-145 (1983).
145. Capella,P., Galli,C. and Fumagalli,R., *Lipids*, **3**, 431-438 (1968).
146. Capella,P. and Zorzut,C.M., *Anal. Chem.*, **40**, 1458-1463 (1968).
147. Carballeira,N.M., Maldonado,L. and Porras,B., *Lipids*, **22**, 767-769 (1987).
148. Carey,E.M., *Lipids*, **17**, 656-661 (1982).
149. Carter,H.E. and Gaver,R.C., *J. Lipid Res.*, **8**, 391-395 (1967).
150. Carter,H.E., Rothfus,J.A. and Gigg,R., *J. Lipid Res.*, **2**, 227-234 (1961).
151. Chadha,R.K., Lawrence,J.F. and Conacher,H.B.S., *J. Chromatogr.*, **356**, 441-444 (1986).
152. Challinor,C.J., Hamilton,R.J. and Simpson,K., *Chem. Phys. Lipids*, **3**, 145-151 (1969).
153. Chapman,D. and Goni,F.M., in *The Lipid Handbook*, pp. 385-447 (1986) (edited by F.D.Gunstone, J.L.Harwood and F.B.Padley, Chapman and Hall, London).
154. Chen,S.L., Stein,R.A. and Mead,J.F., *Chem. Phys. Lipids*, **16**, 161-166 (1976).
155. Chobanov,D., Tarandjiska,R. and Chobanova,R., *J. Am. Oil Chem. Soc.*, **53**, 48-51 (1976).
156. Christiansen,K., Mahadevan,V., Viswanathan,C.V. and Holman,R.T., *Lipids*, **4**, 421-427 (1969).
157. Christie,W.W., *J. Chromatogr.*, **34**, 405-406 (1968).
158. Christie,W.W., *J. Chromatogr.*, **37**, 27-32 (1968).
159. Christie,W.W., in *Topics in Lipid Chemistry. Vol. 1.*, pp. 1-49 (1970) (edited by F.D.Gunstone, Logos Press, London).
160. Christie,W.W., in *Topics in Lipid Chemistry. Vol. 3.*, pp. 171-197 (1972) (edited by F.D.Gunstone, Logos Press, London).
161. Christie,W.W., *Biochim. Biophys. Acta*, **316**, 204-211 (1973).
162. Christie,W.W., in *Lipid Metabolism in Ruminant Animals*, pp. 95-191 (1981) (edited by W.W.Christie, Pergamon Press, Oxford).
163. Christie,W.W., *Lipid Analysis* (Second Edition) (1982) (Pergamon Books, Oxford).
164. Christie,W.W., *J. Lipid Res.*, **23**, 1072-1075 (1982).
165. Christie,W.W., *J. Lipid Res.*, **26**, 507-512 (1985).
166. Christie,W.W., *J. Chromatogr.*, **361**, 396-399 (1986).
167. Christie,W.W., in *Analysis of Oils and Fats*, pp. 313-339 (1986) (edited by R.J. Hamilton and J.B.Rossell, Elsevier Applied Science, London).
168. Christie,W.W., *High-Performance Liquid Chromatography and Lipids* (1987) (Pergamon Books, Oxford).
169. Christie,W.W., *J. High Res. Chromatogr., Chromatogr. Commun.*, **10**, 148-150 (1987).
170. Christie,W.W., *J. Chromatogr.*, **447**, 305-314 (1988).
171. Christie,W.W., Brechany,E.Y., Gunstone,F.D., Lie Ken Jie,M.S.F. and Holman,R.T., *Lipids*, **22**, 664-666 (1987).
172. Christie,W.W., Brechany,E.Y. and Holman,R.T., *Lipids*, **22**, 224-228 (1987).
173. Christie,W.W., Brechany,E.Y., Johnson,S.B. and Holman,R.T., *Lipids*, **21**, 657-661 (1986).
174. Christie,W.W., Brechany,E.Y. and Lie Ken Jie,M.S.F., *Chem. Phys. Lipids*, **46**, 225-229 (1988).

175. Christie,W.W., Brechany,E.Y. and Stefanov,K., *Chem. Phys. Lipids*, **46**, 127-136 (1988).
176. Christie,W.W., Gunstone,F.D., Ismail,I.A. and Wade,L., *Chem. Phys. Lipids*, **2**, 196-202 (1968).
177. Christie,W.W., Gunstone,F.D. and Prentice,H.G., *J. Chem. Soc.*, 5768-5771 (1963).
178. Christie,W.W., Gunstone,F.D., Prentice,H.G. and Sen Gupta,S.C., *J. Chem. Soc.*, 5833-5837 (1964).
179. Christie,W.W. and Holman,R.T., *Lipids*, **1**, 176-182 (1966).
180. Christie,W.W. and Holman,R.T., *Chem. Phys. Lipids*, **1**, 407-423 (1967).
181. Christie,W.W. and Moore,J.H., *J. Sci. Food Agric.*, **22**, 120-124 (1971).
182. Christie,W.W., Noble,R.C. and Moore,J.H., *Analyst (London)*, **95**, 940-944 (1970).
183. Christie,W.W., Rebello,D. and Holman,R.T., *Lipids*, **4**, 229-231 (1969).
184. Christie,W.W. and Stefanov,K., *J. Chromatogr.*, **392**, 259-265 (1987).
185. Christopherson,S.W. and Glass,R.L., *J. Dairy Sci.*, **52**, 1289-1290 (1969).
186. Ciucanu,I. and Kerek,F., *J. Chromatogr.*, **284**, 179-185 (1984).
187. Cohen,A., Hertz,H.S., Mandel,J., Paule,R.C., Schaffer,R., Sniegoski,L.T., Welch,M.J. and White,E., *Clin. Chem.*, **26**, 854-860 (1980).
188. Colborne,A.J. and Laidman,D.L., *Phytochem.*, **14**, 2639-2645 (1975).
189. Conacher,H.B.S., *J. Ass. Off. Anal.Chem.*, **58**, 488-491 (1975).
190. Conacher,H.B.S., *J. Chromatogr. Sci.*, **14**, 405-411 (1976).
191. Conacher,H.B.S. and Gunstone,F.D., *Chem. Phys. Lipids*, **3**, 203-220 (1969).
192. Conacher,H.B.S., Gunstone,F.D., Hornby,G.M. and Padley,F.B., *Lipids*, **5**, 434-441 (1970).
193. Conacher,H.B.S., Iyengar,J.R. and Beare-Rogers,J.L., *J. Assoc. Off. Anal. Chem.*, **60**, 899-903 (1977).
194. Conway,J., Ratnayake,W.M.N. and Ackman,R.G., *J. Am. Oil Chem. Soc.*, **62**, 1340-1343 (1985).
195. Corey,E.J. and Venkateswarlu,A., *J. Am. Chem. Soc.*, **94**, 6190-6191 (1972).
196. Cranwell,P.A., Robinson,N. and Eglinton,G., *Lipids*, **20**, 645-651 (1985).
197. Craske,J.D. and Bannon,C.D., *J. Am. Oil Chem. Soc.*, **64**, 1413-1417 (1987).
198. Craske,J.D. and Edwards,R.A., *J. Chromatogr.*, **53**, 253-261 (1970).
199. Creer,M.H., Pastor,C., Corr,P.B., Gross,R.W. and Sobel,B.E., *Anal. Biochem.*, **144**, 65-74 (1985).
200. Curstedt,T., *Biochim. Biophys. Acta*, **360**, 12-23 (1974).
201. Curstedt,T., *Biochim. Biophys. Acta*, **489**, 79-88 (1977).
202. Curstedt,T., in *Ether Lipids: Biochemical and Biomedical Aspects*, pp. 1-15 (1983) (edited by H.K.Mangold and F.Paltauf, Academic Press, London).
203. Curstedt,T. and Sjovall,J., *Biochim. Biophys. Acta*, **360**, 24-37 (1974).
204. D'Alonzo,R.P., Kozarek,W.J. and Wade,R.L., *J. Am. Oil Chem. Soc.*, **59**, 292-295 (1982).
205. Darbre,A., in *Handbook of Derivatives for Chromatography*, pp. 36-103 (1978) (edited by K.Blau & G.S.King, Heyden & Son, London).
206. Dasgupta,A., Ayanoglu,E. and Djerassi,C., *Lipids*, **19**, 768-776 (1984).
207. Dasgupta,A., Ayanoglu,E., Wegmann-Szente,A., Tomer,K.B. and Djerassi,C., *Chem. Phys. Lipids*, **41**, 335-347 (1986).
208. Davies,J.E.D., Hodge,P., Barve,J.A., Gunstone,F.D. and Ismail,I.A., *J. Chem. Soc. Perkin II*, 1557-1561 (1972).
209. Davies,J.E.D., Hodge,P., Gunstone,F.D. and Lie Ken Jie,M.S.F., *Chem. Phys. Lipids*, **15**, 48-52 (1975).
210. Davies,J.E.D., Lie Ken Jie,M.S.F. and Lam,C.H., *Chem. Phys. Lipids*, **15**, 157-160 (1975).
211. Deeth,H.C., Fitz-Gerald,C.H. and Snow,A.J., *N. Z. J. Dairy Sci. Technol.*, **18**, 13-20 (1983).
212. Deffense,E., *Rev. Franc. Corps Gras*, **31**, 123-129 (1984).
213. De Haas,G.H. and Van Deenen,L.L.M., *Biochim. Biophys. Acta*, **106**, 315-325 (1965).
214. Derks,H.J.G.M., van Heiningen,A. and Koedam,H.C., *Clin. Chem.*, **31**, 691-694 (1985).
215. De Rosa,M. and Gambacorta,A., *Phytochem.*, **14**, 209-210 (1975).
216. De Rosa,M. and Gambacorta,A., *Prog. Lipid Res.*, **27**, 153-175 (1988).
217. Desty,D.H., Goldup,A. and Swanton,W.T., in *Gas Chromatography*, pp. 105-138 (1962)

(edited by N.Brenner, J.E.Callen & M.D.Weiss, Academic Press, New York).
218. Devinat,G., Coustille,J.L., Perrin,J-L. and Prevot,A., *Rev. Franc. Corps Gras*, **30**, 463-468 (1983).
219. Dickens,B.F., Ramesha,C.S. and Thompson,G.A., *Anal. Biochem*., **127**, 37-48 (1982).
220. Dickens,B.F. and Thompson,G.A., *Biochemistry*, **21**, 3604-3611 (1982).
221. Dinh-Nguyen,N., Ryhage,R. and Stallberg-Stenhagen,S., *Arkiv. Kemi*, **15**, 433-438 (1960).
222. Dittmar,K.E.J., Heckers,H. and Melcher,F.W., *Fette Seifen Anstrichm*., **80**, 297-303 (1978).
223. Dix,T.A. and Marnett,L.J., *J. Biol. Chem*., **260**, 5351-5357 (1985).
224. Dolev,A., Rohwedder,W.K. and Dutton,H.J., *Lipids*, **1**, 231-233 (1966).
225. Dommes,V., Wirtz-Peitz,F. and Kunau,W-H., *J. Chromatogr. Sci*., **14**, 360-366 (1976).
226. Dorris,G.M., Douek,M. and Allen,L.H., *J. Am. Oil Chem. Soc*., **59**, 494-500 (1982).
227. Downing,D.T. and Greene,R.S., *Lipids*, **3**, 96-100 (1968).
228. Downing,D.T. and Sharaf,D.M., *Biochim. Biophys. Acta*, **431**, 378-389 (1976).
229. Draffan,G.H., Stillwell,R.N. and McCloskey,J.A., *Org. Mass Spectrom*., **1**, 669-685 (1968).
230. Drozd,J., *J. Chromatogr*., **113**, 303-356 (1975).
231. Dudley,P.A. and Anderson,R.E., *Lipids*, **10**, 113-114 (1975).
232. Duncan,W.R.H., Lough,A.K., Garton,G.A. and Brooks,P., *Lipids*, **9**, 669-673 (1974).
233. Dunkelblum,E., Tan,S.H. and Silk,P.J., *J. Chem. Ecol*., **11**, 265-277 (1985).
234. Dunn,J.A., Holland,K.B. and Jezorek,J.R., *J. Chromatogr*., **394**, 375-381 (1987).
235. Eagles,J., Fenwick,G.R. and Self,R., *Biomed. Mass Spectrom*., **6**, 462-464 (1979).
236. Eckert,W.R., *Fette Seifen Anstrichm*., **79**, 360-362 (1977).
237. Egge,H., in *Ether Lipids: Biochemical and Biomedical Aspects*, pp. 17-47 (1983) (edited by H.K.Mangold and F.Paltauf, Academic Press, London).
238. Egge,H., Murawski,U., Ryhage,R., Gyorgy,P., Chatranon,W. and Zilliken,P., *Chem. Phys. Lipids*, **8**, 42-55 (1972).
239. Eglinton,G. and Hunneman,D.H., *Phytochem*., **7**, 313-322 (1968).
240. Eglinton,G., Hunneman,D.H. and McCormick,A., *Org. Mass Spectrom*., **1**, 593-611 (1968).
241. Eisele,T.A., Libbey,L.M., Pawloski,N.E., Nixon,J.E. and Sinnhuber,R.O., *Chem. Phys. Lipids*, **12**, 316-326 (1974).
242. El-Hamdy,A.H. and Perkins,E.G., *J. Am. Oil Chem. Soc*., **58**, 867-872 (1981).
243. Emken,E.A., *Lipids*, **6**, 686-687 (1971).
244. Emken,E.A., *Lipids*, **7**, 459-466 (1972).
245. Emken,E.A. and Dutton,H.J., *Lipids*, **9**, 272-278 (1974).
246. Emken,E.A., Hartman,J.C. and Turner,C.R., *J. Am. Oil Chem. Soc*., **55**, 561-563 (1978).
247. Evans,N., Games,D.E., Harwood,J.L. and Jackson,A.H., *Biochem. Soc. Trans*., **2**, 1091-1093 (1970).
248. Evershed,R.P. and Goad,L.J., *Biomed. Environ. Mass Spectrom*., **14**, 131-140 (1987).
249. Evershed,R.P., Male,V. and Goad,L.J., *J. Chromatogr*., **400**, 187-205 (1987).
250. Evershed,R.P., Prescott,M.C., Goad,L.J. and Rees,H.H., *Biochem. Soc. Trans*., **15**, 175-179 (1987).
251. Fallon,W.E. and Shimizu,Y., *Lipids*, **12**, 765-768 (1977).
252. Farooqui,A.A., *Adv. Lipid Res*., 18, 159-227 (1981).
253. Fellenberg,A.J., Johnson,D.W., Poulos,A. and Sharp,P., *Biomed. Environ. Mass Spectrom*., **14**, 127-130 (1987).
254. Ferrell,W.J. and Radloff,D.M., *Physiol. Chem. Phys*., **2**, 551-557 (1970).
255. Ferrell,W.J., Radloff,J.F. and Jackiw,A.B., *Lipids*, **4**, 278-282 (1969).
256. Fine,J.B. and Sprecher,H., *J. Lipid Res*., **23**, 660-663 (1982).
257. Firestone,D. and Horowitz,W., *J. Ass. Off. Anal Chem*., **62**, 709-721 (1979).
258. Fisher,G.S. and Schuller,W.H., *J. Am. Oil Chem. Soc*., **58**, 943-946 (1981).
259. Folch,J., Lees,M. and Stanley,G.H.S., *J. Biol. Chem*., **226**, 497-509 (1957).
260. Francesangeli,E., Porcellati,S., Horrocks,L.A. and Goracci,G., *J. Liqu. Chromatogr*., **10**, 2799-2808 (1987).
261. Francis,G.W., *Chem. Phys. Lipids*, **29**, 369-374 (1981).
262. Frankel,E.N., Neff,W.E. and Plattner,R.D., *Lipids*, **21**, 333-337 (1986).
263. Frankel,E.N., Neff,W.E., Rohwedder,W.K., Khambay,B.P.S., Garwood,R.F. and

Weedon,B.C.L., *Lipids*, **12**, 901-907 (1977).
264. Frede,E., *Chromatographia*, **21**, 29-36 (1986).
265. Frede,E. and Thiele,H., *J. Am. Oil Chem. Soc.*, **64**, 521-528 (1987).
266. Freeman,N.K., *J. Am. Oil Chem. Soc.*, **45**, 798-809 (1968).
267. Freeman,R.R. (editor), *High Resolution Gas Chromatography* (Second Edition) (1981) (Hewlett-Packard Inc.).
268. Frost,D.J. and Barzilay,J., *Anal. Chem.*, **43**, 1316-1318 (1971).
269. Frost,D.J. and Gunstone,F.D., *Chem. Phys. Lipids*, **15**, 53-85 (1975).
270. Frost,D.J. and Sies,I., *Chem. Phys. Lipids*, **13**, 173-177 (1974).
271. Funasaki,H. and Gilbertson,J.R., *J. Lipid Res.*, **9**, 766-768 (1968).
272. Gailly,C., Sandra,P., Verzele,M. and Cocito,C., *Eur. J. Biochem.*, **125**, 83-94 (1982).
273. Gambert,P., Lallemant,C., Archambault,A., Maume,B.F. and Padieu,P., *J. Chromatogr.*, **162**, 1-6 (1978).
274. Games,D.E., *Chem. Phys. Lipids*, **21**, 389-402 (1978).
275. Garton,G.A., *Chem. Ind. (London)*, 295-300 (1985).
276. Gaskell,S.J., in *Glass Capillary Chromatography in Clinical Medicine and Pharmacology,* pp. 329-348 (1985) (edited by H. Jaeger, Marcel Dekker, New York).
277. Gaskell,S.J. and Brooks,C.J.W., *J. Chromatogr.*, **122**, 415-424 (1976).
278. Gaskell,S.J. and Brooks,C.J.W., *J. Chromatogr.*, **142**, 469-480 (1977).
279. Gaskell,S.J., Edmonds,C.G. and Brooks,C.J.W., *J. Chromatogr.*, **126**, 591-599 (1976).
280. Gaver,R.C. and Sweeley,C.C., *J. Am. Chem. Soc.*, **88**, 3643-3647 (1966).
281. Gaydou,E.M., Biachini,J.P. and Ralaimanarivo,A., *Anal. Chem.*, **55**, 2313-2317 (1983).
282. Gaydou,E.M., Miralles,J. and Rasoazanakolona,V., *J. Am. Oil Chem. Soc.*, **64**, 997-1000 (1987).
283. Geeraert,E., in *Sample Introduction in Capillary Gas Chromatography, Vol. 1*, pp. 133-157 (1985) (edited by P. Sandra, Huethig, Heidelberg).
284. Geeraert,E., in *Chromatography of Lipids in Biomedical Research and Clinical Diagnosis*, pp. 48-75 (1987) (edited by A.Kuksis, Elsevier, Amsterdam).
285. Geeraert,E. and De Schepper,D., *J. High Res. Chromatogr., Chromatogr. Commun.*, **5**, 80-84 (1982).
286. Geeraert,E., De Schepper,D. and Sandra,P., *J. High Res. Chromatogr., Chromatogr. Commun.*, **6**, 386-387 (1983).
287. Geeraert,E. and Sandra,P., *J. High Res. Chromatogr., Chromatogr. Commun.*, **7**, 431-432 (1984).
288. Geeraert,E. and Sandra,P., *J. High Res. Chromatogr., Chromatogr. Commun.*, **8**, 415-422 (1985).
289. Geeraert,E. and Sandra,P., in *Proc. 6th Int. Symp. Capillary Chromatogr.*, pp. 174-189 (1985) (Heuthig, Heidelberg).
290. Geeraert,E. and Sandra,P., *J. Am. Oil Chem. Soc.*, **64**, 100-105 (1987).
291. Geeraert,E., Sandra,P. and De Schepper,D., *J. Chromatogr.*, **279**, 287-295 (1983).
292. Gensler,W.J. and Marshall,J.P., *J. Org. Chem.*, **42**, 126-129 (1977).
293. Gensler,W.J. and Marshall,J.P., *Chem. Phys. Lipids*, **19**, 128-143 (1977).
294. Gerber,J.G., Barnes,J.S. and Nies,A.S., *J. Lipid Res.*, **20**, 912-914 (1979).
295. Gershfeld,N.L., *Anal. Biochem.*, **116**, 75-79 (1981).
296. Gerson,T., Patel,J.J. and Nixon,L.N., *Lipids*, **10**, 134-139 (1975).
297. Gildenberg,L. and Firestone,D., *J. Assoc. Off. Anal. Chem.*, **68**, 46-51 (1985).
298. Gillan,F.T., *J. Chromatogr. Sci.*, **21**, 293-297 (1983).
299. Glass,C.A. and Dutton,H.J., *Anal. Chem.*, **36**, 2401-2404 (1964).
300. Glass,R.L., Krick,T.P., Olson,D.L. and Thorson,R.L., *Lipids*, **12**, 828-836 (1977).
301. Glass,R.L., Krick,T.P., Sand,D.M., Rahn,C.R. and Schlenk,H., *Lipids*, **10**, 695-702 (1975).
302. Graille,J., Pina,M. and Pioch,D., *J. Am. Oil Chem. Soc.*, **63**, 111-116 (1986).
303. Gray,G.M., in *Lipid Chromatographic Analysis, Second Edition, Vol. 3.*, pp. 897-923 (1975) (edited by G.V.Marinetti, Marcel Dekker, New York).
304. Greenspan,M.D. and Schroeder,E.A., *Anal. Biochem.*, **127**, 441-448 (1982).
305. Grob,K., *J. Chromatogr.*, **178**, 387-392 (1979).
306. Grob,K., *J. Chromatogr.*, **205**, 289-296 (1981).
307. Grob,K., *J. Chromatogr.*, **251**, 235-248 (1982).

308. Grob,K., *J. Chromatogr.*, **279**, 225-232 (1983).
309. Grob,K., *J. Chromatogr.*, **287**, 1-14 (1984).
310. Grob,K. and Bossard,M., *J. Chromatogr.*, **294**, 65-75 (1984).
311. Grob,K. and Laubli,T., *J. Chromatogr.*, **357**, 345-355 (1986).
312. Grob,K. and Neukom,H.P., *J. Chromatogr.*, **189**, 109-117 (1980).
313. Grob,K., Neukom,H.P. and Battaglia,R., *J. Am. Oil Chem. Soc.*, **57**, 282-286 (1980).
314. Grob,R.L. (editor) *Modern Practice of Gas Chromatography* (1977) (J. Wiley & Sons, New York).
315. Grogan,W.M., *Lipids*, **19**, 341-346 (1984).
316. Gross,R.W. and Sobel,B.E., *J. Chromatogr.*, **197**, 79-85 (1980).
317. Gunstone,F.D., *Trends Biochem. Sci.*, **3**, 54N (1978).
318. Gunstone,F.D., in *The Lipid Handbook*, pp. 1-23 (1986) (edited by F.D.Gunstone, J.L.Harwood and F.B.Padley, Chapman and Hall, London).
319. Gunstone,F.D., Harwood,J.L. and Padley,F.B. (editors), *The Lipid Handbook*, (1986) (Chapman and Hall, London).
320. Gunstone,F.D. and Inglis,R.P., in *Topics in Lipid Chemistry. Vol. 2*, pp. 287-307 (1971) (edited by F.D. Gunstone, Logos Press, London).
321. Gunstone,F.D. and Inglis,R.P., *Chem. Phys. Lipids*, **10**, 73-88 (1973).
322. Gunstone,F.D. and Ismail,I.A., *Chem. Phys. Lipids*, **1**, 337-340 (1967).
323. Gunstone,F.D., Ismail,I.A. and Lie Ken Jie,M.S.F., *Chem. Phys. Lipids*, **1**, 376-385 (1967).
324. Gunstone,F.D. and Jacobsberg,F.R., *Chem. Phys. Lipids*, **9**, 26-34 (1972).
325. Gunstone,F.D., Kilcast,D., Powell,R.G. and Taylor,G.M., *Chem. Commun.*, 295 (1967).
326. Gunstone,F.D. and Lie Ken Jie,M.S.F., *Chem. Phys. Lipids*, **4**, 131-138 (1970).
327. Gunstone,F.D. and Lie Ken Jie,M.S.F., *Chem. Phys. Lipids*, **4**, 139-146 (1970).
328. Gunstone,F.D., Lie Ken Jie,M.S.F. and Wall,R.T., *Chem. Phys. Lipids*, **3**, 297-303 (1969).
329. Gunstone,F.D. and Morris,L.J., *J. Chem. Soc.*, 2127-2132 (1959).
330. Gunstone,F.D. and Padley,F.B., *J. Am. Oil Chem. Soc.*, **42**, 957-961 (1965).
331. Gunstone,F.D., Pollard,M.R., Scrimgeour,C.M., Gilman,N.W. and Holland,B.C., *Chem. Phys. Lipids*, **17**, 1-13 (1976).
332. Gunstone,F.D., Pollard,M.R., Scrimgeour,C.M. and Vedanayagam, H.S., *Chem. Phys. Lipids*, **18**, 115-129 (1977).
333. Gunstone,F.D. and Subbarao,R., *Chem. Ind. (London)*, 461-462 (1966).
334. Gunstone,F.D. and Subbarao,R., *Chem. Phys. Lipids*, **1**, 349-359 (1967).
335. Gunstone,F.D., Wijesundera,R.C. and Scrimgeour,C.M., *J. Sci. Food Agric.*, **29**, 539-550 (1978).
336. Haekel,R., Sonntag,O., Kulpman,W.R. and Feldman,U., *J. Clin. Chem. Clin. Biochem.*, **17**, 553-563 (1979).
337. Hallgren,B., Niklasson,A., Stallberg,G. and Thorin,H., *Acta Chem. Scand.*, **B28**, 1035-1040 (1974).
338. Hallgren,B., Ryhage,R. and Stenhagen,E., *Acta Chem. Scand.*, **13**, 845-847 (1959).
339. Hallgren,B. and Stallberg,G., *Acta Chem. Scand.*, **21**, 1519-1529 (1967).
340. Hamberg,M., *Chem. Phys. Lipids*, **6**, 152-158 (1971).
341. Hamberg,M. and Samuelsson,B., *J. Biol. Chem.*, **242**, 5329-5335 (1967).
342. Hamilton,R.J. and Ackman,R.G., *J. Chromatogr. Sci.*, **13**, 474-478 (1975).
343. Hamilton,R.J. and Raie,M.Y., *Chem. Ind. (London)*, 1228-1229 (1971).
344. Hamilton,R.J., Raie,M.Y. and Miwa,T.K., *Chem. Phys. Lipids*, **14**, 92-96 (1975).
345. Hammarstrom,S., *J. Lipid Res.*, **11**, 175-182 (1970).
346. Hammarstrom,S., *Eur. J. Biochem.*, **15**, 581-591 (1970).
347. Hammarstrom,S., *Eur. J. Biochem.*, **21**, 388-392 (1971).
348. Hammarstrom,S., *Methods Enzymol.*, **35**, 326-333 (1975).
349. Hammarstrom,S. and Hamberg,M., *Anal. Biochem.*, **52**, 169-179 (1973).
350. Hammarstrom,S. and Samuelsson,B., *J. Biol. Chem.*, **247**, 1001-1011 (1972).
351. Hammarstrom,S., Samuelsson,B. and Samuelsson,K., *J. Lipid Res.*, **11**, 150-157 (1970).
352. Hanahan,D.J., Ekholm,J. and Jackson,C.M., *Biochemistry*, **2**, 630-641 (1963).
353. Hanahan,D.J. and Kumar,R., *Prog. Lipid Res.*, **26**, 1-28 (1987).
354. Hansen,R.P., *Chem. Ind. (London)*, 1640-1641 (1967).
355. Hara,A. and Radin,N.S., *Anal. Biochem.*, **90**, 420-426 (1978).

356. Hara,A. and Taketomi,T., *J. Biochem. (Tokyo)*, **100**, 415-423 (1986).
357. Harlow,R.D., Litchfield,C. and Reiser,R., *Lipids*, **1**, 216-223 (1966).
358. Haroldsen,P.E., Clay,K.L. and Murphy,R.C., *J. Lipid Res.*, **28**, 42-49 (1987).
359. Harvey,D.J., *Biomed. Mass Spectrom.*, **9**, 33-38 (1982).
360. Harvey,D.J., *Biomed. Mass Spectrom.*, **11**, 187-192 (1984).
361. Harvey,D.J., *Biomed. Mass Spectrom.*, **11**, 340-347 (1984).
362. Harvey,D.J., *Biomed. Environ. Mass Spectrom.*, **14**, 103-110 (1987).
363. Harvey,D.J. and Tiffany,J.M., *J. Chromatogr.*, **301**, 173-188 (1984).
364. Harvey,D.J. and Tiffany,J.M., *Biomed. Mass Spectrom.*, **11**, 353-359 (1984).
365. Harvey,D.J., Tiffany,J.M., Duerden,J.M., Pandher,K.S. and Mengher,L.S., *J. Chromatogr.*, **414**, 253-263 (1987).
366. Harwood,J.L., in *The Biochemistry of Plants. Vol. 4. Lipids: Structure and Function*, pp. 1-55 (1980) (edited by P.K.Stumpf, Academic Press, New York).
367. Harwood,J.L. and Russell,N.J., *Lipids in Plants and Microbes* (1984) (Allen & Unwin, London).
368. Hase,A., Hase,T. and Anderegg,R., *J. Am. Oil Chem. Soc.*, **55**, 407-411 (1978).
369. Hasegawa,K. and Suzuki,T., *Lipids*, **8**, 631-634 (1973).
370. Hasegawa,K. and Suzuki,T., *Lipids*, **10**, 667-672 (1975).
371. Haslbeck,F. and Grosch,W., *Lipids*, **18**, 706-713 (1983).
372. Haslbeck,F., Grosch,W. and Firl,J., *Biochim. Biophys. Acta*, **750**, 185-193 (1983).
373. Hay,J.D. and Morrison,W.R., *Biochim. Biophys. Acta*, **202**, 237-243 (1970).
374. Hayashi,A. and Matsubara,T., *Biochim. Biophys. Acta*, **248**, 306-314 (1971).
375. Hayashi,A., Matsubara,T. and Matsuura,T., *Chem. Phys. Lipids*, **14**, 103-105 (1975).
376. Hayes,L., Lowry,R.R. and Tinsley,I.J., *Lipids*, **6**, 65-66 (1971).
377. Heath,R.R., Bursed,G.E. Tumlinson,J.H. and Doolittle,R.E., *J. Chromatogr.*, **189**, 199-208 (1980).
378. Heath,R.R., Tumlinson,J.H. and Doolittle,R.E., *J. Chromatogr. Sci.*, **15**, 10-13 (1977).
379. Heath,R.R., Tumlinson,J.H., Doolittle,R.E. and Proveaux,A.T., *J. Chromatogr. Sci.*, **13**, 380-382 (1975).
380. Heckers,H., Melcher,F.W. and Schloeder,U., *J. Chromatogr.*, **136**, 311-317 (1987).
381. Heftmann,E., *in Chromatography: Fundamentals and Applications of Chromatographic and Electrophoretic Methods. B. Applications*, pp. B191-B222 (1983) (edited by E.Heftmann, Elsevier, Amsterdam).
382. Heinze,F.J., Linscheid,M. and Heinz,E., *Anal. Biochem.*, **139**, 126-133 (1984).
383. Hinshaw,J.V. and Seferovic,W., *J. High Res. Chromatogr., Chromatogr. Commun.*, **9**, 69-72 (1986).
384. Hinshaw,J.V. and Seferovic,W., *J. High Res. Chromatogr., Chromatogr. Commun.*, **9**, 731-736 (1986).
385. Hites,R.A., *Anal. Chem.*, **42**, 1736-1740 (1970).
386. Hites,R.A., *Methods Enzymol.*, **35**, 348-358 (1975).
387. Hofstetter,H.H., Sen,N. and Holman,R.T., *J. Am. Oil Chem. Soc.*, **42**, 537-540 (1965).
388. Hokin,L.E., *Ann. Rev. Biochem.*, **54**, 205-235 (1985).
389. Holla,K.S. and Cornwell,D.G., *J. Lipid Res.*, **6**, 322-324 (1965).
390. Holla,K.S., Horrocks,L.A. and Cornwell,D.G., *J. Lipid Res.*, **5**, 263-265 (1964).
391. Holloway,P.J., in *Handbook of Chromatography. Vol. I. Lipids*, pp. 321-345 (1984) (edited by H.K.Mangold, CRC Press, Boca Raton).
392. Holman,R.T. and Hofstetter,H.H., *J. Am. Oil Chem. Soc.*, **42**, 540-544 (1965).
393. Holman,R.T. and Rahm,J.J., *Prog. Chem. Fats other Lipids*, **9**, 13-90 (1966).
394. Holmbom,B., *J. Am. Oil Chem. Soc.*, **54**, 289-293 (1977).
395. Holub,B.J. and Kuksis,A., *Adv. Lipid Res.*, **16**, 1-125 (1978).
396. Holub,B.J., Kuksis,A. and Thompson,W., *J. Lipid Res.*, **11**, 558-564 (1970).
397. Holz,G.G., Beach,D.H., Singh,B.N. and Fish,W.R., *Lipids*, **18**, 607-610 (1983).
398. Hooper,N.K. and Law,J.H., *J. Lipid Res.*, **9**, 270-275 (1968).
399. Hopkins,C.Y., *Prog. Chem. Fats other Lipids*, **8**, 213-252 (1966).
400. Hopkins,C.Y., *J. Am. Oil Chem. Soc.*, **45**, 778-783 (1968).
401. Hopkins,C.Y. and Bernstein,H.J., *Can. J. Chem.*, **37**, 775-782 (1959).
402. Horiike,M. and Hirano,C., *Biomed. Environ. Mass Spectrom.*, **14**, 183-186 (1987).
403. Horning,E.C., Karmen,A. and Sweeley,C.C., *Prog. Chem. Fats other Lipids*, **7**, 167-246

(1964).
404. Horning,M.G., Casparrini,G. and Horning,E.C., *J. Chromatogr. Sci.*, **7**, 267-275 (1969).
405. Horning,M.G., Murakami,S. and Horning,E.C., *Am. J. Clin. Nutr.*, **24**, 1086-1096 (1971).
406. Hughes,H., Smith,C.V., Horning,E.C. and Mitchell,J.R., *Anal. Biochem.*, **130**, 431-436 (1983).
407. Hughes,H., Smith,C.V., Tsokos-Kuhn,J.O. and Mitchell,J.R., *Anal. Biochem.*, **152**, 107-112 (1986).
408. Hunter,M.L., Christie,W.W. and Moore,J.H., *Lipids*, **8**, 65-70 (1973).
409. Imai,C., Watanabe,H., Haga,N. and Ii,T., *J. Am. Oil Chem. Soc.*, **51**, 326-330 (1974).
410. Ioannou,P.V. and Golding,B.T., *Prog. Lipid Res.*, **17**, 279-318 (1979).
411. Itabashi,Y. and Takagi,T., *Lipids*, **15**, 205-215 (1980).
412. Itabashi,Y. and Takagi,T., *J. Chromatogr.*, **299**, 351-364 (1984).
413. Itoh,K., Suzuki,A., Kuroki,Y. and Akino,T., *Lipids*, **20**, 611-616 (1985).
414. IUPAC, in *Standard Methods for the Analysis of Oils and Fats (7th Ed.)*, **Method 2.207**, pp. 99-102 (1987) (Pergamon Press, Oxford).
415. IUPAC-IUB Commission on Biochemical Nomenclature, *Eur. J. Biochem.*, **2**, 127-131 (1967); *Biochem. J.*, **105**, 897-902 (1967).
416. IUPAC-IUB Commission on Biochemical Nomenclature, *Hoppe-Seyler's Z. Physiol. Chem.*, **358**, 599-616 (1977); *J. Lipid Res.*, **19**, 114-125 (1978).
417. Iversen,S.A., Cawood,P., Madigan,M.J., Lawson,A.M. and Dormandy,T.L., *FEBS Letts.*, **171**, 320-324 (1984).
418. Iverson,J.L. and Sheppard,A.J., *J. Chromatogr. Sci.*, **13**, 505-508 (1975).
419. Iverson,J.L. and Sheppard,A.J., *J. Assoc. Off. Anal. Chem.*, **60**, 284-288 (1977).
420. Iverson,J.L. and Sheppard,A.J., *Food. Chem.*, **21**, 223-234 (1986).
421. Iverson,J.L. and Weik,R.W., *J. Ass. Off. Anal. Chem.*, **50**, 1111-1118 (1967).
422. Iyengar,R. and Schlenk,H., *Biochemistry*, **6**, 396-402 (1967).
423. Jacob,J., *J. Chromatogr. Sci.*, **13**, 415-422 (1975).
424. Jacob,J., *Hoppe-Seyler's Z. Physiol Chem.*, **357**, 609-611 (1976).
425. Jacob,J., in *Chemistry and Biochemistry of Natural Waxes*, pp. 94-146 (1976) (edited by P.E.Kolattukudy, Elsevier, Amsterdam).
426. Jacob,J. and Poltz,J., *Lipids*, **10**, 1-8 (1975).
427. Jaeger,H. (editor), *Glass Capillary Chromatography in Clinical Medicine and Pharmacology* (1985) (Marcel Dekker, New York).
428. Jaeger,H., Frank,H., Kloer,H-U. and Ditschuneit,H., in *Applications of Glass Capillary Gas Chromatography*, pp. 331-364 (1981) (edited by W.G. Jennings, Marcel Dekker, New York).
429. Jaeger,H., Kloer,H-U. and Ditschuneit,H., *J. Lipid Res.*, **17**, 185-190 (1976).
430. Jaeger,H., Kloer,H-U., Ditschuneit,H. and Frank,H., in *Glass Capillary Chromatography in Clinical Medicine and Pharmacology*, pp. 271-314 (1985) (edited by H.Jaeger, Marcel Dekker, New York).
431. James,A.T., *Spec. Pub. R. Soc. Chem. (Memb. Gas Sep. Enrich.)*, **62**, 175-200 (1986).
432. James,A.T. and Martin,A.J.P., *Biochem. J.*, **50**, 679-690 (1952).
433. James,A.T. and Martin,A.J.P., *Biochem. J.*, **63**, 144-152 (1956).
434. James,A.T. and Webb,J., *Biochem. J.*, **66**, 515-520 (1957).
435. Jamieson,G.R., in *Topics in Lipid Chemistry Vol. 1*, pp. 107-159 (1970) (edited by F.D. Gunstone, Logos Press, London).
436. Jamieson,G.R., *J. Chromatogr. Sci.*, **13**, 491-497 (1975).
437. Jamieson,G.R., McMinn,A.L. and Reid,E.H., *J. Chromatogr.*, **178**, 555-558 (1979).
438. Jamieson,G.R. and Reid,E.H., *J. Chromatogr.*, **26**, 8-16 (1967).
439. Jamieson,G.R. and Reid,E.H., *J. Chromatogr.*, **40**, 160-162 (1969).
440. Jamieson,G.R. and Reid,E.H., *J. Chromatogr.*, **42**, 304-310 (1969).
441. Jamieson,G.R. and Reid,E.H., *J. Chromatogr.*, **128**, 193-195 (1976).
442. Janssen,G. and Parmentier,G., *Biomed. Mass Spectrom.*, **5**, 439-443 (1978).
443. Janssen,G., Parmentier,G., Verhulst,A. and Eyssen,H., *Biomed. Mass Spectrom.*, **12**, 134-138 (1985).
444. Jennings,W.G. (editor), *Applications of Glass Capillary Chromatography* (1981) (Marcel Dekker, New York).
445. Jensen,N.J. and Gross,M.L., *Lipids*, **21**, 362-365 (1986).

446. Jensen,N.J. and Gross,M.L., *Mass Spectrom. Rev.*, **6**, 497-536 (1987).
447. Jensen,N.J., Tomer,K.B. and Gross,M.L., *Anal. Chem.*, **57**, 2018-2021 (1985).
448. Johns,S.R., Leslie,D.R., Willing,R.I. and Bishop,D.G., *Austral. J. Chem.*, **30**, 813-822 (1977).
449. Johnson,A.R., Murray,K.E., Fogerty,A.C., Kennett,B.H., Pearson,J.A. and Shenstone,F.S., *Lipids*, **2**, 316-322 (1967).
450. Johnson,C.B. and Holman,R.T., *Lipids*, **1**, 371-380 (1966).
451. Johnston,A.E., Dutton,H.J., Scholfield,C.R. and Butterfield,R.O., *J. Am. Oil Chem. Soc.*, **55**, 486-490 (1978).
452. Joseph,J.D., *Lipids*, **10**, 395-403 (1975).
453. Juaneda,P. and Rocquelin,G., *Lipids*, **20**, 40-41 (1985).
454. Kadowaki,H., Bremer,E.G., Evans,J.E., Jungalwala,F.B. and McCluer,R.H., *J. Lipid Res.*, **24**, 1389-1397 (1983).
455. Kalo,P., Vaara,K. and Antila,M., *J. Chromatogr.*, **368**, 145-151 (1986).
456. Kamerling,J.P. and Vliegenthart,J.F.G., in *Glass Capillary Chromatography in Clinical Medicine and Pharmacology*, pp. 371-379 (1985) (edited by H.Jaeger, Marcel Dekker, New York).
457. Kaneda,K., Naito,S., Imaizumi,S., Yano,I., Mizuno,S., Tomiyasu,I., Baba,T., Kusunose,E. and Kusunose,M., *J. Clin. Microbiol.*, **24**, 1060-1070 (1986).
458. Kaneda,T., *J. Chromatogr.*, **136**, 323-327 (1977).
459. Kanfer,J.N., *Methods Enzymol.*, **14**, 660-664 (1969).
460. Kanfer,J.N. and Hakomori,S-I. (editors), *Sphingolipid Biochemistry (Handbook of Lipid Research Vol. 3.)* (1983) (Plenum Press, New York).
461. Karlsson,K-A., *Acta Chem. Scand.*, **19**, 2425-2427 (1965).
462. Karlsson,K-A., *Acta Chem. Scand.*, **21**, 2577-2578 (1967).
463. Karlsson,K-A., *Acta Chem. Scand.*, **22**, 3050-3052 (1968).
464. Karlsson,K-A., *Lipids*, **5**, 878-891 (1970).
465. Karlsson,K-A., *Chem. Phys. Lipids*, **5**, 6-41 (1970).
466. Karlsson,K-A. and Martensson,E., *Biochim. Biophys. Acta*, **152**, 230-233 (1968).
467. Karlsson,K-A. and Pascher,I., *J. Lipid Res.*, **12**, 466-472 (1971).
468. Karlsson,K-A. and Pascher,I., *Chem. Phys. Lipids*, **12**, 65-74 (1974).
469. Karlsson,K-A., Samuelsson,B.E. and Steen,G.O., *Biochim. Biophys. Acta*, **316**, 336-362 (1973).
470. Kattner,G. and Fricke,H.S.G., *J. Chromatogr.*, **361**, 263-268 (1986).
471. Kaulen,H.D., *Anal. Biochem.*, **45**, 664-667 (1972).
472. Kawaguchi,A., Kobayashi,Y., Ogawa,Y. and Okuda,S., *Chem. Pharm. Bull.*, **31**, 3228-3232 (1983).
473. Kenner,G.W. and Stenhagen,E., *Acta Chem. Scand.*, **18**, 1551-1552 (1964).
474. Keough,T., Mihelich,E.D. and Eickhoff,D.J., *Anal. Chem.*, **56**, 1849-1852 (1984).
475. Kesselmeier,J. and Heinz,E., *Anal. Biochem.*, **144**, 319-328 (1985).
476. Kim,R.S. and LaBella,F.S., *J. Lipid Res.*, **28**, 1110-1117 (1987).
477. Kino,M., Matsumura,T., Gamo,M. and Saito,K., *Biomed. Mass Spectrom.*, **9**, 363-369 (1982).
478. Kito,M., Ishinaga,M., Nishihara,M., Kato,M., Sawada,S. and Hata,T., *Eur. J. Biochem.*, **54**, 55-63 (1975).
479. Kito,M., Takamura,H., Narita,H. and Urade,R., *J. Biochem. (Tokyo)*, **98**, 327-339 (1985).
480. Kleiman,R., Bohannon,M.B., Gunstone,F.D. and Barve,J.A., *Lipids*, **11**, 599-603 (1976).
481. Kleiman,R. and Spencer,G.F., *J. Am. Oil Chem. Soc.*, **50**, 31-38 (1973).
482. Kleiman,R., Spencer,G.F., Earle,F.R., Nieshlag,H.J. and Barclay,A.S., *Lipids*, **7**, 660-665 (1972).
483. Klein,R.A. and Kemp,P., in *Methods in Membrane Biology. Vol. 8*, pp. 51-217 (1977) (edited by E.D.Korn, Plenum Press, New York).
484. Klein,R.A. and Schmitz,B., *Biomed. Environ. Mass Spectrom.*, **13**, 429-438 (1986).
485. Klopfenstein,W.E., *J. Lipid Res.*, **12**, 773-776 (1971).
486. Klump,B., Melchert,H-U. and Rubach,K., *Fres. Z. Anal. Chem.*, **313**, 553-560 (1982).
487. Knights,B.A., Brown,A.C., Conway,E. and Middleditch,B.S., *Phytochem.*, **9**, 1317-1324 (1970).

488. Knoche,H.W., *Lipids*, **6**, 581-583 (1971).
489. Kobayashi,T., *J. Chromatogr.*, **194**, 404-409 (1980).
490. Kochhar,S.P. and Rossell,J.B., *International Analyst*, Issue 5, 23-26 (1987).
491. Kolattukudy,P.E. (editor), *The Chemistry and Biochemistry of Natural Waxes* (1976) (Elsevier, Amsterdam).
492. Kolattukudy,P.E. and Sawaya,W.N., *Lipids*, **9**, 290-292 (1974).
493. Kramer,J.K.G., Fouchard,R.C. and Jenkins,K.J., *J. Chromatogr. Sci.*, **23**, 54-56 (1985).
494. Kramer,J.K.G. and Hulan,H.W., *J. Lipid Res.*, **17**, 674-676 (1976).
495. Krisnangkura,K. and Sweeley,C.C., *Chem. Phys. Lipids*, **13**, 415-428 (1974).
496. Krohn,K., Eberlein,K. and Gercken,G., *J. Chromatogr.*, **153**, 550-552 (1978).
497. Krupcik,J. and Bohov,P., *J. Chromatogr.*, **346**, 33-42 (1985).
498. Kuksis,A., *Can. J. Biochem.*, **42**, 407-430 (1964).
499. Kuksis,A., *Fette Seifen Anstrichm.*, **73**, 130-138 (1971).
500. Kuksis,A., *Fette Seifen Anstrichm.*, **73**, 332-342 (1971).
501. Kuksis,A., *Can. J. Biochem.*, **49**, 1245-1250 (1971).
502. Kuksis,A., *J. Chromatogr. Sci.*, **10**, 53-56 (1972).
503. Kuksis,A., in *Analysis of Lipids and Lipoproteins*, pp. 36-62 (1975) (edited by E.G.Perkins, American Oil Chemists' Soc., Champaign).
504. Kuksis,A., in *Lipid Chromatographic Analysis. (2nd Edition) Vol. 1*, pp. 215-337 (1976) (edited by G.V.Marinetti, Marcel Dekker, New York).
505. Kuksis,A., *Separation Purification Methods*, **6**, 353-395 (1977).
506. Kuksis,A. (editor), *Handbook of Lipid Research. Vol. 1. Fatty Acids and Glycerides*, (1978) (Plenum Press, New York).
507. Kuksis,A., *in Chromatography: Fundamentals and Applications of Chromatographic and Electrophoretic Methods. B. Applications*, pp. B75-B146 (1983) (edited by E.Heftmann, Elsevier, Amsterdam).
508. Kuksis,A., in *Lipid Research Methodology*, pp. 78-132 (1984) (edited by J.A.Story, A.R.Liss Inc., New York).
509. Kuksis,A., in *Handbook of Chromatography. Vol. I. Lipids*, pp. 381-480 (1984) (edited by H.K.Mangold, CRC Press, Boca Raton).
510. Kuksis,A. (editor), *Chromatography of Lipids in Biomedical Research and Clinical Diagnosis*, (1987) (Elsevier, Amsterdam).
511. Kuksis,A. and Breckenridge,W.C., in *Dairy Lipids and Lipid Metabolism*, pp. 28-98 (1968) (edited by M.F.Brink and D.Kritchevsky, Avi Publishing Co., Westport).
512. Kuksis,A. and Marai,L., *Lipids*, **2**, 217-224 (1967).
513. Kuksis,A., Marai,L. and Gornall,D.A., *J. Lipid Res.*, **8**, 352-358 (1967).
514. Kuksis,A., Marai,L., Myher,J.J., Cerbulis,J. and Farrell,H.M., *Lipids*, **21**, 183-190 (1986).
515. Kuksis,A., Marai,L., Myher,J.J. and Pind,S., in *Chromatography of Lipids in Biomedical Research and Clinical Diagnosis*, pp. 403-440 (1987) (edited by A.Kuksis, Elsevier, Amsterdam).
516. Kuksis,A. and Myher,J.J., *J. Chromatogr.*, **379**, 57-90 (1986).
517. Kuksis,A. and Myher,J.J., in *Chromatography of Lipids in Biomedical Research and Clinical Diagnosis*, pp. 1-47 (1987) (edited by A. Kuksis, Elsevier, Amsterdam).
518. Kuksis,A., Myher,J.J., Geher,K., Breckenridge,W.C., Jones,G.J.L. and Little,J.A., *J. Chromatogr.*, **224**, 1-23 (1981).
519. Kuksis,A., Myher,J.J., Geher,K., Breckenridge,W.C. and Little, J.A., *J. Chromatogr.*, **230**, 231-252 (1981).
520. Kuksis,A., Myher,J.J., Geher,K., Hoffman,A.G.D., Breckenridge, W.C., Jones,G.J.L. and Little,J.A., *J. Chromatogr.*, **146**, 393-412 (1978).
521. Kuksis,A., Myher,J.J., Geher,K., Shaik,N.A., Breckenridge,W.C., Jones,G.J.L. and Little,J.A., *J. Chromatogr.*, **182**, 1-26 (1980).
522. Kuksis,A., Myher,J.J. and Marai,L., *J. Am. Oil Chem. Soc.*, **61**, 1582-1589 (1984).
523. Kuksis,A., Myher,J.J. and Marai,L., *J. Am. Oil Chem. Soc.*, **62**, 762-767 (1985).
524. Kuksis,A., Myher,J.J. and Marai,L., *J. Am. Oil Chem. Soc.*, **62**, 767-773 (1985).
525. Kuksis,A., Myher,J.J., Marai,L. and Geher,K., *J. Chromatogr. Sci.*, **13**, 423-430 (1975).
526. Kuksis,A., Myher,J.J., Marai,L. and Geher,K., *Anal. Biochem.*, **70**, 302-312 (1976).
527. Kuksis,A., Myher,J.J., Marai,L., Little,J.A., McArthur,R.G. and Roncari,D.A.K.,

Lipids, **21**, 371-377 (1986).
528. Kuksis,A., Myher,J.J., Marai,L., Little,J.A., McArthur,R.G. and Roncari,D.A.K., *J. Chromatogr.*, **381**, 1-12 (1986).
529. Kuksis,A., Stachnyk,O. and Holub,B.J., *J. Lipid Res.*, **10**, 660-667 (1969).
530. Kusram,K. and Polgar,N., *Lipids*, **6**, 961-962 (1971).
531. Laine,R.A., Young,N.D., Gerber,J.N. and Sweeley,C.C., *Biomed. Mass Spectrom.*, **1**, 10-14 (1974).
532. Lam,C.H. and Lie Ken Jie,M.S.F., *J. Chromatogr.*, **117**, 365-374 (1976).
533. Lam,C.H. and Lie Ken Jie,M.S.F., *J. Chromatogr.*, **121**, 303-311 (1976).
534. Lankelma,J., Ayanoglu,E. and Djerassi,C., *Lipids*, **18**, 853-858 (1983).
535. Lanza,E. and Slover,H.T., *Lipids*, **16**, 260-267 (1981).
536. Lanza,E., Zyren,J. and Slover,H.T., *J. Agr. Food Chem.*, **28**, 1182-1186 (1980).
537. Lauer,W.M., Aasen,W.M., Graff,G. and Holman,R.T., *Lipids*, **5**, 861-868 (1970).
538. Lawrence,J.F., Chadha,R.K. and Conacher,H.B.S., *J. Ass. Off. Anal. Chem.*, **66**, 1385-1389 (1984).
539. Lehmann,W.D. and Kessler,M., *Fres. Z. Anal. Chem.*, **312**, 311-316 (1982).
540. Leonhardt,B.A. and DeVilbiss,E.D., *J. Chromatogr.*, **322**, 484-490 (1985).
541. Leonhardt,B.A., DeVilbiss,E.D. and Klun,J.A., *Org. Mass Spectrom.*, **18**, 9-11 (1983).
542. Lercker,G., *J. Chromatogr.*, **279**, 543-548 (1983).
543. Lercker,G., Capella,P., Conte,L.S., Ruini,F. and Giordani,G., *Lipids*, **16**, 912-919 (1981).
544. Lercker,G., Cocchi,M., Turchetto,E. and Savioli,S., *J. High Res. Chromatogr., Chromatogr. Commun.*, **7**, 274-276 (1984).
545. Lie Ken Jie,M.S.F., *J. Chromatogr.*, **109**, 81-87 (1975).
546. Lie Ken Jie,M.S.F., *Adv. Chromatogr.*, **18**, 1-57 (1980).
547. Lie Ken Jie,M.S.F., in *Handbook of Chromatography. Vol. I. Lipids*, pp. 277-294 (1984) (edited by H.K.Mangold, CRC Press, Boca Raton).
548. Lie Ken Jie,M.S.F. and Lam,C.H., *J. Chromatogr.*, **138**, 373-380 (1977).
549. Lie Ken Jie,M.S.F. and Lam,C.H., *J. Chromatogr.*, **138**, 446-448 (1977).
550. Limsathayourat,N. and Melchert,H-U., *Fres. Z. Anal. Chem.*, **318**, 410-413 (1984).
551. Lin,K.C., Marchello,M.J. and Fischer,A.G., *J. Food Sci.*, **49**, 1521-1524 (1984).
552. Linstead,R.P. and Whalley,M., *J. Chem. Soc.*, 2987-2989 (1950).
553. Litchfield,C., *Analysis of Triglycerides* (1972) (Academic Press, New York).
554. Litchfield,C. and Ackman,R.G., *J. Chromatogr.*, **75**, 137-140 (1973).
555. Litchfield,C., Ackman,R.G., Sipos,J.C. and Eaton,C.A., *Lipids*, **6**, 674-681 (1971).
556. Litchfield,C., Greenberg,A.J., Ackman,R.G. and Eaton,C.A., *Lipids*, **13**, 860-866 (1978).
557. Litchfield,C., Harlow,R.D. and Reiser,R., *Lipids*, **2**, 363-370 (1967).
558. Litchfield,C., Miller,E., Harlow,R.D. and Reiser,R., *Lipids*, **2**, 345-350 (1967).
559. Litchfield,C., Tyszkiewicz,J., Marcantonio,E.E. and Noto,G., *Lipids*, **14**, 619-622 (1979).
560. Lohninger,A. and Nikiforov,A., *J. Chromatogr.*, **192**, 185-192 (1980).
561. Longone,F.T. and Miller,A.H., *Chem. Commun.*, 447-448 (1967).
562. Lough,A.K., *Prog. Chem. Fats other Lipids*, **14**, 1-48 (1975).
563. Low,M.G. and Finean,J.B., *Biochem. J.*, **167**, 281-284 (1977).
564. Lunazzi,L.. Placucci,G., Grossi,L. and Strocchi,A., *Chem. Phys. Lipids*, **30**, 347-352 (1982).
565. Lund,P. and Jensen,F., *Milchwissenschaft*, **38**, 193-196 (1983).
566. Lusby,W.R., Thompson,M.J. and Kochansky,J., *Lipids*, **19**, 888-901 (1984).
567. Luthra,M.G. and Sheltawy,A., *Biochem. J.*, **126**, 1231-1239 (1972).
568. Maas,R.L., Turk,J., Oates,J.A. and Brash,A.R., *J. Biol. Chem.*, **257**, 7056-7067 (1982).
569. McCloskey,J.A., in *Topics in Lipid Chemistry. Vol. 1*, pp. 369-440 (1970) (edited by F.D.Gunstone, Logos Press, London).
570. McCloskey,J.A., *Methods Enzymol.*, **35**, 340-348 (1975).
571. McCloskey,J.A. and Law,J.H., *Lipids*, **2**, 225-230 (1967).
572. McCloskey,J.A. and McClelland,M.J., *J. Am. Chem. Soc.*, **87**, 5090 5093 (1965).
573. McCluer,R.H., Ullman,M.D. and Jungalwala,F.B., *Adv. Chromatogr.*, **25**, 309-353 (1986).
574. McLafferty,F.W., *Anal. Chem.*, **31**, 82-87 (1959).
575. Maeda,K., Kawaguchi,S., Niwa,T., Ohki,T. and Kobayashi,K., *J. Chromatogr.*, **221**, 199-204 (1980).

576. Mahadevan,V., *Prog. Chem. Fats other Lipids*, **11**, 83-135 (1971).
577. Mahadevan,V., *Prog. Chem. Fats other Lipids*, **15**, 255-299 (1978).
578. Mahadevan,V., in *Fatty Acids*, pp. 527-542 (1979) (edited by E.H.Pryde, American Oil Chemists' Society, Champaign).
579. Mahadevan,V. and Ackman,R.G., in *Handbook of Chromatography. Vol. I. Lipids*, pp. 73-93 (1984) (edited by H.K.Mangold, CRC Press, Boca Raton).
580. Mahadevan,V., Phillips,F. and Lundberg,W.O., *J. Lipid Res.*, **6**, 434-435 (1965).
581. Mahadevan,V., Phillips,F. and Lundberg,W.O., *Lipids*, **1**, 183-187 (1966).
582. Mahadevan,V., Viswanathan,C.V. and Lundberg,W.O., *J. Chromatogr.*, **24**, 357-363 (1966).
583. Mahadevan,V., Viswanathan,C.V. and Lundberg,W.O., *J. Lipid Res.*, **8**, 2-6 (1967).
584. Mahadevappa,V.G. and Holub,B.J., in *Chromatography of Lipids in Biomedical Research and Clinical Diagnosis*, pp. 225-265 (1987) (edited by A.Kuksis, Elsevier, Amsterdam).
585. Manganaro,F., Myher,J.J., Kuksis,A. and Kritchevsky,D., *Lipids*, **16**, 508-517 (1981).
586. Mangold,H.K. and Paltauf,F. (editors), *Ether Lipids: Biochemical and Biomedical Aspects* (1983) (Academic Press, New York).
587. Mangold,H.K. and Spener,F., in *Lipids and Lipid Polymers in Higher Plants*, pp. 85-101 (1977) (edited by M.Tevini and H.K.Lichtenthaler, Springer-Verlag, Berlin).
588. Mangold,H.K. and Totani,N., in *Ether Lipids: Biochemical and Biomedical Aspects*, pp. 377-387 (1983) (edited by H.K.Mangold and F.Paltauf, Academic Press, London).
589. Mantz,A.W., *Ind. Res.*, **19**, 90-94 (1977).
590. Mares,P., in *Chromatography of Lipids in Biomedical Research and Clinical Diagnosis*, pp. 128-162 (1987) (edited by A.Kuksis, Elsevier, Amsterdam).
591. Mares,P., *Prog. Lipid Res.*, **27**, 107-133 (1988).
592. Mares,P. and Husek,P., *J. Chromatogr.*, **350**, 87-103 (1985).
593. Mares,P., Ranny,M., Sedlacek.J. and Skorepa,J., *J. Chromatogr.*, **275**, 295-306 (1983).
594. Mares,P., Skorepa,J., Sindelkova,E. and Tvrzicka,E., *J. Chromatogr.*, **273**, 172-179 (1983).
595. Mares,P., Tvrzicka,E. and Skorepa,J., *J. Chromatogr.*, **164**, 331-343 (1979).
596. Mares,P., Tvrzicka,E. and Tamchyna,V., *J. Chromatogr.*, **146**, 241-252 (1978).
597. Martin,A.J.P. and James,A.T., *Biochem. J.*, **63**, 138-143 (1956).
598. Martin,A.J.P. and Synge,R.L.M., *Biochem. J.*, **35**, 1358-1368 (1941).
599. Martinez-Castro,I., Alonso,L. and Juarez,M., *Chromatographia*, **21**, 37-40 (1986).
600. Matsubara,T., *Chem. Phys. Lipids*, **14**, 247-259 (1975).
601. Matsubara,T. and Hayashi,A., *Biochim. Biophys. Acta*, **296**, 171-178 (1973).
602. Matsui,E., Ogura,K. and Handa,S., *J. Biochem. (Tokyo)*, **101**, 423-432 (1987).
603. Matsuura,F., Matsubara,T. and Hayashi,A., *J. Biochem. (Tokyo)*, **74**, 49-57 (1973).
604. Mayer,B., Moser,R., Leis,H-J. and Gleispach,H., *J. Chromatogr.*, **378**, 430-436 (1986).
605. Meakins,G.D. and Swindells,R., *J. Chem. Soc.*, 1044-1047 (1959).
606. Mecham,D.K. and Mohammad,A., *Cereal Chem.*, **32**, 405-415 (1955).
607. Merritt,C., Vajdi,M., Kayser,S.G., Halliday,J.W. and Bazinet,M.L., *J. Am. Oil Chem. Soc.*, **59**, 422-432 (1982).
608. Michelsen,P. and Odham,G., *J. Chromatogr.*, **331**, 295-302 (1985).
609. Mikolajczak,K.L. and Bagby,M.O., *J. Am. Oil Chem. Soc.*, **41**, 391 (1964).
610. Minnikin,D.E., *Lipids*, **7**, 398-403 (1972).
611. Minnikin,D.E., *Lipids*, **10**, 55-57 (1975).
612. Minnikin,D.E., *Chem. Phys. Lipids*, **21**, 313-347 (1978).
613. Minnikin,D.E., Abley,P., McQuillin,F.J., Kusamran,K., Maskens,K. and Polgar,N., *Lipids*, **9**, 135-140 (1974).
614. Minnikin,D.E. and Smith,S., *J. Chromatogr.*, **103**, 205-207 (1975).
615. Miwa,T.K., Mikolajczak,K.L., Earle,F.R. and Wolff,I.A., *Anal. Chem.*,**32**, 1739-1742 (1960).
616. Monseigny,A., Vigneron,P-Y., Levacq,M. and Swoboda,F., *Rev. Franc. Corps Gras*, **26**, 107-120 (1979).
617. Morley,N.H., Kuksis,A., Buchnea,D. and Myher,J.J., *J. Biol. Chem.*, **250**, 3414-1418 (1975).
618. Morris,L.J., *J. Chromatogr.*, **12**, 321-328 (1963).
619. Morris,L.J., Holman,R.T. and Fontell,K., *J. Lipid Res.*, **1**, 412-420 (1960).

620. Morris,L.J. and Marshall,M.O., *Chem. Ind. (London)*, 460-461 (1966).
621. Morris,L.J. and Wharry,D.M., *J. Chromatogr.*, **20**, 27-37 (1965).
622. Morris,L.J., Wharry,D.M. and Hammond,E.W., *J. Chromatogr.*, **31**, 69-76 (1967).
623. Morris,L.J., Wharry,D.M. and Hammond,E.W., *J. Chromatogr.*, **33**, 471-479 (1968).
624. Morrison,W.R., *Biochim. Biophys. Acta,* **176**, 537-546 (1969).
625. Morrison,W.R. and Hay,J.D., *J. Chromatogr.*, **202**, 460-467 (1970).
626. Morrison,W.R. and Smith,L.M., *J. Lipid Res.*, **5**, 600-608 (1964).
627. Morrison,W.R., Tan,S.L. and Hargin,K.D., *J. Sci. Food Agric.*, **31**, 329-340 (1980).
628. Motta,L., Brianza,M., Stanga,F. and Amelotti,G., *Riv. Ital. Sostanze Grasse*, **60**, 625-633 (1983).
629. Muramatsu,T. and Schmid,H.H.O., *Chem. Phys. Lipids*, **9**, 123-132 (1972).
630. Murata,T., *Anal. Chem.*, **49**, 2209-2213 (1977).
631. Murata,T., *J. Lipid Res.*, **19**, 166-171 (1978).
632. Murata,T., Ariga,T. and Araki,E., *J. Lipid Res.*, **19**, 172-176 (1978).
633. Murata,T., Ariga,T., Oshima,M. and Miyatake,T., *J. Lipid Res.*, **19**, 370-374 (1978).
634. Murata,T. and Takahashi,S., *Anal. Chem.*, **45**, 1816-1823 (1973).
635. Murata,T. and Takahashi,S., *Anal. Chem.*, **49**, 728-731 (1977).
636. Murata,T., Takahashi,S. and Takeda,T., *Anal. Chem.*, **47**, 577-580 (1975).
637. Murawski,U. and Egge,H., *J. Chromatogr. Sci.*, **13**, 497-505 (1975).
638. Murawski,U. and Jost,U., *Chem. Phys. Lipids*, **13**, 155-158 (1974).
639. Murray,K.E., *Austral. J. Chem.*, **12**, 657-670 (1959).
640. Murray,K.E. and Schulten,H.R., *Chem. Phys. Lipids*, **29**, 11-21 (1981).
641. Muuse,B.G. and van der Kamp,H.J., *Neth. Milk Dairy J.*, **39**, 1-13 (1985).
642. Myher,J.J., in *Handbook of Lipid Research. Vol. 1. Fatty Acids and Glycerides*, pp. 123-196 (1978) (edited by A. Kuksis, Plenum Press, New York).
643. Myher,J.J. and Kuksis,A., *Lipids*, **9**, 382-390 (1974).
644. Myher,J.J. and Kuksis,A., *J. Chromatogr. Sci.*, **13**, 138-145 (1975).
645. Myher,J.J. and Kuksis,A., *Can. J. Biochem.*, **57**, 117-124 (1979).
646. Myher,J.J. and Kuksis,A., *Can. J. Biochem.*, **60**, 638-650 (1982).
647. Myher,J.J. and Kuksis,A., *J. Biochem. Biophys. Methods*, **10**, 13-23 (1984).
648. Myher,J.J. and Kuksis,A., *Can. J. Biochem. Cell Biol.*, **62**, 352-362 (1984).
649. Myher,J.J., Kuksis,A., Breckenridge,W.C. and Little,J.A., *Can. J. Biochem.*, **59**, 626-636 (1981).
650. Myher,J.J., Kuksis,A., Breckenridge,W.C. and Little,J.A., *Lipids*, **19**, 683-691 (1984).
651. Myher,J.J., Kuksis,A., Breckenridge,W.C., McGuire,V. and Little,J.A., *Lipids*, **20**, 90-101 (1985).
652. Myher,J.J., Kuksis,A., Marai,L. and Cerbulis,J., *Lipids*, **21**, 309-314 (1986).
653. Myher,J.J., Kuksis,A., Marai,L. and Yeung,S.K.F., *Anal. Chem.*, **50**, 557-561 (1978).
654. Myher,J.J., Kuksis,A., Yang,L-Y. and Marai,L., *Biochem. Cell Biol.*, **65**, 811-821 (1987).
655. Myher,J.J., Marai,L. and Kuksis,A., *Anal. Biochem.*, **62**, 188-203 (1974).
656. Myher,J.J., Marai,L. and Kuksis,A., *J. Lipid Res.*, **15**, 586-592 (1974).
657. Naccarato,W.F., Gelaman,R.A., Kawalek,J.C. and Gilbertson,J.R., *Lipids*, **7**, 275-281 (1972).
658. Nadenicek,J.D. and Privett,O.S., *Chem. Phys. Lipids*, **2**, 409-414 (1968).
659. Naito,H.K. and David,J.A. in *Lipid Research Methodology*, pp. 1-76 (1984) (edited by J.A.Story, A.R.Liss Inc., New York).
660. Nakagawa,Y. and Horrocks,L.A., *J. Lipid Res.*, **24**, 1268-1275 (1983).
661. Nakagawa,Y. and Waku,K., in *Chromatography of Lipids in Biomedical Research and Clinical Diagnosis*, pp. 163-190 (1987) (edited by A.Kuksis, Elsevier, Amsterdam).
662. Natarajan,V. and Schmid,H.H.O., *Lipids*, **12**, 128-130 (1977).
663. Ng,K.J., Andresen,B.D., Hilty,M.D. and Bianchine,J.R., *J. Chromatogr.*, **276**, 1-10 (1983).
664. Nichols,B.W., *Biochim. Biophys. Acta*, **70**, 417-422 (1963).
665. Nichols,B.W., in *New Biochemical Separations*, pp. 321-337 (1964) (edited by A.T.James and L.J.Morris, Van Norstrand, New York).
666. Nichols,B.W. and James,A.T., *Fette Seifen Anstrichm.*, **66**, 1003-1006 (1964).
667. Nichols,P.D., Guckert,J.B. and White,D.C., *J. Microbiol. Methods*, **5**, 49-55 (1986).
668. Nicolaides,N., Apon,J.M.B. and Wong,D.H., *Lipids*, **11**, 781-790 (1976).

669. Nicolaides,N. and Fu,H.C., *Lipids*, **4**, 83-86 (1969).
670. Nicolaides,N., Soukup,V.G. and Ruth,E.C., *Biomed. Mass Spectrom.*, **10**, 441-449 (1983).
671. Niehaus,W.G. and Ryhage,R., *Anal. Chem.*, **40**, 1840-1847 (1968).
672. Nishihara,J., Ishibashi,T., Imai,Y. and Muramatsu,T., *Lipids*, **19**, 907-910 (1984).
673. Nishihara,M., Kimura,K., Izui,K., Ishinaga,M., Kato,M. and Kito,M., *Biochim. Biophys. Acta*, **409**, 212-217 (1975).
674. Norman,H.A. and Thompson,G.A., *Arch. Biochem. Biophys.*, **242**, 168-175 (1985).
675. Novotny,M., Segura,R. and Zlatkis,A., *Anal. Chem.*, **44**, 9-13 (1972).
676. Nurmela,K.V.V. and Satama,L.E., *J. Chromatogr.*, **435**, 139-148 (1988).
677. Oehlenschlager,J. and Gercken,G., *Lipids*, **13**, 557-562 (1978).
678. Oehlenschlager,J. and Gercken,G., *J. Chromatogr.*, **176**, 126-128 (1979).
679. Oesterhelt,G., Marugg,P., Rueher,R. and Germann,A., *J. Chromatogr.*, **234**, 99-106 (1982).
680. Ojanpera,S.H., *J. Am. Oil Chem. Soc.*, **55**, 290-292 (1978).
681. Oshima,M., Ariga,T. and Murata,T., *Chem. Phys. Lipids*, **19**, 289-299 (1977).
682. Ottenstein,D.M., Bartley,D.A. and Supina,W.R., *J. Chromatogr.*, **119**, 401-407 (1976).
683. Ottenstein,D.M., Witting,L.A., Silvis,P.H., Hometchko,D.J. and Pelick,N., *J. Am. Oil Chem. Soc.*, **61**, 390-394 (1981).
684. Ottolenghi,A.C., *Methods Enzymol.*, **14**, 188-197 (1969).
685. Ozcimder,M. and Hammers,W.E., *J. Chromatogr.*, **187**, 307-317 (1980).
686. Padley,F.B., Gunstone,F.D. and Harwood,J.L., in *The Lipid Handbook*, pp. 49-170 (1986) (edited by F.D.Gunstone, J.L.Harwood and F.B.Padley, Chapman & Hall, London).
687. Padley,F.B. and Timms,R.E., *J. Am. Oil Chem. Soc.*, **57**, 286-292 (1980).
688. Paltauf,F., *Biochim. Biophys. Acta*, **239**, 38-46 (1971).
689. Panganamala,R.V., Horrocks,L.A., Geer,J.C. and Cornwell,D.G., *Chem. Phys. Lipids*, **6**, 97-102 (1971).
690. Park,D.K., Terao,J. and Matsushita,S., *Agric. Biol. Chem.*, **45**, 2443-2448 (1981).
691. Park,S.W. and Addis,P.B., *Anal. Biochem.*, **149**, 275-283 (1985).
692. Parsons,H., Emken,E.A., Marai,L. and Kuksis,A., *Lipids*, **21**, 247-251 (1986).
693. Patton,G.M., Cann,S., Brunengraber,H. and Lowenstein,J.M., *Methods Enzymol.*, **72**, 8-20 (1981).
694. Patton,G.M., Fasulo,J.M. and Robins,S.J., J. Lipid Res., 23, 190-196 (1982).
695. Patton,G.M. and Lowenstein,J.M., in *Handbook of Chromatography. Vol. II. Lipids*, pp. 225-239 (1984) (edited by H.K.Mangold, CRC Press, Boca Raton).
696. Patton,G.M. and Robins,S.J., in *Chromatography of Lipids in Biomedical Research and Clinical Diagnosis*, pp. 311-347 (1987) (edited by A.Kuksis, Elsevier, Amsterdam).
697. Patton,S. and Jensen,R.G., *Prog. Chem. Fats other Lipids*, **14**, 163-277 (1975).
698. Pawlowski,N.E., Eisele,T.A., Lee,D.J., Nixon,J.E. and Sinnhuber,R.O., *Chem. Phys. Lipids*, **13**, 164-172 (1974).
699. Pawlowski,N.E., Nixon,J.E. and Sinnhuber,R.O., *J. Am. Oil Chem. Soc.*, **49**, 387-392 (1972).
700. Peake,D.A. and Gross,M.L., *Anal. Chem.*, **57**, 115-120 (1985).
701. Pechine,J.M., Perez,F., Antony,C. and Jallon,J.M., *Anal. Biochem.*, **145**, 177-182 (1985).
702. Peers,K.E. and Coxon,D.T., *J. Food Technol.*, **21**, 463-469 (1986).
703. Pelletier,O., Wright,L.A. and Breckenridge,W.C., *Clin. Chem.*, **33**, 1403-1411 (1987).
704. Perkins,E.G., in *Analysis of Lipids and Lipoproteins*, pp. 183-203 (1975) (edited by E.G.Perkins, American Oil Chemists' Soc., Champaign).
705. Perkins,E.G. and Iwaoka,W.T., *J. Am. Oil Chem. Soc.*, **50**, 44-49 (1973).
706. Perkins,E.G., McCarthy,T.P., O'Brien,M.A. and Kummerow,F.A., *J. Am. Oil Chem. Soc.*, **54**, 279-281 (1977).
707. Pfeffer,P.E., in *Analysis of Lipids and Lipoproteins*, pp. 153-169 (1975) (edited by E.G.Perkins, American Oil Chemists' Soc., Champaign).
708. Pfeffer,P.E., Luddy,F.E., Unruh,J. and Schoolery,J.N., *J. Am. Oil Chem. Soc.*, **54**, 380-386 (1977).
709. Phillipou,G., Bigham,D.A. and Seamark,R.F., *Lipids*, **10**, 714-716 (1975).
710. Phillipou,G. and Poulos,A., *Chem. Phys. Lipids*, **22**, 51-54 (1978).
711. Phillips,F.C., Erdahl,W.L., Nadenicek,J.D., Nutter,L.J., Schmit,J.A. and Privett,O.S.,

Lipids, **19**, 142-150 (1984).
712. Phillips,F.C., Erdahl,W.L., Schmit,J.A. and Privett,O.S., *Lipids*, **19**, 880-887 (1984).
713. Pike,J.E. and Morton,D.R. (editors), *Advances in Prostaglandin, Thromboxane and Leukotriene Research*, Vol. 14 (1985) (Raven Press, New York).
714. Pina,M., Pioch,D. and Graille,J., *Lipids*, **22**, 358-361 (1987).
715. Piretti,M. and Pagliuca,G., *Rev. Franc. Corps Gras*, **34**, 26-27 (1987).
716. Plattner,R.D. and Gardner,H.W., *Lipids*, **20**, 126-131 (1985).
717. Plattner,R.D., Gardner,H.W. and Kleiman,R., *J. Am. Oil Chem. Soc.*, **60**, 1298-1303 (1983).
718. Plattner,R.D. and Spencer,G.F., *Lipids*, **18**, 68-73 (1983).
719. Plattner,R.D., Spencer,G.F. and Kleiman,R., *Lipids*, **11**, 222-227 (1976).
720. Plattner,R.D., Spencer,G.F. and Kleiman,R., *J. Am. Oil Chem. Soc.*, **54**, 511-515 (1977).
721. Pocklington,W.D. and Hautfenne,A., *Pure & Appl. Chem.*, **57**, 1515-1522 (1985).
722. Pohl,P., Glasl,H. and Wagner,H., *J. Chromatogr.*, **49**, 488-492 (1970).
723. Polgar,N., in *Topics in Lipid Chemistry. Vol. 2.*, pp. 207-246 (1971) (edited by F.D.Gunstone, Logos Press, London).
724. Polito,A.J., Akita,T. and Sweeley,C.C., *Biochemistry*, **7**, 2609-2614 (1968).
725. Pollard,M., in *Analysis of Oils and Fats*, pp. 401-434 (1986) (edited by R.J. Hamilton & J.B. Rossell, Elsevier Applied Science, London).
726. Pompella,A., Maellaro,E., Casini,A.F., Ferrali,M., Ciccoli,L. and Comporti,M., *Lipids*, **22**, 206-211 (1987).
727. Poole, C.F., in *Handbook of Derivatives for Chromatography*, pp. 152-200 (1978) (edited by K.Blau and G.S.King, Heyden & Sons, London).
728. Poole,C.F. and Schuette,S.A. (editors) *Contemporary Practice of Chromatography* (1984) (Elsevier, Amsterdam).
729. Poole,C.F. and Zlatkis,A., *J. Chromatogr. Sci.*, **17**, 115-123 (1979).
730. Poole,C.F. and Zlatkis,A., *J. Chromatogr.*, **184**, 99-183 (1980).
731. Porschmann,J., Welsh,T., Engelwald,W. and Vigh,G., *J. High Res. Chromatogr., Chromatogr. Commun.*, **7**, 509-514 (1984).
732. Poulos,A., Johnson,D.W., Beckman,K., White,A.G. and Easton,C., *Biochem. J.*, **248**, 961-964 (1987).
733. Poulos,A., Sharp,P., Singh,H., Johnson,D., Fellenberg,A. and Pollard,A., *Biochem. J.*, **235**, 607-610 (1986).
734. Powell,R.G. and Smith,C.R., *Biochemistry*, **5**, 625-631 (1966).
735. Powell,R.G., Smith,C.R. and Wolff,I.A., *Lipids*, **2**, 172-177 (1967).
736. Powell,W.S., *Methods Enzymol.*, **86**, 530-543 (1982).
737. Privett,O.S., *Prog. Chem. Fats other Lipids*, **9**, 91-117 (1966).
738. Privett,O.S. and Nickell,E.C., *Lipids*, **1**, 98-103 (1966).
739. Privett,O.S. and Nutter,L.J., *Lipids*, **2**, 149-154 (1967).
740. Proot,M., Sandra,P. and Geeraert,E., *J. High Res. Chromatogr., Chromatogr. Commun.*, **9**, 189-192 (1986).
741. Purnell,J.H., *J. Chem. Soc.*, 1268-1274 (1960).
742. Qureshi,N., Takayama,K. and Schnoes,H.K., *J. Biol. Chem.*, **255**, 182-189 (1980).
743. Radin,N.S., *Methods Enzymol.*, **72**, 5-7 (1981).
744. Radwan,S.S., in *Handbook of Chromatography. Vol. I. Lipids*, pp. 481-508 (1984) (edited by H.K.Mangold, CRC Press, Boca Raton).
745. Rahn,C.H. and Schlenk,H., *Lipids*, **8**, 612-616 (1973).
746. Raju,P.K. and Reiser,R., *Lipids*, **1**, 10-15 (1966).
747. Rakoff,H. and Emken,E.A., *Chem. Phys. Lipids*, **31**, 215-225 (1982).
748. Rakoff,H., Wiesleder,D. and Emken,E.A., *Lipids*, **14**, 81-83 (1979).
749. Ramachandran,S., Panganamala,R.V. and Cornwell,D.G., *J. Lipid Res.*, **10**, 465-467 (1969).
750. Ramachandran,S., Sprecher,H.W. and Cornwell,D.G., *Lipids*, **3**, 511-518 (1968).
751. Ramachandran,S., Venkata Rao,P. and Cornwell,D.G., *J. Lipid Res.*,**9**, 137-139 (1968).
752. Ramesha,C.S., Dickens,B.F. and Thompson,G.A., *Biochemistry*, **21**, 3618-3622 (1982).
753. Ramesha,C.S. and Pickett,W.C., *Biomed. Envir. Mass Spectrom.*, **13**, 107-111 (1986).
754. Rankin,P.C., *Lipids*, **5**, 825-831 (1970).
755. Rapport,M.M. and Alonzo,N., *J. Biol. Chem.*, **217**, 199-204 (1955).

756. Ratnayake,W.M.N. and Ackman,R.G., *Lipids*, **14**, 580-584 (1979).
757. Ratnayake,W.M.N. and Ackman,R.G., *Lipids*, **14**, 795-803 (1979).
758. Ratnayake,W.M.N., Timmins,A., Ohshima,T. and Ackman,R.G., *Lipids*, **21**, 518-524 (1986).
759. Recourt,J.H., Jurriens,G. and Schmitz,M., *J. Chromatogr.*, **30**, 35-42 (1967).
760. Reddy,P.V., Natarajan,P.V. and Sastry,P.S., *Chem. Phys. Lipids*, **17**, 373-377 (1976).
761. Reers,M., Schmid,P.C., Erdahl,W.L. and Pfeiffer,D.R., *Chem. Phys. Lipids*, **42**, 315-321 (1986).
762. Renkonen,O., *J. Am. Oil Chem. Soc.*, **42**, 298-304 (1965).
763. Renkonen,O., *Biochim. Biophys. Acta*, **125**, 288-309 (1966).
764. Renkonen,O., *Biochim. Biophys. Acta*, **137**, 575-577 (1967).
765. Renkonen,O., *Biochim. Biophys. Acta,* **152**, 114-135 (1968).
766. Renkonen,O., *Lipids*, **3**, 191-192 (1968).
767. Renkonen,O. and Luukkonen,A., in *Lipid Chromatographic Analysis, 2nd Edition, Vol.1.*, pp. 1-58 (1976) (edited by G.V.Marinetti, Marcel Dekker, New York).
768. Rezanka,T., Mares,P., Husek,P. and Podojil,M., *J. Chromatogr.*, **355**, 265-271 (1986).
769. Rezanka,T. and Podojil,M., *J. Chromatogr.*, **362**, 399-406 (1986).
770. Rezanka,T., Vokoun,J., Slavicek,J. and Podojil,M., *J. Chromatogr.*, **268**, 71-78 (1983).
771. Ries,S.K., Wert,V., Sweeley,C.C. and Leavitt,R.A., *Science*, **195**, 1339-1341 (1977).
772. Riva,M., Daghetta,A. and Galli,M., *Riv. Ital. Sostanze Grasse*, **58**, 432-443 (1981).
773. Robinson,P.G., *J. Lipid Res.*, **23**, 1251-1253 (1982).
774. Rock,C.O. and Snyder,F., *Arch. Biochem. Biophys.*, **171**, 631-636 (1975).
775. Roehm,J.N. and Privett,O.S., *Lipids*, **5**, 353-358 (1970).
776. Rossi,R., Carpita,A., Quirici,M.G. and Veracini,C.A., *Tetrahedron*, **38**, 639-644 (1982).
777. Roughan,P.G. and Batt,R.D., *Phytochem.*, **8**, 363-369 (1969).
778. Rouser,G., Kritchevsky,G. and Yamamoto,A., in *Lipid Chromatographic Analysis, Vol. 1*, pp. 99-162 (1967) (edited by G.V.Marinetti, Edward Arnold Ltd, London).
779. Rustow,B., Rabe,H. and Kunze,D., in *Chromatography of Lipids in Biomedical Research and Clinical Diagnosis*, pp. 191-224 (1987) (edited by A.Kuksis, Elsevier, Amsterdam).
780. Ryhage,R., Stallberg-Stenhagen,S. and Stenhagen,E., *Arkiv. Kemi*, **18**, 179-186 (1961).
781. Ryhage,R. and Stenhagen,E., *Arkiv. Kemi*, **13**, 523-542 (1959).
782. Ryhage,R. and Stenhagen,E., *Arkiv. Kemi*, **14**, 497-509 (1959).
783. Ryhage,R. and Stenhagen,E., *Arkiv. Kemi*, **15**, 545-574 (1960).
784. Saito,K., Satouchi,K., Kino,M., Ogino,H. and Oda,M., in *Chromatography of Lipids in Biomedical Research and Clinical Diagnosis*, pp. 378-402 (1987) (edited by A.Kuksis, Elsevier, Amsterdam).
785. Sampugna,J., Pallansch,L.A., Enig,M.G. and Keeney,M., *J. Chromatogr.*, **249**, 245-255 (1982).
786. Samuelsson,B. and Samuelsson,K., *J. Lipid Res.*, **10**, 41-46 (1969).
787. Samuelsson,B. and Samuelsson,K., *J. Lipid Res.*, **10**, 47-55 (1969).
788. Samuelsson,K. and Samuelsson,B., *Biochem. Biophys. Res. Commun.*, **37**, 15-21 (1969).
789. Samuelsson,K. and Samuelsson,B., *Chem. Phys. Lipids*, **5**, 44-79 (1970).
790. Sand,D.M., Schlenk,H., Thoma,H. and Spiteller,G., *Biochim. Biophys. Acta*, **751**, 455-461 (1983).
791. Sandra,P. (editor), *Sample Introduction in Capillary Gas Chromatography. Vol. 1.* (1985) (Huethig, Heidelberg).
792. Sarmientos,F., Schwarzmann,G. and Sandhoff,K., *Eur. J. Biochem.*, **146**, 59-64 (1985).
793. Satouchi,K., Oda,M., Yasunaga,K. and Saito,K., *J. Biochem. (Tokyo)*, **94**, 2067-2070 (1983).
794. Satouchi,K., Pinckard,R.N., McManus,L.M. and Hanahan,D.J., *J. Biol. Chem.*, **256**, 4425-4432 (1981).
795. Satouchi,K. and Saito,K., *Biomed. Mass Spectrom.*, **3**, 122-126 (1976).
796. Satouchi,K. and Saito,K., *Biomed. Mass Spectrom.*, **4**, 107-112 (1977).
797. Satouchi,K. and Saito,K., *Biomed. Mass Spectrom.*,**6**, 396-402 (1979).
798. Satouchi,K., Saito,K. and Kates,M., *Biomed. Mass Spectrom.*, **5**, 87-88 (1978).
799. Saunders,R.D. and Horrocks,L.A., *Anal. Biochem.*, **143**, 71-75 (1984).
800. Schlenk,H. and Gellerman,J.L., *Anal. Chem.*, **32**, 1412-1414 (1960).
801. Schmid,H.H.O., Bandi,P.C. and Su,K.L., *J. Chromatogr. Sci.*, **13**, 478-486 (1975).

802. Schmid,H.H.O., Baumann,W.J. and Mangold,H.K., *J. Am. Chem. Soc.*, **89**, 4797-4798 (1967).
803. Schmid,H.H.O., Jones,L.L. and Mangold,H.K., *J. Lipid Res.*, **8**, 692-693 (1967).
804. Schmid,H.H.O. and Mangold,H.K., *Biochim. Biophys. Acta*, **125**, 182-184 (1966).
805. Schmid,H.H.O. and Takahashi,T., *Hoppe-Seyler's Z. Physiol. Chem.*, **349**, 1673-1676 (1968).
806. Schmid,P.P., Muller,M.D. and Simon,W., *J. High Res. Chromatogr., Chromatogr. Commun.*, **2**, 675-676 (1979).
807. Schmidt,N., Gercken,G. and Konig,W.A., *J. Chromatogr.*, **410**, 458-462 (1987).
808. Schmitz,B. and Egge,H., *Chem. Phys. Lipids*, **25**, 287-298 (1979).
809. Schmitz,B. and Klein,R.A., *Chem. Phys. Lipids*, **39**, 285-311 (1986).
810. Schmitz,B., Murawski,U., Pfluger,M. and Egge,H., *Lipids*, **12**, 307-313 (1977).
811. Schmitz,F.J. and McDonald,F.J., *J. Lipid Res.*, **15**, 158-164 (1974).
812. Schogt,J.C.M. and Haverkamp-Begemann,P., *J. Lipid Res.*, **6**, 466-470 (1965).
813. Scholfield,C.R., in *Geometrical and Positional Fatty Acid Isomers*, pp. 17-52 (1978) (edited by E.A.Emken & H.J.Dutton, American Oil Chemists' Society, Champaign).
814. Scholfield,C.R., *J. Am. Oil Chem. Soc.*, **56**, 510-511 (1979).
815. Scholfield,C.R., *J. Am. Oil Chem. Soc.*, **58**, 662-663 (1981).
816. Scholfield,C.R. and Dutton,H.J., *J. Am. Oil Chem. Soc.*, **48**, 228-231 (1971).
817. Schulte,E., *Fette Seifen Anstrichm.*, **83**, 289-291 (1981).
818. Schulte,E., Hohn,M. and Rapp,U., *Fres. Z. Anal. Chem.*, **307**, 115-119 (1981).
819. Schulte,K.E. and Rucker,G., *J. Chromatogr.*, **49**, 317-322 (1970).
820. Schulten,H.R., Murray,K.E. and Simmleit,N., *Z. Naturforsch.*, **42C**, 178-190 (1987).
821. Schulten,H.R., Simmleit,N. and Murray,K., *Fres. Z. Anal. Chem.*, **327**, 235-238 (1987).
822. Schulten,H.R., Simmleit,N. and Rump,H.H., *Chem. Phys. Lipids*, **41**, 209-224 (1986).
823. Schwartz,D.P., *Anal. Biochem.*, **71**, 24-28 (1976).
824. Scorepa,J., *Molecular Species of Triglycerides in Biological Systems* (1975) (Universita Karlova, Prague).
825. Scrimgeour,C.M., *J. Am. Oil Chem. Soc.*, **54**, 210-211 (1977).
826. Sebedio,J-L. and Ackman,R.G., *Can. J. Chem.*, **56**, 2480-2485 (1978).
827. Sebedio,J-L. and Ackman,R.G., *Lipids*, **16**, 461-467 (1981).
828. Sebedio,J-L. and Ackman,R.G., *J. Chromatogr. Sci.*, **20**, 231-234 (1982).
829. Sebedio,J-L. and Ackman,R.G., *J. Am. Oil Chem. Soc.*, **60**, 1986-1991 (1983).
830. Sebedio,J-L., Farquharson,T.E. and Ackman,R.G., *Lipids*, **17**, 469-475 (1982).
831. Sebedio,J-L., Le Quere,J.L., Semon,E., Morin,O., Prevost,J. and Grandgirard,A., *J. Am. Oil Chem. Soc.*, **64**, 1324-1333 (1987).
832. Serdarevich,B. and Carroll,K.K., *J. Lipid Res.*, **7**, 277-284 (1966).
833. Sessa,D.J., Gardner,H.W., Kleiman,R. and Weisleder,D., *Lipids*, **12**, 613-619 (1977).
834. Shantha,N.C. and Kaimal,T.N.B., *Lipids*, **19**, 971-974 (1984).
835. Shapiro,I.L. and Kritchevsky,D., *J. Chromatogr.*, **18**, 599-601 (1965).
836. Sharkey,A.G., Friedel,R.A. and Langer,S.H., *Anal. Chem.*, **29**, 770-776 (1957).
837. Sharkey,A.G., Shultz,J.L. and Friedel,R.A., *Anal. Chem.*, **31**, 87-94 (1959).
838. Sharp,P., Poulos,A., Fellenberg,A. and Johnson,D., *Biochem. J.*, **248**, 61-67 (1987).
839. Sheppard,A.J. and Iverson,J.L., *J. Chromatogr. Sci.*, **13**, 448-452 (1975).
840. Sheppard,A.J., Waltking,A.E., Zmachinski,H. and Jones,S.T., *J. Assoc. Off. Anal. Chem.*, **61**, 1419-1423 (1978).
841. Shibahara,A., Yamamoto,K., Nakayama,T. and Kajimoto,G., *Yukagaku*, **34**, 618-625 (1985).
842. Shukla,V.K.S., Abdel-Moety,M.E., Larsen,E. and Egsgaard,H., *Chem. Phys. Lipids*, **23**, 285-290 (1979).
843. Shukla,V.K.S., Clausen,J., Egsgaard,H. and Larsen,E., *Fette Seifen Anstrichm.*, **82**, 193-199 (1980).
844. Shukla,V.K.S., Nielsen,W.S. and Batsberg,W., *Fette Seifen Anstrichm.*, **85**, 274-278 (1983).
845. Sisfontes,L., Nyborg,G., Svensson,L. and Blomstrand,R., *J. Chromatogr.*, **216**, 115-125 (1981).
846. Skipski,V.P. and Barclay,M., *Methods Enzymol.*, **14**, 530-598 (1969).
847. Skipski,V.P., Peterson,R.F. and Barclay,M., *Biochem. J.*, **90**, 374-378 (1964).

848. Skorepa,J., Kahudova,V., Kotrlikova,E., Mares,P. and Todorovicova,H., *J. Chromatogr.*, **273**, 180-186 (1983).
849. Skorepa,J., Mares,P., Rublicova,J. and Vinogradov,S., *J. Chromatogr.*, **162**, 177-184 (1979).
850. Slomiany,B.L., Murty,V.L.N., Liau,Y.H. and Slomiany,A., *Prog. Lipid Res.*, **26**, 29-51 (1987).
851. Slover,H.T. and Lanza,E., *J. Am. Oil Chem. Soc.*, **56**, 933-943 (1979).
852. Smith,A. and Calder,A.G., *Biomed. Mass Spectrom.*, **6**, 347-349 (1979).
853. Smith,A. and Duncan,W.R.H., *Lipids*, **14**, 350-355 (1979).
854. Smith,A. and Lough,A.K., *J. Chromatogr. Sci.*, **13**, 486-490 (1975).
855. Smith,C.R., *Lipids*, **1**, 268-273 (1966).
856. Smith,C.R., *Prog. Chem. Fats other Lipids*, **11**, 137-177 (1970).
857. Smith,C.R., in *Topics in Lipid Chemistry*, Vol. 3. pp. 89-124 (1972) (edited by F.D.Gunstone, Logos Press, London).
858. Smith,C.R., *J. Chromatogr. Sci.*, **14**, 36-40 (1976).
859. Smith,G.M. and Djerassi,C., *Lipids*, **22**, 236-240 (1987).
860. Smith,L.A., Norman,H.A., Cho,S.H. and Thompson,G.A., *J. Chromatogr.*, **346**, 291-299 (1985).
861. Smith,N.B., *Lipids*, **17**, 464-468 (1982).
862. Smith,N.B., *J. Chromatogr.*, **249**, 57-63 (1982).
863. Smith,N.B., *J. Chromatogr.*, **254**, 195-202 (1983).
864. Smith,R.M. (editor), *Supercritical Fluid Chromatography* (1988) (Royal Society of Chemistry, London).
865. Snyder,F., in *Lipid Chromatographic Analysis (2nd Edition) Vol. 1.*, pp. 111-148 (1976) (edited by G.V.Marinetti, Marcel Dekker, New York).
866. Snyder,F. (editor), *Platelet-Activating Factor and Related Lipid Mediators* (1987) (Plenum Press, New York).
867. Snyder,F., Blank,M.L. and Wykle,R.L., *J. Biol. Chem.*, **246**, 3639-3645 (1971).
868. Snyder,F., Rainey,W.T., Blank,M.L. and Christie,W.H., *J. Biol. Chem.*, **245**, 5853-5856 (1970).
869. Sonesson,A., Larsson,L. and Jimenez,J., *J. Chromatogr.*, **417**, 366-370 (1987).
870. Spark,A.A. and Ziervogel,M., *J. High Res. Chromatogr., Chromatogr. Commun.*, **5**, 206-207 (1982).
871. Spencer,G.F., *Phytochem.*, **16**, 282-284 (1977).
872. Spencer,G.F., *J. Am. Oil Chem. Soc.*, **56**, 642-646 (1979).
873. Spencer,G,F. and Plattner,R.D., *J. Am. Oil Chem. Soc.*, **61**, 90-94 (1984).
874. Spencer,G.F., Plattner,R.D. and Miwa,T., *J. Am. Oil Chem. Soc.*, **54**, 187-189 (1977).
875. Spener,F., in *Handbook of Chromatography. Vol. I. Lipids*, pp. 509-544 (1984) (edited by H.K.Mangold, CRC Press, Boca Raton).
876. Spener,F. and Mangold,H.K., *Biochemistry*, **13**, 2241-2248 (1974).
877. Sprecher,H.W., Maier,R., Barber,M. and Holman,R.T., *Biochemistry*, **4**, 1856-1863 (1965).
878. Stan,H-J. and Scheutwinkel-Reich,M., *Fres. Z. Anal. Chem.*, **296**, 400-405 (1979).
879. Stan,H-J. and Scheutwinkel-Reich,M., *Lipids*, **15**, 1044-1054 (1980).
880. Stein,R.A. and Nicolaides,N., *J. Lipid Res.*, **3**, 476-478 (1962).
881. Stein,R.A. and Slawson,V., *J. Chromatogr.*, **25**, 204-212 (1966).
882. Strocchi,A., *Riv. Ital. Sostanze Grasse*, **63**, 99-117 (1986).
883. Strocchi,A. and Bonaga,G., *Chem. Phys. Lipids*, **15**, 87-94 (1975).
884. Strocchi,A. and Holman,R.T., *Riv. Ital. Sostanze Grasse*, **48**, 617-622 (1971).
885. Strocchi,A., Mariani,C., Camurati,F., Fedeli,E., Baragli,S., Gamba,P., Giro,L. and Motta,L., *Riv. Ital. Sostanze Grasse*, **61**, 499-506 (1984).
886. Su,K.L. and Schmid,H.H.O., *Lipids*, **9**, 208-213 (1974).
887. Sugita,M., *J. Biochem. (Tokyo)*, **82**, 1307-1312 (1977).
888. Sugita,M., Itasaka,O. and Hori,T., *Chem. Phys. Lipids*, **16**, 1-8 (1976).
889. Sun,G.Y. and Horrocks,L.A., *J. Lipid Res.*, **10**, 153-157 (1969).
890. Sundler,R. and Akesson,G.A.E., *J. Chromatogr.*, **80**, 233-240 (1973).
891. *Supelco Bulletin*, No. 822 (1986) (Supelco Inc., Bellefonte, U.S.A.).
892. Suzuki,A., Handa,S. and Yamakawa,T., *J. Biochem. (Tokyo)*, **80**, 1181-1183 (1976).

893. Suzuki,M., Ariga,T., Sekine,M., Araki,E. and Miyatake,T., *Anal. Chem.*, **53**, 985-988 (1981).
894. Svensson,L., Sisfontes,L., Nyborg,G. and Blomstrand,R., *Lipids*, **17**, 50-59 (1982).
895. Sweeley,C.C. and Moscatelli,E.A., *J. Lipid Res.*, **1**, 40-47 (1960).
896. Sweeley,C.C. and Nunez,H.A., *Ann. Rev. Biochem.*, **54**, 765-801 (1985).
897. Taguchi,R., Asahi,Y. and Ikezawa,H., *Biochim. Biophys. Acta*, **619**, 48-57 (1980).
898. Takagi,T. and Itabashi,Y., *Lipids*, **12**, 1062-1068 (1977).
899. Takagi,T. and Itabashi,Y., *J. Chromatogr. Sci.*, **15**, 121-124 (1977).
900. Takagi,T. and Itabashi,Y., *Lipids*, **16**, 546-551 (1981).
901. Takagi,T., Itabashi,Y., Ota,T. and Hayashi,K., *Lipids*, **11**, 354-356 (1976).
902. Takagi,T., Sakai,A., Itabashi,Y. and Hayashi,K., *Lipids*, **12**, 228-234 (1977).
903. Takahashi,T. and Schmid,H.H.O., *Chem. Phys. Lipids*, **2**, 220-222 (1968).
904. Takahashi,T. and Schmid,H.H.O., *Chem. Phys. Lipids*, **4**, 243-246 (1970).
905. Takamura,H., Narita,H., Urade,R. and Kito,M., *Lipids*, **21**, 356-361 (1986).
906. Takeuchi,G., Weiss,M. and Harrison,A.G., *Anal. Chem.*, **59**, 918-921 (1987).
907. Tallent,W.H., Harris,J., Spencer,G.F. and Wolff,I.A., *Lipids*, **3**, 425-430 (1968).
908. Tam,S.P. and Breckenridge,W.C., *J. Lipid Res.*, **24**, 1343-1357 (1983).
909. Taylor,J., Parkes,R.J., Wahle,K.W.J. and Graham,A., *Lipids*, **20**, 393-397 (1985).
910. Terao,J., Asano,I. and Matsushita,S., *Arch. Biochem. Biophys.*, **235**, 326-333 (1984).
911. Terao,J., Inoue,T., Yamagata,S., Murakami,H. and Matsushita,S., *Agric. Biol. Chem.*, **48**, 1735-1743 (1984).
912. Termonia,M., Munari,F. and Sandra,P., *J. High Res. Chromatogr., Chromatogr. Commun.*, **10**, 263-268 (1987).
913. Thenot,J-P., Horning,E.C., Stafford,M. and Horning,M.G., *Anal. Letts.*, **5**, 217-223 (1972).
914. Thompson,G.A. and Lee,P., *Biochim. Biophys. Acta,* **98**, 151-159 (1965).
915. Thompson,W., *Biochim. Biophys. Acta*, **187**, 150-153 (1969).
916. Tokumura,A., Suzuki,T., Takauchi,K. and Tsukatani,H., *J. Chromatogr.*, **343**, 138-142 (1985).
917. Tokumura,A., Yoshioka,Y., Tsukatani,H. and Handa,Y., *Biomed. Envir. Mass Spectrom.*, **13**, 175-180 (1986).
918. Tomer,K.B., Crow,F.W. and Gross,M.L., *J. Am. Chem. Soc.*, **105**, 5487-5488 (1983).
919. Tomiyasu,I., Toriyama,S., Yano,I. and Masui,M., *Chem. Phys. Lipids*, **28**, 41-54 (1981).
920. Totani,N., *Fette Seifen Anstrichm.*, **84**, 70-73 (1982).
921. Totani,N. and Mangold,H.K., *Mikrochim. Acta*, **1**, 73-78 (1981).
922. Traitler,H., *Prog. Lipid Res.*, **26**, 257-280 (1987).
923. Traitler,H. and Prevot,A., *J. High Res. Chromatogr., Chromatogr. Commun.*, **4**, 109-114 (1981).
924. Traitler,H. and Prevot,A., *Rev. Franc. Corps Gras*, **28**, 263-268 (1981).
925. Tserng,K-Y., Kliegman,R.M., Miettinen,E.L. and Kalhan,S.C., *J. Lipid Res.*, **22**, 852-858 (1981).
926. Tulloch,A.P., *J. Am. Oil Chem. Soc.*, **41**, 833-836 (1964).
927. Tulloch,A.P., *J. Am. Oil Chem. Soc.*, **43**, 670-674 (1966).
928. Tulloch,A.P., *J. Chromatogr. Sci.*, **13**, 403-407 (1975).
929. Tulloch,A.P., in *Chemistry and Biochemistry of Natural Waxes*, pp. 236-287 (1976) (edited by P.E.Kolattukudy, Elsevier, Amsterdam).
930. Tulloch,A.P., *Can. J. Chem.*, **55**, 1135-1142 (1977).
931. Tulloch,A.P., *Org. Mag. Res.*, **11**, 109-115 (1978).
932. Tulloch,A.P., *Lipids*, **15**, 881-888 (1980).
933. Tulloch,A.P., *Lipids*, **17**, 544-549 (1982).
934. Tulloch,A.P., *Prog. Lipid Res.*, **22**, 235-256 (1983).
935. Tulloch,A.P., *Lipids*, **20**, 404-411 (1985).
936. Tulloch,A.P., *Lipids*, **20**, 652-663 (1985).
937. Tulloch,A.P. and Bergter,L., *Lipids*, **14**, 996-1002 (1979).
938. Tulloch,A.P. Heinz,E. and Fischer, *Hoppe-Seyler's Z. Physiol. Chem.*, **354**, 879-889 (1973).
939. Tulloch,A.P. and Mazurek,M., *Lipids*, **11**, 228-234 (1976).
940. Ulsaker,G.A. and Teien,G., *Analyst (London)*, **111**, 559-562 (1986).

941. Valicenti,A.J., Chapman,C.J., Holman,R.T. and Chipault,J.R., *Lipids*, **13**, 190-194 (1978).
942. Valicenti,A.J., Heimermann,W.H. and Holman,R.T., *J. Org. Chem.*, **44**, 1068-1073 (1979).
943. Vance,D.E. and Ridgway,N.E., *Prog. Lipid Res.*, **27**, 61-79 (1988).
944. Vandenheuvel,W.J.A., Gardiner,W.L. and Horning,E.C., *J. Chromatogr.*, **19**, 263-276 (1965).
945. Van Gorkom,M. and Hall,G.E., *Spectrochim. Acta*, **22**, 990-992 (1966).
946. Van Kuijk,F.J.G.M., Thomas,D.W., Stephens,R.J. and Dratz,E.A., *J. Free Radicals Biol. Med.*, **1**, 215-225 (1985).
947. Van Os,C.P.A., Rijke-Schilder,G.P.M., Kammerling,J.P., Gerwig,G.J. and Vliegenthart,J.F.G., *Biochim. Biophys. Acta*, **620**, 326-331 (1980).
948. VanRollins,M. and Murphy,R.C., *J. Lipid Res.*, **25**, 507-517 (1984).
949. Van Vleet,E.S. and Quinn,J.G., *J. Chromatogr.*, **151**, 396-400 (1978).
950. Vetter,W. and Meister,W., *Org. Mass Spectrom.*, **16**, 118-122 (1981).
951. Vigneron,P.Y., Henon,G., Monseigny,A., Levacq,M., Stoclin,B. and Delvoye,P., *Rev. Franc. Corps Gras*, **29**, 423-435 (1982).
952. Vincenti,M., Guglieilmetti,G., Cassani,G. and Tonini,C., *Anal. Chem.*, **59**, 694-699 (1987).
953. Vine,J., *J. Chromatogr.*, **196**, 415-424 (1980).
954. Vioque,E., in *Handbook of Chromatography. Vol. I. Lipids*, pp. 295-320 (1984) (edited by H.K.Mangold, CRC Press, Boca Raton).
955. Vioque,E. and Holman,R.T., *J. Am. Oil Chem. Soc.*, **39**, 63-66 (1962).
956. Viswanathan,C.V., *J. Chromatogr.*, **98**, 105-128 (1974).
957. Viswanathan,C.V., *J. Chromatogr.*, **98**, 129-155 (1974).
958. Von Rudloff,E., *J. Am. Oil Chem. Soc.*, **33**, 126-128 (1956).
959. Von Rudloff,E., *Can. J. Chem.*, **34**, 1413-1418 (1956).
960. Wait,R. and Hudson,M.J., *Lett. Appl. Microbiol.*, **1**, 95-99 (1985).
961. Wakeham,S.G. and Frew,N.M., *Lipids*, **17**, 831-843 (1982).
962. Waku,K. and Nakazawa,Y., *J. Biochem. (Tokyo)*, **72**, 149-155 (1972).
963. Walker,M.A., Roberts,D.R. and Dumbroff,E.B., *J. Chromatogr.*, **241**, 390-391 (1982).
964. Walton,T.J. and Kolattukudy,P.E., *Biochemistry*, **11**, 1885-1897 (1972).
965. Watts,R. and Dils,R., *J. Lipid Res.*,**9**, 40-51 (1968).
966. Watts,R. and Dils,R., *J. Lipid Res.*, **10**, 33-40 (1969).
967. Ways,P. and Hanahan,D.J., *J. Lipid Res.*, **5**, 318-328 (1964).
968. Weihrauch,J.L., Brewington,C.R. and Schwartz,D.P., *Lipids*, **9**, 883-890 (1974).
969. White,C.M. and Houck,R.K., *J. High Res. Chromatogr., Chromatogr. Commun.*, **8**, 293-296 (1985).
970. White,D.A., in *Form and Function of Phospholipids*, pp. 441-482 (1973) (edited by G.B.Ansell, J.N.Hawthorne and R.M.C.Dawson, Elsevier, Amsterdam).
971. White,D.C., Tucker,A.N. and Sweeley,C.C., *Biochim. Biophys. Acta*, **187**, 527-533 (1969).
972. White,H.B., *J. Chromatogr.*, **21**, 213-222 (1966).
973. White,H.B. and Powell,S.S., *J. Chromatogr.*, **32**, 451-457 (1968).
974. Wiegandt,H. (editor), *Glycolipids. New Comprehensive Biochemistry, Vol. 10*, (1985) (Elsevier, New York).
975. Wijekoon,W.M.D., Ayanoglu,E. and Djerassi,C., *Tetrahedron Letts.*, **25**, 3285-3288 (1984).
976. Williams,J.P., Watson,G.R., Khan,M., Leung,S., Kuksis,A., Stachnyk,O. and Myher,J.J., *Anal. Biochem.*, **66**, 110-122 (1975).
977. Wineberg,J.P. and Swern,D., *J. Am. Oil Chem. Soc.*, **50**, 142-146 (1973).
978. Wineberg,J.P. and Swern,D., *J. Am. Oil Chem. Soc.*, **51**, 528-533 (1974).
979. Wing,D.R., Harvey,D.J., La Droitte,P., Robinson,K. and Belcher,S., *J. Chromatogr.*, **368**, 103-111 (1986).
980. Winterfeld,M. and Debuch,H., *Hoppe Seyler's Z. Physiol. Chem.*, **345**, 11-21 (1966).
981. Witting,L.A. (editor), *Glycolipid Methodology* (1976) (American Oil Chemists' Society, Champaign).
982. Wolff,I.A. and Miwa,T.K., *J. Am. Oil Chem. Soc.*, **42**, 208-215 (1965).

983. Wolthers,B.G., Hindriks,F.R., Muskiet,F.A.J. and Groen,A., *Clin. Chim. Acta*, **103**, 305-315 (1980).
984. Wong,W-S.D. and Hammond,E.G., *Lipids*, **12**, 475-479 (1977).
985. Wood,R., *Lipids*, **2**, 199-203 (1967).
986. Wood,R., *J. Chromatogr.*, **287**, 202-208 (1984).
987. Wood,R., *Biochem. Arch.*, **2**, 63-71 (1986).
988. Wood,R., Bever,E.L. and Snyder,F., *Lipids*, **1**, 399-408 (1966).
989. Wood,R. and Harlow,R.D., *Arch. Biochem. Biophys.*, **131**, 495-501 (1969).
990. Wood,R. and Harlow,R.D., *J. Lipid Res.*, **10**, 463-465 (1969).
991. Wood,R. and Harlow,R.D., *Arch. Biochem. Biophys.*, **135**, 272-281 (1969).
992. Wood,R. and Healy,K., *Lipids*, **5**, 661-663 (1970).
993. Wood,R. and Lee,T., *Lipids*, **15**, 876-879 (1981).
994. Wood,R. and Lee,T., *J. Chromatogr.*, **254**, 237-246 (1983).
995. Wood,R., Raju,P.K. and Reiser,R., *J. Am. Oil Chem. Soc.*, **42**, 81-85 (1965).
996. Wood,R., Raju,P.K. and Reiser,R., *J. Am. Oil Chem. Soc.*, **42**, 161-165 (1965).
997. Wood,R. and Reiser,R., *J. Am. Oil Chem. Soc.*, **42**, 315-320 (1965).
998. Wood,R. and Snyder,F., *J. Am. Oil Chem. Soc.*, **43**, 53-54 (1966).
999. Wood,R. and Snyder,F., *Lipids*, **1**, 62-72 (1966).
1000. Wood,R. and Snyder,F., *J. Lipid Res.*, **8**, 494-500 (1967).
1001. Wood,R. and Snyder,F., *Lipids,* **3**, 129-135 (1968).
1002. Wood,R. and Snyder,F., *Arch. Biochem. Biophys.*, **131**, 478-494 (1969).
1003. Woodford,F.P. and Van Gent,C.M., *J. Lipid Res.*, **1**, 188-190 (1960).
1004. Woollard,P.M., *Biomed. Mass Spectrom.*, **10**, 143-154 (1983).
1005. Woollard,P.M. and Mallet,A.I., *J. Chromatogr.*, **306**, 1-21 (1984).
1006. Yamaoka,R., Tokoro,M. and Hayashiya,K., *J. Chromatogr.*, **399**, 259-267 (1987).
1007. Yao,J.K. and Rastetter,G.M., *Anal. Biochem.*, **150**, 111-116 (1985).
1008. Yasugi,E., Kasama,T. and Seyama,Y., *J. Biochem. (Tokyo)*, **102**, 1477-1482 (1987).
1009. Yeung,S.K.F., Kuksis,A., Marai,L. and Myher,J.J., *Lipids*, **12**, 529-537 (1977).
1010. Youngs,C.C., *J. Am. Oil Chem. Soc.*, **61**, 576-581 (1984).
1011. Zak,B., *Lipids*, **15**, 698-704 (1980).
1012. Zeman,I. and Pokorny,J., *J. Chromatogr.*, **10**, 15-20 (1963).
1013. Zeman,I., Ranny,M. and Winterova,L., *J. Chromatogr.*, **354**, 283-292 (1986).
1014. Zinkel,D.F. and Rowe,J.W., *Anal. Chem.*, **36**, 1160-1161 (1964).

Index